Bridging Human-Machine Interaction for Energy Infrastructure

Mohamed Naoui
https://orcid.org/0000-0001-9117-3510
National Engineering School of Gabes, Tunisia

Lassaad Sbita
https://orcid.org/0000-0002-6589-3976
National Engineering School of Gabes, Tunisia

IGI Global
Scientific Publishing
Publishing Tomorrow's Research Today

Vice President of Editorial Melissa Wagner
Director of Acquisitions Mikaela Felty
Director of Book Development Jocelynn Hessler
Production Manager Mike Brehm
Cover Design Jose Rosado

Published in the United States of America by
IGI Global Scientific Publishing
701 East Chocolate Avenue
Hershey, PA, 17033, USA
Tel: 717-533-8845 | Fax: 717-533-7115
Website: https://www.igi-global.com E-mail: cust@igi-global.com

Library of Congress Cataloging-in-Publication Data

LCCN: 2025033227 (CIP Data Pending)
ISBN13: 9798337356211
Isbn13Softcover: 9798337356228
EISBN13: 9798337356235
British Cataloguing in Publication Data
A Cataloguing in Publication record for this book is available from the British Library.

Table of Contents

Detailed Table of Contents

This chapter explores Permanent Magnet Synchronous Machines, which are favored in electric vehicle traction for their efficiency, power density, and dynamic performance. Despite their advantages, PMSMs can experience faults that affect both operational safety and reliability. In this chapter, we begin by classifying common faults into two main categories: rotor-related faults and stator-related faults which are represented as electrical faults. Moreover, we introduce a detailed mathematical model of PMSMs using the direct-quadrature (dq) frame, which simulates both healthy and faulty conditions, enabling analysis of fault impacts on currents, voltages, and torque. We further explore Field Oriented Control, and explain its key components such as PI controllers and coordinate transformations. Finally, a comparative study based on standard driving cycles (FTP, NEDC, WLTP) is carried out to illustrate how faults impact system performance, providing insights into early fault detection and supporting the development of robust monitoring and fault-tolerant systems.

In recent decades, most countries rely more and more on renewable energy sources such as solar, wind, biomass and geothermal energies, etc. Solar and wind energy, in particular, are widely used because they are naturally abundant and environmentally friendly. Electricity generation based on renewable energy sources is increasing as a core component of modern energy infrastructure. To increase the output power of photovoltaic solar power generation, it is extremely important to force the system to operate at the highest possible efficiency. Maximum Power Point Tracking (MPPT) algorithms play a crucial role in optimizing the performances of the photovoltaic (PV) power generation system. This paper contributes to the modeling and development of the complete PV pumping system(Youssef,#6). Additionally, this research seeks to improve human-machine communication in energy systems by incorporating industrial oversight instruments. This communication happens through monitoring and controlling the system using STEP 7 software, which helps create a smooth link between the operator and the automated process, while comparing three Maximum Power Point Tracking (MPPT) methods: the widely used Perturb and Observe (P&O) and Incremental Conductance (INC) techniques, along with a method based on Artificial Neural Networks (ANN). The use of the ANN method demonstrates the benefits of smart control techniques. especially where their use is potential with rapidly varying irradiance. While the INC method has an appeal to its simplicity, and minimal installation costs, a comparative evaluation made reference to important metrics like voltage, current, and power production in different environments.

In this chapter adopted a critical view of the evolution of Human Machine Interfaces (HMIs) in energy infrastructures from analog controls to AI augmented systems, particularly with regard to how digitalisation, automation and the integration of renewables have changed operator working practices and how systems are designed. Opportunity and development of HMIs have appeared in many forms, including SCADA, DCS, predictive analytics and the incorporation of explainable AI. Using frameworks including ISO 9241, NASA TLX and SAGAT the chapter evaluates the effectiveness of HMIs for usability, safety and trust. Using examples including smart grids and nuclear controls, HMI design is argued to also be human centred to allow energy systems to be resilient and sustainable whilst enabling operator cognition.

This chapter presents an adaptive Maximum Power Point Tracking (MPPT) control framework for photovoltaic systems featuring real-time human-driven algorithm selection via an intuitive Human-Machine Interface (HMI). The platform integrates four classical algorithms—Perturb and Observe (P&O), Incremental Conductance (INC), Constant Voltage (CV), and enhanced P&O (dP-P&O)—implemented on dSPACE DS1104 hardware. Experimental evaluation under standardized staircase irradiance profiles reveals INC achieves highest efficiency (99.946%), followed by P&O (99.884%) and dP-P&O (99.780%), while CV offers computational simplicity despite lower efficiency (95.324%). This human-in-the-loop approach enables context-aware algorithm optimization without interrupting power generation, demonstrating superior adaptability compared to fixed MPPT implementations. The modular architecture supports future integration of AI-based techniques and comprehensive PV diagnostics.

This chapter addresses the critical role of supervisory human–machine interaction (HMI) in the effective energy management of hybrid renewable energy systems. It begins by presenting a comprehensive modeling framework of a hybrid system combining photovoltaic panels, wind turbines, batteries, and electrical loads. A hybrid control and energy management strategy is then developed to optimize power flow and enhance system stability. Central to this work is the design and implementation of a user-friendly HMI built using MATLAB App Designer and integrated with a Simulink model. The interface allows real-time monitoring of power outputs, DC bus voltage, and battery state of charge, while providing interactive controls for load adjustment and simulation management. The seamless integration between the HMI and the control system facilitates improved decision-making and operational flexibility. Finally, the chapter highlights the importance of such supervisory interfaces in advancing the reliability and efficiency of renewable energy systems, and suggests pathways for future enhancements involving adaptive and remote control features.

The rapid evolution of energy infrastructure towards decentralization and automation has amplified the need for secure, transparent, and efficient human-machine interactions. In emerging smart grids, autonomous energy trading platforms, and distributed generation systems, transactions between human operators, IoT-enabled devices, and AI-driven systems must be safeguarded against cyber threats, fraud, and operational discrepancies. This chapter examines the transformative role of blockchain technology in enabling secure human-machine transactions within modern energy markets. It highlights how blockchain's decentralized ledger, cryptographic validation, and immutable record-keeping can enhance trust, traceability, and accountability in both peer-to-peer (P2P) and machine-to-machine (M2M) energy exchanges. Through case studies and implementation models, the chapter explores blockchain-enabled microgrids, dynamic pricing systems, and automated settlement mechanisms that facilitate real-time, tamper-proof transactions.

The rapid penetration of renewable energy into smart grid has decreased grid inertia due to the substitution of synchronous generators with inverter-based resources. Due to their lack of rotational inertia, renewable energy create a crucial 'inertia gap'. In response to this challenge, virtual inertia control (VIC) has emerged as an important solution, evolving from primitive emulation methods to advanced adaptive and artificial intelligence-based methods. This work investigates the evolution of virtual inertia emulation methods as important countermeasures for smart grids, from their initial fixed-parameter methods through the development into real-time adaptive systems and intelligent control structures. The integration of Human-Machine Interaction (HMI) interfaces is emphasized to enable operators to monitor, interpret, and intervene in inertia management systems. These HMI systems support decision-making by visualizing real-time data, enabling intuitive control of adaptive VIC algorithms, and ensuring human oversight in safety-critical scenarios within low-inertia power grids.

In this work, a high-efficiency photovoltaic water pumping system is conceived, modelled, implemented and validated in Matlab/Simulink environment. The system is made up of a triple junction photovoltaic generator, a centrifugal pumping system, and a three-phase permanent magnet synchronous motor (PMSM) supplied with a three-phase inverter chosen as a pump driver. As a motor drive decoupling strategy, the field oriented control approach is adopted. Three control loops make up the traditional vector control strategy: a speed loop control, two loops for the stator current components of the PMSM. In most cases, the reference quadrature stator flux is obtained from the speed controller. However, the reference direct stator flux is typically set to zero values in many applications. However, issues arose if the application required the high-speed mode. Consequently, a vector control technique and a field weakening region algorithm are developed in this study. The designed system and all control algorithms are implemented in Matlab/Simulink environment. At various speeds, including the two working modes at standard test conditions, significant results are presented and analyzed.

The global transition from conventional vehicles to electric mobility is accelerating, driven by environmental concerns, regulatory frameworks, and advances in energy infrastructure. As electric vehicles (EVs) become increasingly embedded within smart grid, intelligent energy management systems are required to address variable demand, battery aging, charging variability, and grid stability. Artificial Intelligence (AI) has become a key enabler by providing efficiency and reliability. This chapter reviews AI-driven strategies for battery charging and bidirectional energy exchange, focusing on Computational Intelligence (CI) and Machine Learning (ML). Their applications in state estimation, charging control, optimization, anomaly detection, and smart grid coordination are analyzed. Particular attention is given to V2G and G2V architectures and the critical role of Human-Machine Interaction (HMI), which translates AI-generated insights into understandable, user centered charging decisions. By comparing CI and ML approaches and identifying open challenges, this chapter offers a structured perspective on intelligent and HMI integrated EV charging systems.

The development of underwater vehicles is a vital tool for exploring, monitoring, and responding to marine environments that are typically inaccessible or frightening to human beings. The intervention of underwater robots makes cross-oceans more manageable. As part of this work, the chapter focus on the fundamental aspects of underwater robot modeling and control. The second part of this chapter deals with the stabilization problem of the robot. The use of a control system is essential for underwater robots to ensure precise trajectory tracking, position stability, and attitude stabilization. The author offer in this section an explicit comparison between PID and FOPID Controllers. Finally, the study emphasizes the importance of Human-Machine Interaction (HMI) in robot control. An intuitive Dashboard (Simulink) was developed for real-time trajectory adjustment. Results validate the robot's ability to follow complex 3D paths and the HMI's effectiveness, although further controller optimization is required to improve performance during dynamic movements.

This chapter presents a comprehensive modeling and control framework for a grid-connected wind energy conversion system (WECS) based on a Permanent Magnet Synchronous Generator (PMSG). The proposed system integrates aerodynamic and electromechanical models, including wind profile, turbine, shaft dynamics, PMSG, and full-scale back-to-back converters. A Tip Speed Ratio (TSR)-based Maximum Power Point Tracking (MPPT) method is implemented to ensure optimal energy extraction, while pitch angle control maintains rated power under high wind conditions. Vector control is applied to both machine-side and grid-side converters to regulate torque, flux, DC-link voltage, and manage active/reactive power injection into the grid. A Phase-Locked Loop (PLL) ensures synchronization. MATLAB/Simulink

Preface

INTRODUCTION

The global transition toward intelligent, decentralized, and sustainable energy systems has accelerated unprecedented transformations in how humans interact with the technological infrastructures that power modern life. From renewable energy conversion, electric mobility, industrial automation, and hybrid microgrids to digital twins, robotics, artificial intelligence, and blockchain enabled energy markets, Human–Machine Interaction (HMI) has become a defining pillar of energy innovation. As energy networks grow more complex interconnected through sensors, data platforms, algorithms, and cyber-physical systems the human operator remains indispensable. The challenge is no longer simply to automate machines, but to harmonize human reasoning with artificial intelligence, decision support tools, and advanced control systems. This book, Bridging Human–Machine Interaction for Energy Infrastructure, brings together multidisciplinary contributions that explore this evolving landscape from technical, operational, and human-centered perspectives.

Human–machine collaboration now extends far beyond traditional SCADA and analog control rooms. Modern energy systems incorporate predictive analytics, machine learning, adaptive controllers, immersive dashboards, gesture- or voice-driven interfaces, and supervisory platforms that must remain intuitive, reliable, and secure. As renewable penetration increases and energy infrastructures become distributed and intelligent, the importance of designing systems that are both technologically advanced and operationally transparent has never been greater. This book aims to fill this critical knowledge gap by presenting research that binds energy engineering with digital interaction, user-centered design, intelligent control, and cybersecurity.

OBJECTIVES

The overarching objective of this book is to create a unified reference that highlights the essential role of Human–Machine Interaction (HMI) in modern and future energy infrastructures, presenting how intuitive interfaces, supervisory systems, AI-driven decision tools, and interactive platforms enhance reliability, safety, sustainability, and adaptability across a wide spectrum of energy applications. It aims to bridge energy engineering with human-centered interface design by emphasizing the impact of dashboards, monitoring tools, intuitive controls, and visualization platforms; to demonstrate advanced control and artificial intelligence approaches in which human supervision complements automation for achieving fault tolerance, optimal performance, and efficient energy management; and to showcase emerging paradigms such as blockchain, virtual inertia, fractional control, cybersecurity frameworks, and smart interaction techniques tailored for increasingly autonomous systems. The book further seeks to integrate renewable energy technologies with HMI innovations by illustrating the importance of interactive supervision in photovoltaic, wind, hybrid, and microgrid systems, while providing a multidisciplinary resource that brings together electrical engineering, digital technologies, HMI, AI, and industrial applications. Ultimately, the motivation is to offer both conceptual and practical insights into how humans and machines co-operate, co-adapt, and co-evolve within highly autonomous and data-driven environments, contributing to the development of safe, efficient, resilient, and human-aware energy infrastructures.

TARGET AUDIENCE

The overarching objective of this book is to create a unified reference that highlights the essential role of Human–Machine Interaction (HMI) in modern and future energy infrastructures, presenting how intuitive interfaces, supervisory systems, AI-driven decision tools, and interactive platforms enhance reliability, safety, sustainability, and adaptability across a wide spectrum of energy applications. It aims to bridge energy engineering with human-centered interface design by emphasizing the impact of dashboards, monitoring tools, intuitive controls, and visualization platforms; to demonstrate advanced control and artificial intelligence approaches in which human supervision complements automation for achieving fault tolerance, optimal performance, and efficient energy management; and to showcase emerging paradigms such as blockchain, virtual inertia, fractional control, cybersecurity frameworks, and smart interaction techniques tailored for increasingly autonomous systems. The book further seeks to integrate renewable energy technologies with HMI innovations by illustrating the importance of interactive supervision in photovoltaic, wind, hybrid, and microgrid systems, while providing a multidisciplinary resource that brings together electrical engineering, digital technologies, HMI, AI, and industrial applications. Ultimately, the motivation is to offer both conceptual and practical insights into how humans and machines co-operate, co-adapt, and co-evolve within highly autonomous and data-driven environments, contributing to the development of safe, efficient, resilient, and human-aware energy infrastructures.

THE CHALLENGES

The development of modern energy systems that seamlessly integrate human–machine interaction presents a unique set of challenges that arise from increasing system complexity, rapid digitalization, and the growing reliance on intelligent automation. As renewable energy, electric mobility, decentralized grids, and cyber-physical infrastructures expand, operators must supervise systems that are more dynamic, data-intensive, and automated than ever before. One major challenge is ensuring that human operators remain capable of understanding and managing highly complex control architectures, especially as artificial intelligence and adaptive algorithms make decisions that may not always be transparent. Designing user interfaces that are intuitive, cognitively efficient, and capable of translating complex technical information into actionable insights is another significant difficulty. Additionally, as inverter-based renewable systems introduce issues such as low inertia, fluctuating energy supply, and fast transient dynamics, HMIs must support real-time situational awareness and rapid human intervention. The human role in system oversight is further challenged by cybersecurity risks: modern HMI, SCADA, and IoT-enabled platforms expand the attack surface, exposing the energy sector to malicious intrusions that can target both technical systems and human decision processes. Ensuring resilience, trust, and safe interaction within human-in-the-loop and machine-driven systems requires advanced cybersecurity, robust decision-support tools, and careful interface design. Moreover, achieving interoperability across heterogeneous technologies ranging from drones, underwater robots, and electric vehicles to hybrid renewable systems and blockchain-enabled energy markets demands standardized communication frameworks that still allow flexible human control. Balancing automation with human authority, ensuring transparency in AI-assisted decisions, preventing operator overload, and maintaining system stability under uncertainty all represent crucial challenges that must be addressed for future energy infrastructures to remain safe, efficient, and human centered.

OVERVIEW OF THE BOOK

This book " Bridging Human-Machine Interaction for Energy Infrastructure" contains 16 chapters covering different themes related Human-Machine Interaction for Energy Infrastructure, here is a summary of the likely chapter topics:

Chapter 1 bridges Human–Machine Interaction for Energy Infrastructure by examining electrical faults in Permanent Magnet Synchronous Machines used in electric vehicles and demonstrating how dq-modeling, fault analysis, and real-time monitoring through HMI platforms strengthen system observability. By evaluating machine behavior under standard driving cycles, the chapter underscores how human-centered monitoring improves safety, reliability, and early fault detection, making HMI an indispensable tool for modern electric mobility.

Chapter 2 contributes to bridging HMI with energy infrastructure by integrating ANN-based, P&O, and INC algorithms within supervisory platforms that enhance photovoltaic MPPT performance. It highlights how HMI-driven oversight allows operators to compare algorithms, adjust parameters, and validate real-time responses, ultimately fostering interactive, efficient, and human-aware solar energy environments.

Chapter 3 advances the theme by tracing the evolution of HMI from analog systems to AI-enhanced platforms used in contemporary energy control rooms. Through ISO and ergonomics-based evaluation, the chapter connects usability, cognitive load, and operator resilience to the design of future human-centered interfaces, illustrating how HMI becomes the bridge between complex digital operations and effective human decision-making.

Chapter 4 reinforces Human–Machine Interaction for energy systems by implementing MPPT algorithms on dSPACE DS1104 and enabling real-time interactive switching via an HMI interface. This human-in-the-loop capability illustrates how operator involvement enhances adaptability and responsiveness in photovoltaic systems, leading to more efficient and context-aware energy harvesting.

Chapter 5 directly supports the book's theme by developing an advanced supervisory interface in MATLAB App Designer for controlling hybrid PV–wind–battery systems. Through real-time monitoring, load management, and energy flow optimization, this chapter exemplifies how HMI enhances modern microgrid operation and fosters improved human–machine collaboration in distributed renewable systems.

Chapter 6 bridges HMI with secure machine interactions in energy markets by analyzing blockchain-based applications for trustworthy and autonomous transactions. The chapter explains how decentralized ledgers strengthen human–machine cooperation, reduce fraud, and protect P2P and M2M energy trading an essential component of future interactive infrastructures.

Chapter 7 advances the integration of HMI with control engineering by addressing inertia loss in inverter-dominated grids and showing how intuitive interfaces facilitate operator interaction with virtual inertia algorithms. By linking adaptive control with real-time decision support, the chapter demonstrates how HMI serves as a vital mediator for maintaining frequency stability.

Chapter 8 connects HMI principles to renewable-driven water pumping systems by modeling and optimizing a high efficiency PV water pumping solution equipped with vector control and field weakening. Interactive supervision tools highlighted in this chapter show how human operators can enhance performance, diagnostic issues, and tune system behavior in real-world agricultural applications.

Chapter 9 bridges HMI with emerging mobility infrastructures by reviewing AI-powered smart charging strategies for electric vehicles. Through human-centered decision interfaces, the chapter explores V2G, G2V, state estimation, and machine learning techniques, emphasizing how operator interaction strengthens reliability, transparency, and user trust in future smart charging ecosystems.

Chapter 10 strengthens the theme by exploring underwater robotics for offshore energy operations, integrating PID/FOPID comparisons with real-time HMI dashboards for 3D trajectory tracking. This interactive control approach demonstrates the essential role of HMI in extreme marine environments where automation and human supervision must work seamlessly together.

Chapter 11 contributes to the theme by modeling a complete wind energy conversion system covering aerodynamic, electromechanical, and grid-integration elements to establish the foundations for intelligent HMI supervision in wind farms. The chapter demonstrates how interactive platforms enhance monitoring, fault detection, and performance optimization in large-scale renewable installations.

Chapter 12 concludes the sequence by addressing cyber threats targeting HMI and human-in-the-loop systems in critical energy infrastructures. By offering mitigation strategies, it underscores the need for secure, resilient, and human-aware digital platforms, reinforcing the central thesis that bridging Human–Machine Interaction is essential for the safe evolution of modern energy systems.

AUDIENCE

This book is intended for a broad yet specialized audience engaged in the development, operation, and study of modern energy systems and human–machine interaction technologies, serving readers from both academic and industrial communities who seek to understand how intelligent interfaces, digital tools, artificial intelligence, and advanced control systems are shaping the next generation of energy infrastructures. The primary target audience includes researchers, scientists, and academics working in renewable energy systems, power electronics, robotics, electrical engineering, artificial intelligence, and cyber-physical systems who require an interdisciplinary reference that integrates HMI with advanced energy technologies, as well as graduate and postgraduate students specializing in energy engineering, automation, control systems, smart grids, and digital technologies, who can use this book as a learning resource for coursework, thesis research, and practical experimentation. It is equally vital for engineers, practitioners, and system operators responsible for designing, monitoring, and maintaining energy infrastructures, including photovoltaic plants, wind farms, electric vehicle charging systems, hybrid microgrids, and industrial automation platforms, as it provides practical insights into applying HMI tools for improved decision-making, reliability, and operational efficiency. Furthermore, the book addresses HMI designers, interface developers, and software engineers involved in building control dashboards, monitoring tools, and supervisory platforms for technical infrastructures, offering them guidance on integrating user-centered design with real-time control and data-driven algorithms, while also proving valuable for policy makers, energy planners, and technology analysts interested in understanding how human–machine interaction, cybersecurity, and intelligent systems influence the future of sustainable, decentralized, and resilient energy networks. By bridging advanced engineering with human-centered interaction, this book offers valuable knowledge for anyone involved in shaping the future of intelligent energy systems from conceptual design and experimental research to industrial deployment and strategic policy development.

Mohamed Naoui
National Engineering school of Gabes, University of Gabès, Tunisia

Lassaad Sbita
National Engineering school of Gabes, University of Gabès, Tunisia

Acknowledgment

As the editors, we hope that the chapters in this book will stimulate further research in modeling, analyzing, and studying modern energy systems and human–machine interaction technologies. We hope that this book, covering so many different aspects, will be valuable for all readers. We also want to thank the reviewers for their diligence in reviewing the chapters. Special thanks go to the IGI-Global publisher, especially the book editorial team.

Chapter 1
Impact of Electrical Faults on PMSM Performance in Electric Vehicles: HMI Monitoring and Modeling – FOC-Based Evaluation Under Standard Driving Cycles

Mabrouka Romdhane
https://orcid.org/0009-0005-5779-6518
Faculty of Sciences of Gafsa, University of Gafsa, Tunisia

Mohamed Naoui
https://orcid.org/0000-0001-9117-3510
Research Unit of Energy Processes Environment and Electrical Systems, National Engineering School of Gabes, Tunisia

Ali Mansouri
Higher School of Applied Sciences and Technology of Gafsa, Tunisia

ABSTRACT

This chapter explores Permanent Magnet Synchronous Machines, which are favored in electric vehicle traction for their efficiency, power density, and dynamic performance. Despite their advantages, PMSMs can experience faults that affect both operational safety and reliability. In this chapter, we begin by classifying common faults into two main categories: rotor-related faults and stator-related faults which are represented as electrical faults. Moreover, we introduce a detailed mathematical model of PMSMs using the direct-quadrature (dq) frame, which simulates both healthy and faulty conditions, enabling analysis of fault impacts on currents, voltages, and

DOI: 10.4018/979-8-3373-5621-1.ch001

torque. We further explore Field Oriented Control, and explain its key components such as PI controllers and coordinate transformations. Finally, a comparative study based on standard driving cycles (FTP, NEDC, WLTP) is carried out to illustrate how faults impact system performance, providing insights into early fault detection and supporting the development of robust monitoring and fault-tolerant systems.

INTRODUCTION

The 21st century has been characterized by profound technological advancements, particularly in the field of electric vehicles (EVs), which have rapidly evolved from a niche innovation into a cornerstone of sustainable transportation (Korneev et al., 2024). This transformation is driven by two major forces: the accelerated pace of technological progress in energy storage, power electronics, and control systems, and the global imperative to mitigate climate change through the reduction of greenhouse gas emissions. In this context, EVs are no longer regarded merely as alternatives to conventional internal combustion engine vehicles but are increasingly positioned as key enablers of the energy transition and the sustainable transformation of the mobility sector (EFEMWENKIEKIE et al., 2020; Kumar, 2024; Lau et al., 2022).

Automobile manufacturers, in response to both regulatory pressures and market demand, continue to invest heavily in the research, development, and commercialization of a wide spectrum of EV technologies. These efforts encompass improvements in battery chemistry to enhance energy density and charging efficiency, the integration of advanced control and connectivity features to enable higher degrees of autonomy, and the diversification of motorization strategies to balance performance, cost, and sustainability.

The growing emphasis on electric vehicles thus reflects a broader paradigm shift within the transportation industry one that is redefining design priorities, manufacturing processes, and energy infrastructures. EVs are not only reshaping the future of mobility but are also serving as a critical intersection between the automotive sector, renewable energy integration, and smart grid technologies, thereby solidifying their role as central actors in the global pursuit of a cleaner and more sustainable future.

Regarding electric propulsion systems, the continuous pursuit of efficiency, compactness, and reliability has established the permanent magnet synchronous motor (PMSM) as one of the most widely adopted solutions in modern electric vehicles (Özçiflikçi et al., 2024).

This preference is primarily due to its high power density, excellent efficiency, and superior torque-to-inertia ratio, which make it particularly well-suited for applications requiring dynamic performance and energy savings (Pravin Sankhwar, 2024). Despite these advantages, PMSMs are not immune to failures and are especially vulnerable to inter-turn short-circuit (ITSC) faults within the stator windings. These faults can arise from several factors, including mechanical stresses, insulation degradation due to excessive heating, partial discharges, and electrical overloads (Naoui et al., 2021).

Statistical studies confirm that ITSC faults represent a major reliability concern, accounting for nearly 38% of all motor failures in electrical machines (Wakefield,1998; Carlson,1988). The consequences of such faults extend beyond minor performance degradation; they can lead to severe electromagnetic and mechanical disturbances (Allal et al., 2024). For instance, impulsive short circuits generate strong magnetic counter-fields that may accelerate the demagnetization process of the permanent magnets. Furthermore, the abrupt onset of a short circuit produces a large alternating component in the electromagnetic torque, which not only degrades the smoothness of motor operation but can also induce mechanical stresses capable of damaging couplings, bearings, or even the rotor shaft (Kirsch, 1995).

Given the potential risks associated with ITSC faults, their early detection and accurate diagnosis are of paramount importance to guarantee the safe, reliable, and uninterrupted operation of electric vehicles. Robust fault detection strategies not only prevent catastrophic failures but also contribute to extending the lifespan of the motor, reducing maintenance costs, and enhancing overall system dependability (Zhao et al., 2021). Consequently, research in this field has increasingly focused on developing advanced monitoring and diagnostic techniques that can identify early indicators of ITSC faults, even under dynamic driving conditions, thereby ensuring that PMSMs continue to meet the stringent reliability requirements of electric mobility (Chen et al., 2022).

In order to address the problem of detecting inter-turn short-circuit (ITSC) faults, it is essential to first develop an accurate model of the PMSM under faulty conditions. Such a model serves as a prerequisite for understanding the complex interactions between the fault and the electrical as well as mechanical behavior of the machine. By reproducing the asymmetries and disturbances introduced by ITSC faults, the model provides valuable insights into how these defects manifest themselves in observable signals such as currents, voltages, torque, and speed.

This modeling step is crucial because it allows the creation of a virtual monitoring platform capable of simulating both healthy and faulty scenarios. Through this approach, it becomes possible to identify distinctive signatures of ITSC faults, which can later be exploited by diagnostic and monitoring algorithms. Moreover, modeling not only highlights the degradation of performance but also enables the evaluation of the impact of different fault severities and locations on the machine's dynamic response (Ahmed Baba et al., 2023).

Therefore, the implementation of a PMSM fault model constitutes the foundation for any advanced fault detection strategy. It bridges the gap between theoretical analysis and practical diagnosis by providing a controlled environment in which the effect of faults can be studied in detail without the risks and costs associated with experimental testing on real machines. This constitutes the main objective of the present chapter: to establish a robust modelling framework that captures the behaviour of the PMSM under ITSC faults, thereby laying the groundwork for the development of reliable monitoring and detection techniques.

This chapter is organized into four main sections, each designed to build a comprehensive understanding of fault behavior in permanent magnet synchronous motors (PMSMs) and their impact on control and performance.

The first section presents a detailed classification of electrical faults occurring in PMSMs, with a focus on faults affecting both the stator and the rotor. This includes common issues such as inter-turn short-circuits in stator windings, as well as demagnetization and rotor eccentricity. Each fault type is described in terms of its physical origin, typical manifestations, and potential impact on the motor's electromagnetic behavior and operational safety.

The second section introduces the mathematical modeling of the PMSM in the direct–quadrature (dq) reference frame. Here, both healthy and faulty operating conditions are represented.

The third section focuses on the principles and implementation of Field Oriented Control (FOC). As a widely adopted technique in modern PMSM drives, FOC ensures decoupled regulation of torque and flux, making it particularly effective for high-performance applications. In this part, we detail the structure of the FOC algorithm and its advantages.

Finally, the fourth section highlights the importance of MATLAB/Simulink as a human–machine interface (IHM), used not only for implementing control algorithms but also for monitoring and visualization. Within this environment, a comparative analysis of standardized driving cycles (FTP, NEDC, and WLTP) is performed under both healthy and inter-turn short-circuit (ITSC) conditions. This allows us to assess the impact of faults on key performance indicators such as speed regulation, torque ripple, and current distortion, while underlining the relevance of MATLAB/Simulink as a practical tool for supervision, fault analysis, and early detection.

Through this structure, the chapter establishes a logical progression from fault classification and modeling to control and performance evaluation, thereby providing a solid basis for subsequent research on diagnostic techniques and resilient control solutions for PMSM-based electric drives.

PMSM FAILURES CLASSIFICATION

1. Rotor failures

The rotating part of the motor contains permanent magnets that create a fixed magnetic field. Bearings are used to support the rotor and allow it to rotate smoothly within the stator. Different types of bearings, such as ball or roller bearings, are used, and ultimately, the shaft connects the rotor to the external load, which are the wheels and the vehicle chassis.

One of the most significant magnetic failures is demagnetization. A magnetic failure can significantly affect the performance and efficiency of the motor. In the typical operation of the PMSM, the electric current from the stator winding generates a magnetic field that opposes the residual induction of the permanent magnets. This recurring process leads to a demagnetization of the permanent magnets (Riba Ruiz et al., 2009).

The phenomenon of demagnetization primarily results from an induced reaction, particularly in situations requiring high torque. High temperatures, overcurrent, or exposure to powerful external magnetic fields are some of the many factors that can cause demagnetization (Zhao et al., 2022). When the motor is demagnetized, it loses its ability to generate the necessary magnetic field and reduces torque and efficiency (Usman & Rajpurohit, 2020). In the case where all the permanent magnets of the motor exhibit the same degree and type of demagnetization, it is referred to as uniform demagnetization. This occurs when the magnetization of the permanent magnets decreases over time. However, for partial demagnetization, only a set of magnets is affected by the defect (Krichen et al., 2020).

Eccentricity is another important failure can occur on PMSM rotor, it is defined as a non-uniform distribution of the air gap between the stator and rotor due to various deformations causing substantial damage to their components, due to an unbalanced load, poor alignment, incorrect installation, or a shaft that is not straight (Faiz & Nejadi-Koti, 2019). It is possible to infer three categories of eccentricity; either the axis of the rotor and the stator are not superimposed, and the rotor performs a rotation around its own axis called static eccentricity, or the case of dynamic eccentricity where the axis of the rotor and the stator do not coincide, and the rotor induces a simultaneous rotation of the axis of the stator and the rotor. Or even a combination of both static and dynamic eccentricity. In this situation, the rotor rotates around a third point that differs from both the axis of the stator and the axis of the rotor (Liu et al., 2021; Karami et al., 2021).

In addition to these latter defects, it is worth mentioning the existence of mechanical failures that can occur at the rotor level, such as bearing defects and unbalance and misalignment defects (Alqunun et al., 2025).

The bearings provide essential support to the high-speed rotating shaft of the PMSM motor. They are composed of the outer ring, the inner ring, and the balls held in a cage or rollers (Rosero et al., 2009). Potential factors contributing to bearing failures include high temperatures, inadequate lubrication, corrosion, foreign contaminants, improper installation, and the appearance of fluting (Ullah et al., 2020). Bearing defects induce disturbances in the rotational movement of the bearing. These anomalies generate non-uniform forces and lead to a significant increase in vibrational amplitudes. As a result, irregular friction and suboptimal mechanical interactions between the bearing elements lead to increased acoustic noise production. This correlation between bearing defects, vibrations, and acoustic noise constitutes a crucial aspect of the analysis of mechanical failures in PMSMs.

Rotor unbalance corresponds to a non-uniform distribution of mass around the axis of rotation. It can result from manufacturing defects, particle deposits, wear phenomena, or even improper assembly. This type of failure generates a periodic centrifugal force, responsible for radial vibrations proportional to the rotational speed, appearing mainly at the fundamental frequency (Li et al., 2021). Meanwhile, the Rotary misalignment arises when the rotor axis is not properly aligned with the axis of the driven load or coupling. It is generally classified into two types; parallel and angular and commonly results from improper assembly, mechanical deformation, or installation errors. Such misalignment imposes additional mechanical stresses on bearings and couplings, leading to complex vibration patterns characterized by harmonic components at twice the rotational frequency (Kumar et al., 2023).

2. Stator failures

PMSM electrical failures mainly consist of failures in the stator windings, which are considered the most common problems encountered in alternating current motors. According to statistical data from IEEE and EPRI, failures in the stator winding account for a range of 36% to 66% of all electric motor failures, with variations based on the type and size of the machine (Pietrzak & Wolkiewicz, 2022; Wireko-Brobby et al., 2023).

Stator winding electrical failures are referred to as inter-turn short circuits (ITSC), which are inherently challenging to diagnose and possess the potential to propagate to adjacent winding turns, thereby exacerbating fault severity and compromising machine reliability. If left untreated, this ITSC can quickly damage the subsequent coils and lead to more severe short circuits, such as phase-phase short circuits, or in the most severe case, an open circuit (Usman et al., 2019).

The predominant origin of insulation failures is often linked to abrasion induced by mechanical stress or to overheating of the windings, which may arise from excessive load demand or harsh environmental conditions (Fang et al., 2019). In this context, the underlying causes of short-circuit faults can be systematically classified into three main categories. Electrical factors include thermal overload, typically resulting from voltage amplitude variations, as well as partial discharge phenomena that progressively deteriorate insulation integrity. Mechanical factors encompass defects such as rotor eccentricity, malfunction or obstruction within the ventilation system, and operation under heavy-load starting conditions, all of which contribute to localized stress on the winding structure. Finally, environmental factors primarily elevated temperature and humidity accelerate insulation aging and further aggravate the probability of short-circuit initiation. This structured categorization highlights the multifaceted nature of fault mechanisms have been methodically categorized and synthesized in table (1).

Table 1. Factors of electrical stator failure

Electrical factors	Mechanical factors	Environmental factors
• Thermal overload due to voltage amplitude variation • Partial discharge	• Eccentricity defect • Exhaust in the ventilation system • Starting under heavy load	• Temperature • Humidity

Failure to detect and diagnose these faults immediately can lead to serious consequences, including increased current in the affected coil, causing localized overheating and potential damage to the insulation level. Given the precision required in the operation of PMSMs, the presence of an ITSC fault can compromise the performance, efficiency, and reliability of the motor. Rapid detection and diagnosis of ITSC faults in the PMSM stator winding is essential to implement corrective measures, prevent further damage, and ensure the continued optimal operation of these advanced electrical machines.

PMSM MATHEMATICAL MODELLING

1. Three-phase frame (abc) reference equations

The three-phase stator voltages, flux and currents are respectively written with the following vector notations: $[V_s],[R_s]$ and $[I_s]$. Figure 1 gives the representation of the windings for a three-phase permanent magnet synchronous machine in the electrical space in the three-phase frame (abc) reference and the two-phase frame (dq) reference (Khov, 2009):

Figure 1. PMSM stator winding representation

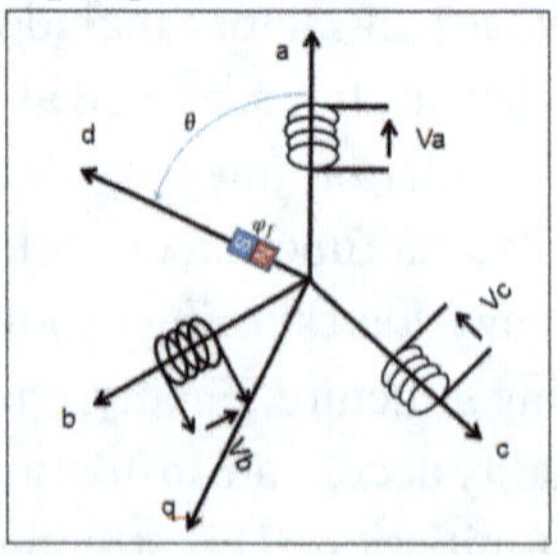

The voltage in the three phase reference frame is written in the equation (1) as:

$$[V_s] = [R_s][I_s] + \frac{d}{dt}[\varphi_s] \tag{1}$$

Where the stator phase voltage vector is presented by the equation (2) and the stator phase resistance is presented in equation (3).

$$[V_s] = [V_a \;\; V_b \;\; V_c]^T \tag{2}$$

$$[R_s] = \begin{bmatrix} R_s & 0 & 0 \\ 0 & R_s & 0 \\ 0 & 0 & R_s \end{bmatrix} \tag{3}$$

The current vector and the flux vector of stator phases are respectively expressed as given in equations (4) and (5).

$$[I_s] = [I_a \;\; I_b \;\; I_c]^T \tag{4}$$

$$[\varphi_s] = [\varphi_a \;\; \varphi_b \;\; \varphi_c]^T \tag{5}$$

The stator and rotor fluxes can be expressed by the magnetic equations (7) and (8) in terms of stator inductance matrix, absolute rotor position and flux peak value as:

$$[\varphi_s] = [L_{ss}][I_{ss}] + [\varphi_f] \tag{6}$$

$$\begin{cases} [\varphi_f] = \begin{bmatrix} \cos(\theta) \\ \cos(\theta - \frac{2\pi}{3}) \\ \cos(\theta - \frac{4\pi}{3}) \end{bmatrix} \\ [L_{ss}] = [L_{s0}] + [L_{s2}(\theta)] \end{cases} \tag{7}$$

Introducing equation (6) into the electrical equation became:

$$[V_s] = [R_s][I_s] + \frac{d}{dt}\left([L_{ss}][I_s] + [\varphi_f]\right) \tag{8}$$

And the mechanical equation is presented as flow:

$$C_{em} - C_r = J.\frac{d}{dt}\omega_r + F.\omega_r \tag{9}$$

The resulting differential equations are nonlinear with time-varying coefficients. To address this complexity, we resort to the use of orthogonal models for the machine, employing two distinct models that position the mutually coupled windings on two orthogonal axes. These two models include: The Park (dq) model and the Concordia (α, β) model.

2. (Dq) model

The Park transformation is a mathematical tool consists of transforming the balanced three-phase PMSM representation into an equivalent two-phase representation characterized by two d-q axes conserving the magnetomotive forces and instantaneous power(Azom et al., 2025).

This transformation aligns the d-axis with the rotor flux or synchronous reference, while the q-axis is orthogonal to it.

The equations that translates the transition from the three-phase system to the two-phase system (dq) is called the direct Park transformation is given by equations (10) and (11).

$$[X_{dq}] = [P(\theta)][X_{abc}]^T \tag{10}$$

$$[P(\theta)] = \sqrt{\frac{2}{3}} \begin{bmatrix} \cos(\theta) & \cos\left(\theta - \frac{2\pi}{3}\right) & \cos\left(\theta + \frac{2\pi}{3}\right) \\ -\sin(\theta) & -\sin\left(\theta - \frac{2\pi}{3}\right) & -\sin\left(\theta + \frac{2\pi}{3}\right) \\ \sqrt{2}/2 & \sqrt{2}/2 & \sqrt{2}/2 \end{bmatrix} \tag{11}$$

The equation (12) describing the transformation from the two-phase frame (dq) reference to the three-phase system (abc) which is the invers Park transformation.

$$[X_{abc}] = [P^{-1}(\theta)][X_{dq}]^T \tag{12}$$

And $[P^{-1}(\theta)]$ is the inverse of Park's transition matrix$[P(\theta)]$.

We also recall that the Concordia transformation is employed to convert the three-phase system (abc) into an orthogonal frame (α,β,0) reference (Naoui et al., 2021) . This transformation is commonly applied to voltages, currents, and fluxes in order to simplify their mathematical representation. Considering $\theta = 0$ in equation (12) yields the formulation of the Concordia transformation direct matrix:

$$\begin{bmatrix} X_\alpha \\ X_\beta \\ X_0 \end{bmatrix} = [C] \begin{bmatrix} X_a \\ X_b \\ X_c \end{bmatrix} \tag{13}$$

$$[C] = \sqrt{\frac{2}{3}} \begin{bmatrix} 1 & -\frac{1}{2} & -\frac{1}{2} \\ 0 & \frac{\sqrt{3}}{2} & -\frac{\sqrt{3}}{2} \\ \frac{1}{\sqrt{2}} & \frac{1}{\sqrt{2}} & \frac{1}{\sqrt{2}} \end{bmatrix} \tag{14}$$

The inverse Concordia transformation is defined by the equation (15), Where $[C]^{-1}$is the invers matrix of $[C]$.

$$\begin{bmatrix} X_a \\ X_b \\ X_c \end{bmatrix} = [C]^{-1} \begin{bmatrix} X_\alpha \\ X_\beta \\ X_0 \end{bmatrix} \tag{15}$$

The equation (16) defines the mathematical transformation linking the (d,q) and (α,β) reference frames.

$$\begin{bmatrix} X_d \\ X_q \end{bmatrix} = [R(\theta)] \begin{bmatrix} X_\alpha \\ X_\beta \end{bmatrix} \tag{16}$$

Where R(θ) is the transformation matrix expressed by the equation (17).

$$[R(\theta)] = \begin{bmatrix} \cos(\theta) & -\sin(\theta) \\ \sin(\theta) & \cos(\theta) \end{bmatrix} \tag{17}$$

Finally, by applying the Park transformation matrix, the PMSM equations can be reformulated in the (d,q) frame reference, as expressed in equation (18):

$$\begin{cases} V_d = R_s I_d + \frac{d\varphi_d}{dt} - p.\omega_r.\varphi_q \\ V_q = R_s I_q + \frac{d\varphi_q}{dt} - p.\omega_r.\varphi_d \end{cases} \tag{18}$$

And the flux is expressed by:

$$\begin{cases} \varphi_d = L_d I_d + \varphi_f \\ \varphi_q = L_q I_q \end{cases} \tag{19}$$

The electromagnetic equation becomes:

$$C_{em} = p.\left[\left(L_d - L_q\right) I_d.I_q + \varphi_f.I_q\right] \tag{20}$$

3. State space representation

a. Healthy PMSM representation

In order to establish the state representation of the permanent magnet synchronous machine (PMSM) in the (d,q) frame, the Park transformation was applied. Given that the permanent magnet synchronous machine (PMSM) is voltage-fed, the voltages $\left(V_d, V_q\right)$ as well as the excitation flux φ_f have been chosen as control variables, while the stator currents $\left(I_d, I_q\right)$ have been selected as state variables.

The state space representation of our system is written under the general structure of the following state equation:

$$\begin{cases} \frac{d[X]}{dt} = [A][X] + [B][U] \\ [Y] = [C][X] \end{cases} \tag{21}$$

The specific model considered in this work is represented by:

$$\begin{cases} \frac{d}{dt}\begin{bmatrix} I_d \\ I_q \end{bmatrix} = \begin{bmatrix} -\frac{R_s}{L_d} & p\omega_r\frac{L_q}{L_d} \\ -p\omega_r\frac{L_q}{L_d} & -\frac{R_s}{L_d} \end{bmatrix}\begin{bmatrix} I_d \\ I_q \end{bmatrix} + \begin{bmatrix} \frac{1}{L_d} & 0 & 0 \\ 0 & \frac{1}{L_q} & 0 \\ 0 & 0 & -p\frac{\omega_r}{L_q} \end{bmatrix}\begin{bmatrix} V_d \\ V_q \\ \varphi_f \end{bmatrix} \\ \begin{bmatrix} I_d \\ I_q \end{bmatrix} = \begin{bmatrix} 1 & 0 \\ 0 & 1 \end{bmatrix}\begin{bmatrix} I_d \\ I_q \end{bmatrix} \end{cases} \quad (22)$$

b. PMSM modelling under Inter turn short circuit fault

Figure 2 illustrates the modelled representation of the PMSM under fault conditions, where an inter-turn short circuit is assumed to occur in phase A. In this model, the fault is emulated by incorporating a short-circuit resistance R_f into the equivalent circuit (Romdhane et al., 2023).

Figure 2. equivalent circuit of the inter-turn short circuit in phase A

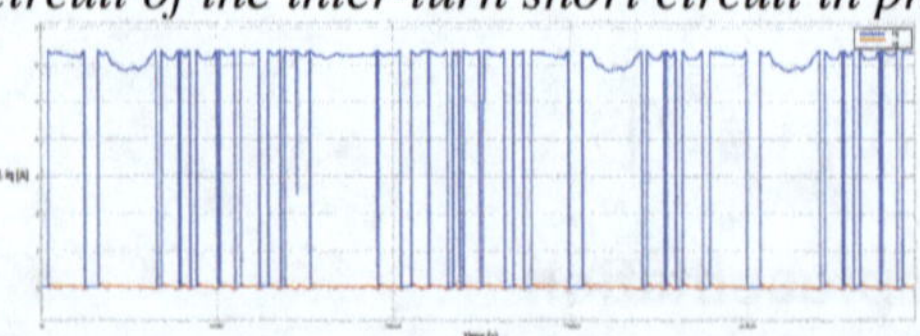

The fault resistance R_f serves as a critical parameter that quantifies the severity of the insulation defect within the stator winding. In a healthy phase, the insulation is intact, effectively resulting in an infinitely high resistance that prevents any current leakage between turns. However, when an inter-turn short circuit occurs, this insulation breaks down locally, causing R_f to drop drastically towards zero. This drastic reduction in resistance facilitates the flow of unintended fault currents between the shorted turns, leading to significant disturbances in the motor's electromagnetic behavior. The variation of R_f from infinite to near zero thus provides a practical means to model different fault severities, ranging from minor insulation degradation to a complete short circuit. Understanding and accurately representing this resistance in the equivalent circuit is essential for simulating fault effects and developing effective fault detection and mitigation strategies.

The winding affected by the fault is consequently subdivided into two distinct sections (J.Zang et al., 2019): a_1 denotes the healthy portion of the winding, while a_2 corresponds to the short-circuited subsection. The parameters R_{a1} and L_{a1} represent the resistance and inductance of the healthy sub-winding, whereas R_{a2} and L_{a2} denote those of the faulty sub-winding. Additionally, the mutual inductance values between the sub-winding of phase A and that of phase Y are characterized by the parameters M_{xy}.

This subdivision is essential for accurately capturing the electrical behavior of the motor under fault conditions. By distinguishing between the healthy and defective parts of the winding, the model can simulate the uneven distribution of currents and magnetic coupling effects caused by the inter-turn short circuit. This approach enables a more precise representation of the fault's impact on the motor's performance, facilitating the development of effective fault detection and mitigation strategies.

To characterize the extent of the inter-turn short circuit, we introduce the coefficient μ in equation (23), which expresses the ratio between the number of short-circuited turns and the total number of turns per phase. In this context, N_f refers to the number of turns in the sub-winding a_2 involved in the fault, while N_s corresponds to the total number of turns in one phase.

$$\mu = \frac{N_f}{N_s} \tag{23}$$

As illustrated in Figure 2, the equality $M_{a2b} = M_{a2c}$ arises from the fact that the axes of sub-windings a_1 and a_2 are aligned with the main axis A. Under this consideration, the complete voltage equations of the faulty PMSM can be formulated in matrix form as:

$$\begin{bmatrix} V_a \\ V_b \\ V_c \\ 0 \end{bmatrix} = [R]\begin{bmatrix} i_a \\ i_b \\ i_c \\ i_f \end{bmatrix} + [L]\frac{d}{dt}\begin{bmatrix} i_a \\ i_b \\ i_c \\ i_f \end{bmatrix} + \begin{bmatrix} e_a \\ e_b \\ e_c \\ -e_{a_2} \end{bmatrix} \tag{24}$$

Where;

$$[R] = \begin{bmatrix} R_s & 0 & 0 & -R_{a_2} \\ 0 & R_s & 0 & 0 \\ 0 & 0 & R_s & 0 \\ -R_{a_2} & 0 & 0 & R_{a_2} + r_f \end{bmatrix} \tag{25}$$

$$[L] = \begin{bmatrix} L_s & 0 & 0 & -L_{a_s} - M_{a_1a_2} \\ 0 & L_s & 0 & -M_{a_2b} \\ 0 & 0 & L_s & -M_{a_2b} \\ -(L_{a_s} + M_{a_1a_2} - M_{a_2b}) & 0 & 0 & L_{a_2} \end{bmatrix} \tag{26}$$

In this formulation, i_a, i_b and i_c denote the instantaneous stator phase currents, while V_a, V_b and V_c correspond to the respective stator phase voltages applied to the PMSM. The parameter R_s represents the stator winding resistance, accounting for the ohmic losses in each phase, whereas L_s denotes the mutual inductance between the stator windings, which governs the magnetic coupling among the phases. The terms e_a, e_b and e_c refer to the electromotive forces (EMFs) induced in each phase as a result of the rotor's motion within the magnetic field. Finally, if designates the circulating fault current in the short-circuited portion of the winding, whose magnitude is directly influenced by the short-circuit fault resistance and, consequently, by the severity of the inter-turn short-circuit (ITSC) defect. These definitions collectively enable a precise mathematical representation of the PMSM's electrical behavior under fault conditions, serving as the basis for the voltage equation matrix shown in Figure 2.

The expression for the electromagnetic torque C_{em}, which represents the mechanical output generated by the interaction between the stator's magnetic field and the rotor's magnetic poles, can be derived from the machine's electromechanical energy conversion principles. This torque, together with the associated mechanical motion of the rotor, is governed by the fundamental mechanical dynamic equation of the PMSM. These relationships, which link the electrical and mechanical domains, are formulated in equation (27):

$$\begin{cases} C_{em} = \dfrac{e_a.i_a + e_b.i_b + e_c.i_c - e_f.i_f}{\omega_r} \\ J.\dfrac{d\omega_r}{dt} - F.\omega_r = C_{em} - C_r \end{cases} \tag{27}$$

Considering that the studied machine adopts a star connection for its three-phase stator windings, and that the neutral point is not externally connected, the sum of the instantaneous phase currents must be equal to zero according to Kirchhoff's current law. This configuration imposes a specific constraint on the phase currents, reducing the number of independent electrical variables in the system and thereby influencing the formulation of the voltage equations.

Mathematically, this condition can be expressed as:

$$i_a + i_b + i_c = 0 \tag{28}$$

This relation is systematically applied in the subsequent modeling steps to simplify the state-space representation of the PMSM under both healthy and fault conditions.

By expressing the stator currents initially defined in the three-phase (a, b, c) reference frame in terms of their corresponding components within the stationary two-axis (α, β) reference frame, it enables the system equations to be reformulated in a more concise and tractable form. This coordinate transformation facilitates the derivation of a comprehensive matrix model for the Permanent Magnet Synchronous Motor (PMSM) that explicitly incorporates the effects of an inter-turn short-circuit fault. Consequently, the dynamic behavior of the faulty PMSM can be succinctly represented by the matrix expression in equation (29):

$$\begin{bmatrix} V_\alpha \\ V_\beta \\ 0 \end{bmatrix} = \begin{bmatrix} R_s & 0 & -R'_{a_2} \\ 0 & R_s & 0 \\ -R'_{a_2} & 0 & R'_f \end{bmatrix} \begin{bmatrix} i_\alpha \\ i_\beta \\ i_f \end{bmatrix} + \begin{bmatrix} L_s & 0 & M_{f_a} \\ 0 & L_s & 0 \\ M_{f_a} & 0 & L_s \end{bmatrix} \frac{d}{dt} \begin{bmatrix} i_\alpha \\ i_\beta \\ i_f \end{bmatrix} + \begin{bmatrix} e_\alpha \\ e_\beta \\ -e_f \end{bmatrix} \tag{29}$$

Where,

$$\begin{cases} R'_{a_2} = \sqrt{\frac{2}{3}} R_{a_2}, R'_f = R_{a_2} + r_f, e_f = e_{a_2} \\ M_{f_a} = -\sqrt{\frac{2}{3}} \left(L_{a_2} + M_{a_1 a_2} - M_{a_2 b} \right) \\ M_{f_\beta} = 0 \end{cases} \tag{30}$$

Based on the proposed model, the electromagnetic torque is given by the equation (31):

$$C_{em} = \frac{e_\alpha . i_\alpha + e_\beta . i_\beta - e_f . i_f}{\omega_r} \tag{31}$$

Given that the analyzed PMSM includes a notch per pole per phase, the equality $M_{a2b} = M_{a2c}$ holds, enabling the simplification of equation (21) as follows:

$$\frac{d}{dt} \begin{bmatrix} i_\alpha \\ i_\beta \\ i_f \end{bmatrix} = \begin{bmatrix} L_s & 0 & M_{f_a} \\ 0 & L_s & 0 \\ M_{f_a} & 0 & L_{a_2} \end{bmatrix}^{-1} \left(-\begin{bmatrix} R_s & 0 & -R'_{a_2} \\ 0 & R_s & 0 \\ -R'_{a_2} & 0 & R'_f \end{bmatrix} \begin{bmatrix} i_\alpha \\ i_\beta \\ i_f \end{bmatrix} + \begin{bmatrix} V_\alpha \\ V_\beta \\ 0 \end{bmatrix} - \begin{bmatrix} e_\alpha \\ e_\beta \\ -e_f \end{bmatrix} \right) \tag{32}$$

The subsequent analysis focuses on the formulation of the state vector for the faulty Permanent Magnet Synchronous Motor (PMSM). Building upon the previously established general framework, the dynamic behavior of the system under fault conditions is represented within the two-phase (dq) reference frame, which facilitates the decoupling of electromagnetic variables and thereby simplifies the analytical process. In accordance with the structure delineated by equation (18), the vectors comprising the state representation of the system in the (dq) frame are expressed in the equations (33) and (34) as follows:

$$[X] = \begin{bmatrix} i_\alpha \\ i_\beta \\ i_f \end{bmatrix} \tag{33}$$

$$[U] = \begin{bmatrix} V_\alpha \\ V_\beta \\ 0 \end{bmatrix} - \begin{bmatrix} e_\alpha \\ e_\beta \\ e_f \end{bmatrix} \tag{34}$$

And

$$A = -\begin{bmatrix} L_s & 0 & M_{f_a} \\ 0 & L_s & 0 \\ M_{f_a} & 0 & L_s \end{bmatrix}^{-1} \begin{bmatrix} R_s & 0 & -R'_{a_2} \\ 0 & R_s & 0 \\ -R'_{a_2} & 0 & R'_f \end{bmatrix} \tag{35}$$

$$B = \begin{bmatrix} L_s & 0 & M_{f_a} \\ 0 & L_s & 0 \\ M_{f_a} & 0 & L_{a_2} \end{bmatrix}^{-1} \tag{36}$$

FILED ORIENTED CONTROL MODEL

1. FOC principal

The flux orientation control (FOC) strategy also known as vector control of the PMSM provides dynamic control of rotor speed and torque due to its simplicity and regulation efficiency (Djelamda & Bouchareb, 2022; Wireko-Brobby et al., 2023). This control method involves the measurement of stator phase currents and motor speed. The following figure represents a general description of the FOC control associated with the PMSM machine under study.

Figure 3. General structure of the Vector Control

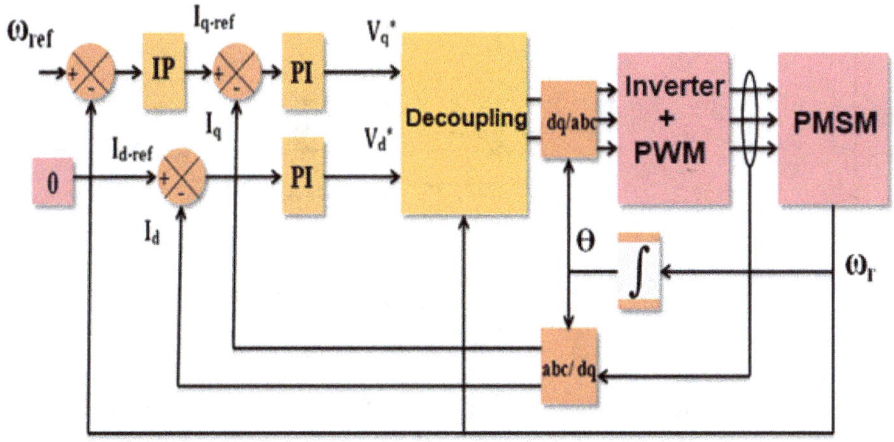

The FOC control of the MSAP is carried out in the (d,q) frame where I_{d_ref} is the current reference and is kept at zero. While I_{q_ref} is limited to the maximum current which is used as the torque reference C_{em}. The system is equipped with an IP speed control loop to produce the I_{q_ref} current, then PI regulators are used to ensure the proximity of the measured values to the reference values.

The outputs of the current regulation Id^* and Iq^*, after decoupling and passing into the three-phase (abc) frame reference, serve as voltage references and then determine the switching of the power supply devices in the voltage source inverter (Lammouchi et al., 2025).

2. Inverter control

The speed regulation of PMSM requires the use of a voltage inverter to convert the direct current U_0 into an alternating current. The inverter in our study consists of three arms, each arm involves two IGBT power switches with diodes in antiparallel.

Series of openings and closings of these switches cause commutations at the level of the current of the phases of the PMSM in order to return to a three-phase system where the voltages and currents are controllable.

The following figure shows an overall illustration of the inverter-machine assembly where the PMSM machine is connected in star.

Figure 4. The voltage Inverter structure

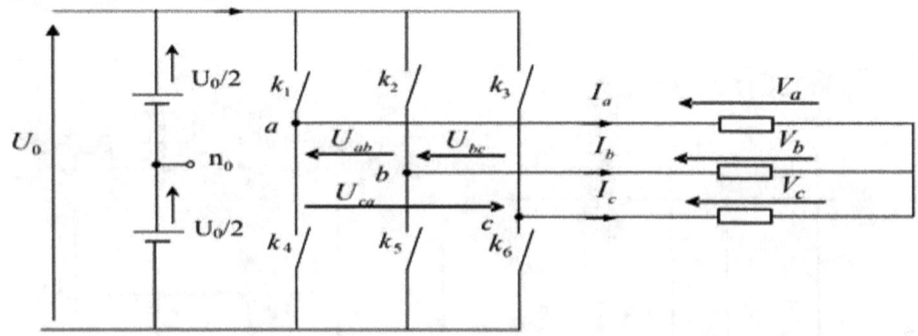

Each arm thus presents two distinct states. These two states can be perceived as a Boolean variable (1 for closed switch and 0 for open switch) and point n_0 represents the potential reference. The definition of the switching function is expressed in equation (37):

$$\begin{cases} k_4 = 1 - k_1 \\ k_5 = 1 - k_2 \\ k_6 = 1 - k_3 \end{cases} \tag{37}$$

The principle can be summarized in the equation (38):

$$if \begin{cases} K_1 = 1 \text{ and } K_2 = 0 \text{ then } V_{a0} = \frac{U_0}{2} \\ K_1 = 0 \text{ and } K_2 = 1 \text{ then } V_{a0} = -\frac{U_0}{2} \end{cases} \tag{38}$$

Applying the same logic for V_{b0} and V_{c0} we obtain the inverter input equations expressed by:

$$\begin{cases} V_{a0} = (2k_1 - 1).\frac{V_0}{2} \\ V_{b0} = (2k_2 - 1).\frac{V_0}{2} \\ V_{c0} = (2k_3 - 1).\frac{V_0}{2} \end{cases} \tag{39}$$

The compound voltages as a function of the input voltages are expressed in equation (40).

$$\begin{cases} V_{ab} = V_{a0} - V_{b0} = (k_1 - k_2).U_0 \\ V_{bc} = V_{b0} - V_{c0} = (k_2 - k_3).U_0 \\ V_{ca} = V_{c0} - V_{a0} = (k_3 - k_1).U_0 \end{cases} \quad (40)$$

And the balanced tensions are written:

$$\begin{bmatrix} V_a \\ V_b \\ V_c \end{bmatrix} = \frac{1}{3}U_0 \begin{bmatrix} 2 & -1 & -1 \\ -1 & 2 & -1 \\ -1 & -1 & 2 \end{bmatrix} \begin{bmatrix} k_1 \\ k_2 \\ k_3 \end{bmatrix} \quad (41)$$

The inverter's switching management is carried out using the pulse width modulation (PWM) technique, allowing the pulse width to be regulated to adjust the system output while allowing the variation of the output voltage frequencies.

This technique is based on the comparison between the modulating signal which is the reference signal in the form of a sine wave at fundamental frequency and the carrier which is a high frequency triangular wave. The switching times of the inverter switches are identified by the intersection points between the carrier and the modulation, while the switching frequency of the switches is established by the carrier frequency (Chethan & Kodeeswara kumaran, 2019; Khalid et al., 2020). At any given moment, one of the two switches on each arm is on and the other is off. At any given moment, one of the two switches on each arm is on and the other is off.

Compensation aims to decouple the d and q components, which facilitates the expression of the machine equations and the regulation part in a simplified manner, thus facilitating the calculation of the regulator coefficients (Khlaief, 2012).

3. Current regulator model

PID (Proportional, Integral, Derivative) regulators represent essential components in the creation of regulators as they have the advantage of being simple and easily executable.

For our system, we have chosen a P.I (proportional, integral) regulator, where the integral action aims to minimize the difference between the setpoint and the regulated value. In parallel, the proportional action allows for adjusting the system's responsiveness.

The figure below illustrates the block diagram of the regulation of I_d and I_q:

Figure 5. Block diagram of direct and quadratic current regulation

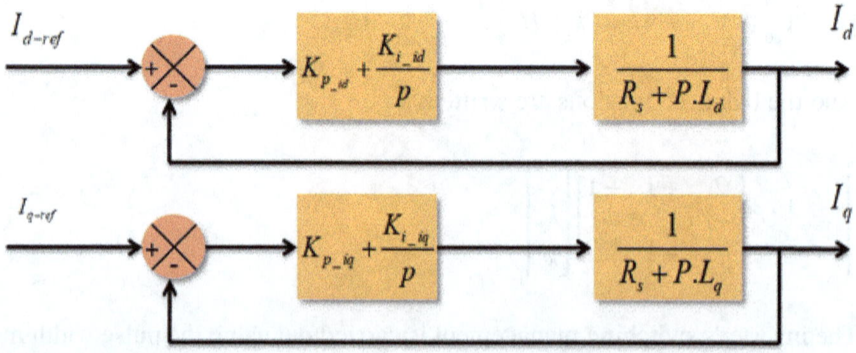

To determine the coefficients of the current regulators K_{p_id}, K_{i_id}, K_{p_iq} and K_{i_iq}, the continuous linear models and the pole compensation method of the closed-loop transfer function are used.

4. Speed regulator model

Rotor speed control is ensured by an IP type regulator whose gains are respectively $K_{i\_\omega r}$ and $K_{p\_\omega r}$. The following figure represents the block diagram of the speed regulation.

Figure 6. Speed regulation block diagram

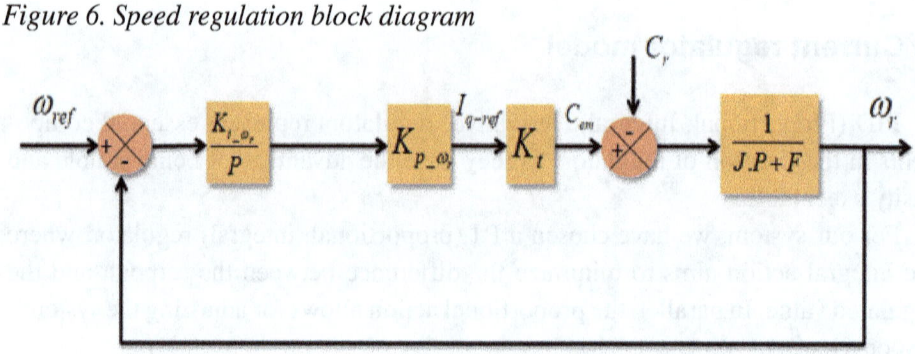

According to the functional diagram of the speed regulation, we observe that:

$$\omega_r = \frac{1}{Jp + F}\left(C_{em} - C_r\right) \tag{42}$$

Where,

$$\begin{cases} C_{em} = K_t . I_{q-ref} \\ K_t = p . \varphi_f \end{cases} \quad (43)$$

The closed-loop transfer function of the IP speed regulator gives a second order dynamic expressed by the equation (44):

$$TF_{v-bf} = \frac{\omega_r}{\omega_{ref}} = \frac{{\omega_0}^2}{P^2 + 2\xi\omega_0 P + {\omega_0}^2} \quad (44)$$

Where,

$$\begin{cases} \omega_0 = \sqrt{\dfrac{K_{p\_\omega_r} . K_{i\_\omega_r} . K_t}{J}} \\ 2\xi\omega_0 = \dfrac{F + K_{p\_\omega_r} . K_t}{J} \end{cases} \quad (45)$$

The speed regulator coefficients $K_{i\_\omega r}$ and $K_{p\_\omega r}$ can be calculated, a damping coefficient of 0.7 is chosen and the pulsation of the oscillations is set according to the desired dynamics.

SIMULATION BASED EVALUATION USING MATLAB/ SIMULINK

1. Matlab/ Simulink as Human-Machine Interface for supervision and simulation

The exploitation of the simplicity of the field-oriented control (FOC) strategy applied to the permanent magnet synchronous motor (PMSM) has been validated through simulations conducted under MATLAB/Simulink in this chapter. Beyond its main role as a numerical modeling and simulation platform, MATLAB/Simulink has been used as a real human-machine interface (HMI) allowing the monitoring, interaction and intuitive visualization of the different phenomena studied (Youssef et al., 2023).

This dual role is particularly relevant: on the one hand, it ensures the efficient implementation of the FOC control algorithm as well as the automatic and accurate calculation of the associated parameters; on the other hand, it offers the user the possibility to monitor in real time the dynamic responses of the PMSM, the behaviors in the presence of faults, and the system performances under various operating conditions. Thus, the simulation environment becomes not only a theoretical evaluation tool, but also a space for interactive experimentation where the user can modify the test conditions, introduce disturbances or analyze the influence of different load profiles. One of the strengths of MATLAB/Simulink is its ability to provide advanced graphical interfaces in the form of interactive dashboards, scopes (virtual oscilloscopes) and customizable visualization blocks. These tools facilitate the reading of electrical and mechanical quantities (currents, voltages, flux, speed, torque), frequency spectral analysis, as well as the evaluation of control robustness. The user thus has a user-friendly HMI that simplifies the understanding of complex results and promotes better decision-making during performance analysis.

Furthermore, integrating HMI into the simulation environment bridges the gap between theoretical modeling and practical evaluation. It allows emulating the role of a virtual experimental test bench, where system responses are observed in a near-real way, reducing the dependence on expensive prototypes or restrictive physical experiments. This approach also contributes to training and research, as it provides clear and interactive didactic support, facilitating the understanding of vector control concepts by students and researchers(Eom & Lee, 2022; Paul, 2021).

Thus, MATLAB/Simulink, as an HMI, is not limited to being a simple simulation tool, but becomes a true digital supervision system that integrates calculation, control and visualization, guaranteeing an exhaustive analysis of the performance of the PMSM under FOC control.

2. Drive cycles test

Standardized driving cycles are essential tools for assessing the energy and environmental performance of vehicles, as they allow for objective comparisons of fuel consumption and pollutant emissions under reproducible conditions. Among the most widely used are the Federal Test Procedure (FTP), the New European Driving Cycle (NEDC), and the Worldwide Harmonized Light Vehicle Test Procedure (WLTP). Although they share a common objective: testing energy efficiency and emissions, each of these cycles has distinct characteristics that influence the results obtained. The difference between these driving cycles lies in the realism of the cycles. NEDC is criticized for not accurately representing real-life driving conditions, while FTP and WLTP aim to provide more representative test scenarios (Martyushev et al., 2023; Singirikonda & Yeddula Pedda, 2023).

The main difference between these protocols lies in their degree of realism compared to everyday driving conditions. The NEDC, developed in the 1980s, has long been the benchmark in Europe, but it has been criticized for its simplicity and lack of representativeness: accelerations are low, speeds relatively low, and driving phases vary little, leading to an underestimation of real-life consumption and emissions. In contrast, the FTP cycle, mainly used in the United States, offers a driving profile closer to urban conditions, incorporating multiple stops, starts, and speed variations. Finally, the WLTP, introduced more recently internationally, aims to harmonize test procedures across countries and more accurately reflect modern driving habits, characterized by higher speeds, more dynamic acceleration, and a variety of road conditions.

Another differentiating aspect concerns the duration of the tests. The WLTP lasts approximately 30 minutes, which is significantly longer than the FTP cycle (23 minutes) and the NEDC (20 minutes). This increased duration makes it possible to cover a wider range of driving situations and to obtain a more complete and realistic assessment of a vehicle's energy performance. Furthermore, the maximum and average speeds achieved in the WLTP are higher than those of the NEDC and FTP, better reflecting the demands of modern traffic, particularly on motorways (Sandrini et al., 2024; Ng & Tong, 2024).

The representation of the simulation results of the permanent magnet synchronous machine, controlled by the Field-Oriented Control (FOC) strategy, is illustrated respectively in Figures 7, 8 and 9, corresponding to the FTP, NEDC and WLTP driving cycles. These representations highlight the adaptation of our model to the different standardized driving profiles and demonstrate the ability of the control system to satisfactorily follow the dynamic loads imposed by each cycle. We thus observe a significant correlation between the evolution of the reference speeds and torques resulting from the cycles and the actual responses of the simulated machine, which confirms the robustness and relevance of the vector control approach used.

Figure 7. FTP drive cycle test

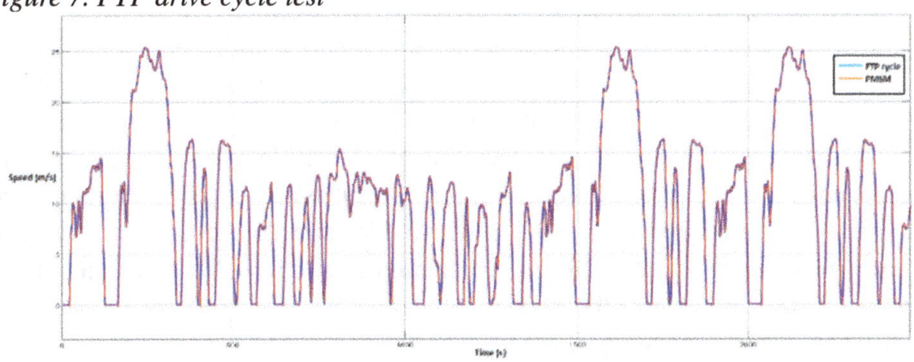

Figure 8. NEDC drive cycle test

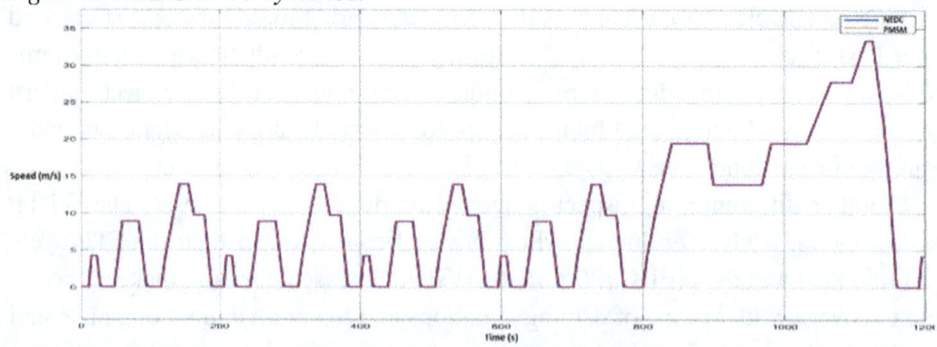

Figure 9. WLTP drive cycle test

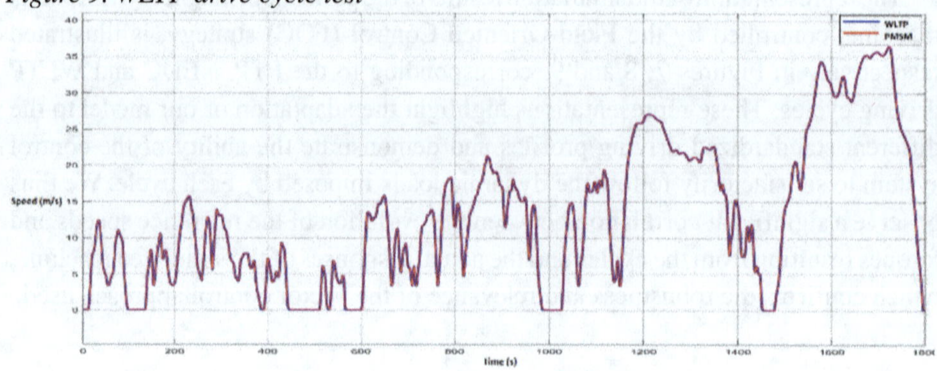

Furthermore, figures 10, 11 and 12 illustrate in detail the behavior of the stator currents, and more specifically that of the I_d component. It is clearly observed that this direct current remains maintained at a zero value throughout the simulation, independently of the applied driving cycle. This result is particularly important since it reflects compliance with the FOC strategy which aims to decouple the magnetic flux from the electromagnetic torque. Maintaining $I_d = 0$implies that all the stator current is used to generate the torque via the quadrature component I_q, thus optimizing the energy efficiency of the machine and minimizing losses.

Figure 10. FTP Current test

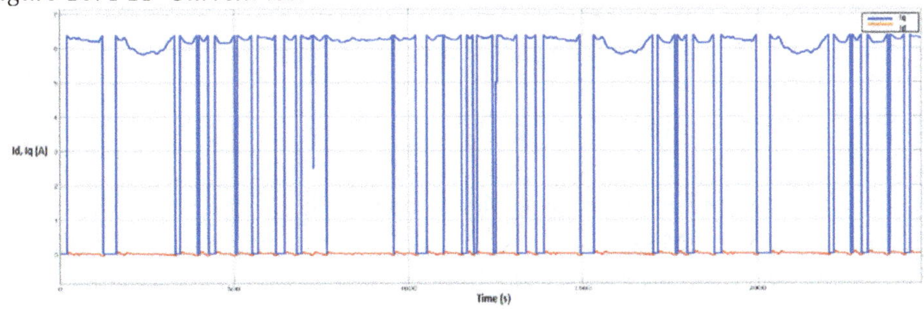

Figure 11. NEDC Current test

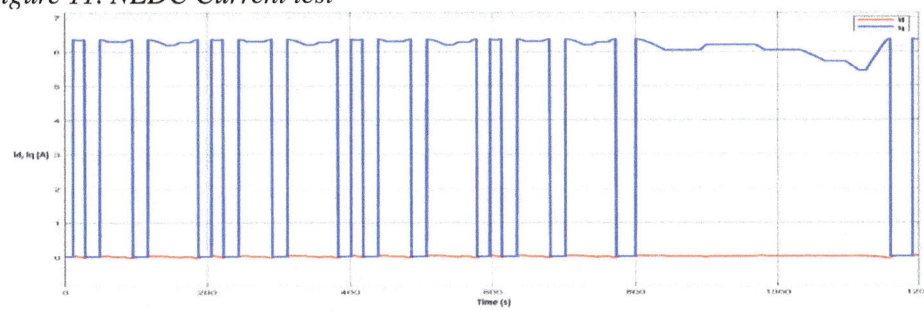

Figure 12. WLTP Current test

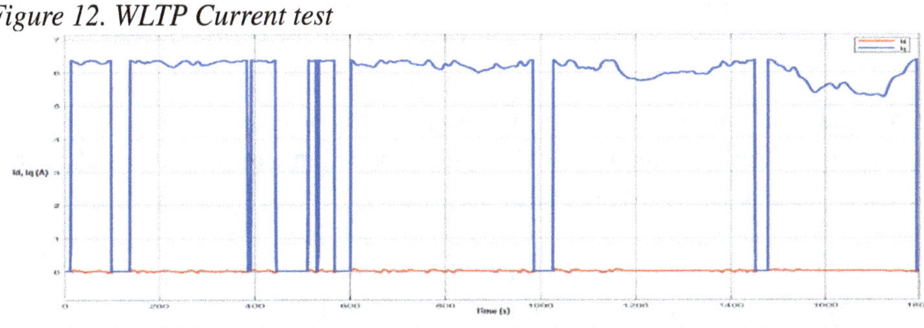

These results highlight not only the compliance of the control with the FOC theory, but also its ability to ensure constant and reliable performance even in very different driving scenarios: urban conditions with frequent stops (FTP), standardized but simplified cycles (NEDC), and finally modern and dynamic cycles that better represent reality (WLTP). In short, the simulation confirms that the FOC strategy applied to our machine guarantees efficient torque regulation and optimal use of stator currents, regardless of load and driving profile variations.

3. ITSC presence test

The simulation presented in this section focuses on the studied permanent magnet synchronous motor (PMSM), which is powered by a pulse-width modulated (PWM) voltage inverter. This power supply configuration is chosen due to its ability to provide a perfectly modulated three-phase voltage, thus ensuring precise control and better quality of the current wave injected into the machine. To manage the PMSM operation, a field-oriented vector control (FOC) strategy is implemented. This technique is widely recognized for its ability to decouple flux and torque control, resulting in fast dynamics, efficient speed regulation, and is introduced at phase A of the machine at time t=0.2 s. The speed setpoint is fixed at 106 rpm, and the simulation lasts for a duration of 0.5 s. optimized energy efficiency. The machine is subjected to a load of 1 N.m from the start, and a short circuit

In this simulation scenario, the machine is subjected from start-up to a constant mechanical load of 1 N m, applied to its shaft. This initial constraint makes it possible to evaluate the robustness of the control strategy in the face of torque stresses, while testing the stability of the dynamic response during start-up. Furthermore, in order to analyze the behavior of the system in the presence of electrical faults, a single-phase short-circuit is deliberately introduced at phase A of the machine at time t = 0.2 s. The introduction of this disturbance makes it possible to examine the transient response of the MSAP and to highlight the effectiveness of the FOC control in the face of a malfunction that alters the symmetry of the three-phase power supply.

The machine's rotation speed is regulated around a setpoint of 106 rpm. This value, chosen as a reference, allows us to precisely observe the regulator's ability to maintain the setpoint speed despite the presence of mechanical and electrical disturbances. The total duration of the simulation is limited to 0.5 s, which is sufficient to observe three distinct regimes:

- Initial start-up with immediate application of load torque,
- Healthy steady-state operation before the fault occurs,
- the post-fault phase, marked by the introduction of the short circuit in phase A and the dynamic evolution of the machine under this abnormal condition.

Thus, this simulation scenario constitutes a complete analysis basis allowing both to validate the effectiveness of the FOC strategy under nominal conditions and to evaluate its capacity to maintain the stability and performance of the MSAP in the presence of electrical faults.

Figure 13 illustrates that, following the occurrence of a short circuit in stator phase A, the rotational speed exhibits noticeable oscillations around an average value corresponding to the synchronous speed. This behavior can be attributed to the unbalanced electromagnetic conditions introduced by the fault, which generate periodic torque disturbances. These disturbances, in turn, induce speed fluctuations despite the presence of the control system, reflecting its limited capability to fully compensate for the asymmetrical fault-induced dynamics. The persistence of oscillations, even when the mean speed remains close to the synchronous value, underscores the degradation in dynamic performance and stability of the machine under faulty conditions.

Figure 13. Healthy VS faulty PMSM speed profile

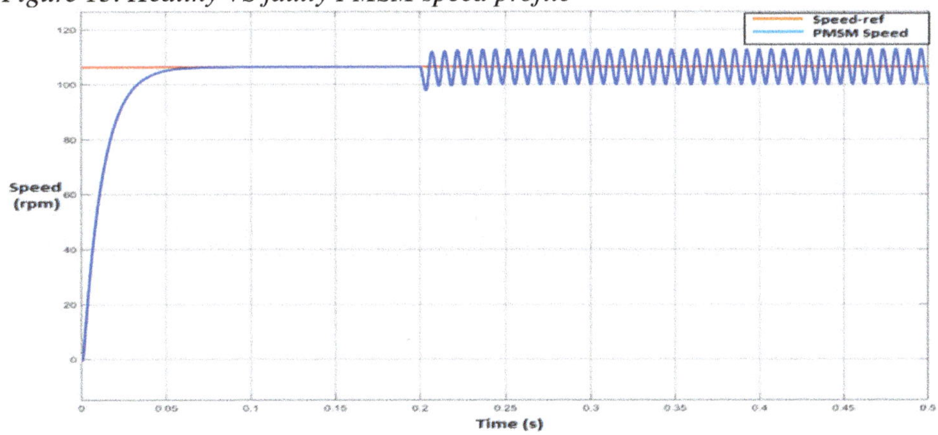

Figures 14 and 15 indicate that the onset of the fault induces substantial harmonic distortion in the stator currents, causing a clear deviation from the ideal sinusoidal waveform. This distortion is evident not only in the defective phase A but also propagates to the healthy phases B and C, likely due to electromagnetic coupling and the redistribution of currents within the machine.

Figure 14. Stator currents of the PMSM under inter-turn short-circuit fault

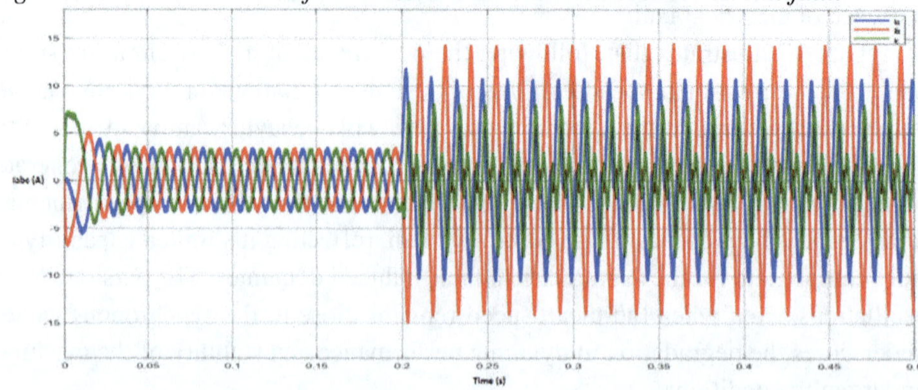

Figure 15. Deffective phase

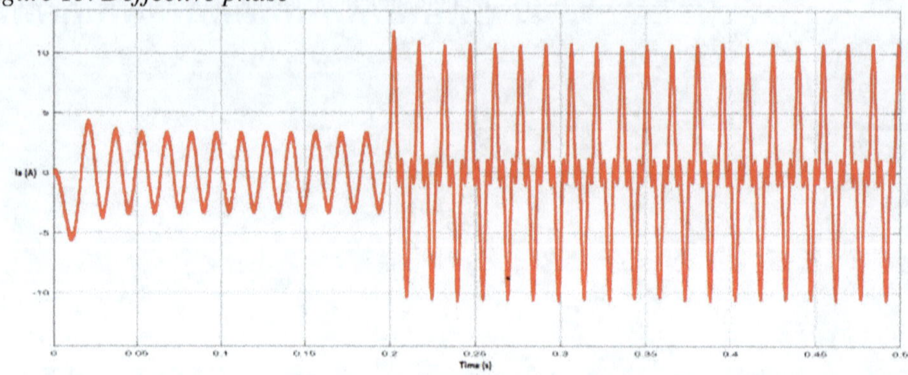

According to the figure 16, a noticeable increase in amplitude, accompanied by pronounced oscillations, can be observed in the direct-axis current (Id). Such behavior reflects a degradation in the performance of the Field-Oriented Control (FOC) strategy, indicating its reduced effectiveness in maintaining current regulation under the given faulty conditions.

Figure 16. Dq-axis currents of healthy PMSM compared to ITSC-faulted PMSM

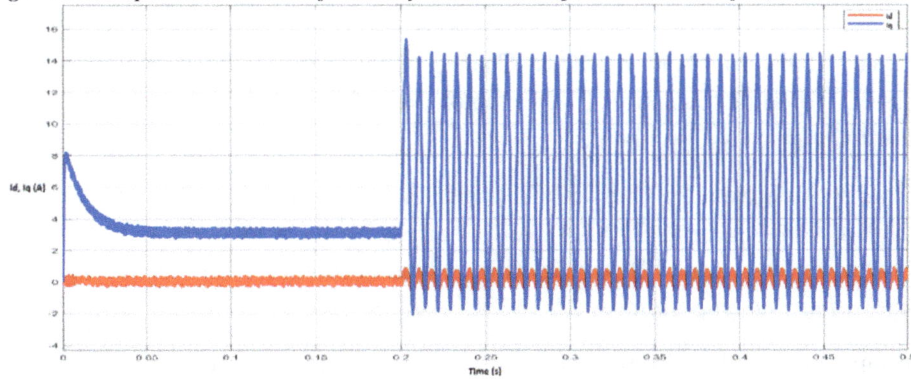

CONCLUSION

In this chapter, we addressed the classification of the most common faults affecting Permanent Magnet Synchronous Machines, with a particular focus on rotor- and stator-related electrical faults. A mathematical model in the direct–quadrature (dq) reference frame was developed to simulate both healthy and faulty operating conditions, enabling a detailed analysis of their impact on key electrical and mechanical parameters. The principles of Field-Oriented Control (FOC) were reviewed to ensure accurate torque and flux decoupling, even under fault conditions. Furthermore, MATLAB/Simulink was employed as a powerful Human–Machine Interface (HMI) platform for implementing the model, simulating standard driving cycles (FTP, NEDC, WLTP), and reproducing inter-turn short-circuit (ITSC) fault scenarios. The comparative performance analysis revealed significant variations in dynamic behavior, including torque ripple, speed regulation accuracy, and current waveform distortion, highlighting the importance of early fault detection and robust monitoring strategies. The insights gained from these simulations lay the groundwork for developing advanced fault-tolerant control systems, ultimately enhancing the reliability and safety of electric vehicle drive systems.

REFERENCES

Ahmed Baba, M., Naoui, M., & Cherkaoui, M. (2023). Fault-tolerant control strategy for Hall sensors in BLDC motor drive for electric vehicle applications. *Sustainability, 15*(13), 10430.

Allal, A., Lammouchi, Z., Khechekhouche, A., Mohamed, N., Alqunun, K., Alghadhban, A., & Guesmi, T. (2024). Advanced Residual Optimal Mapping Approach for Precise Detection of Stator Faults in Induction Motors. *IEEE Access: Practical Innovations, Open Solutions, 12*, 143515–143530.

Alqunun, K., Bechiri, M. B., Naoui, M., Khechekhouche, A., Marouani, I., Guesmi, T., & Allal, A. (2025). An efficient bearing fault detection strategy based on a hybrid machine learning technique. *Scientific Reports, 15*(1), 18739. PMID: 40437009

Azom, M. A., Hossain, M. S., & Khan, M. Y. A. (2025). Review of Electrical and Thermal Modeling Techniques for Three-Phase PMSM Drives. *Control Systems and Optimization Letters, 3*(1), 75–83.

Carlson, W. B. (1988). Thomas Edison as a manager of R&D: The case of the alkaline storage battery, 1898-1915. *IEEE Technology and Society Magazine*, *7*(4), 4–12.

Chen, X., Qin, P., Chen, Y., Zhao, J., Li, W., Mao, Y., & Zhao, T. (2022). Inter-turn short circuit fault diagnosis of PMSM. *Electronics (Basel), 11*(10), 1576.

Chethan, G. N. (2019, August). Performance analysis of PMSM drive with Space-vector PWM and sinusoidal PWM fed VSI. *In 2019 International Conference on Power Electronics Applications and Technology in Present Energy Scenario (PETPES)* (pp. 1-6). IEEE.

Djelamda, I., & Bouchareb, I. (2022). Field-oriented control based on adaptive neuro-fuzzy inference system for PMSM dedicated to electric vehicle. *Bulletin of Electrical Engineering and Informatics, 11*(4), 1892–1901.

EFEMWENKIEKIE, U. K., BABALOLA, O. P., DIRISU, O., & Oyedepo, O. S. (2020). IMPACT OF ELECTRIC VEHICLE IN 21 ST CENTURY DEVELOPING COUNTRIES. [IJMPERD]. *International Journal of Mechanical and Production Engineering Research and Development, 10*(3), 11797–11806.

Eom, H., & Lee, S. H. (2022). Mode confusion of human–machine interfaces for automated vehicles. *Journal of Computational Design and Engineering, 9*(5), 1995–2009.

Faiz, J., & Nejadi-Koti, H. (2019). Eccentricity fault diagnosis indices for permanent magnet machines: State-of-the-art. *IET Electric Power Applications, 13*(9), 1241–1254.

Fang, J., Sun, Y., Wang, Y., Wei, B., & Hang, J. (2019). Improved ZSVC-based fault detection technique for incipient stage inter-turn fault in PMSM. *IET Electric Power Applications, 13*(12), 2015–2026.

Karami, M., Mariun, N. B., Ab-Kadir, M. Z. A., Misron, N., & Mohd Radzi, M. A. (2021). Motor current signature analysis-based non-invasive recognition of mixed eccentricity fault in line start permanent magnet synchronous motor. *Electric Power Components and Systems, 49*(1-2), 133–145.

Khalid, M., & Mohan, A. (2020, October). Performance analysis of vector controlled PMSM drive modulated with sinusoidal PWM and space vector PWM. *In 2020 IEEE International Power and Renewable Energy Conference* (pp. 1-6). IEEE.

Khlaief, A. (2012). Contribution à la commande vectorielle sans capteur mécanique des machines synchrones à aimants permanents (MSAP) (Doctoral dissertation, Aix-Marseille Université).

Khov, M. (2009). Surveillance et diagnostic des machines synchrones à aimants permanents: Détection des courts-circuits par suivi paramétrique (Doctoral dissertation, Institut National Polytechnique de Toulouse-INPT).

Kirsch, D. A. (1995). History of the Electric Automobile: Battery-Only Powered Cars by Ernest H. Wakefield. *Technology and Culture, 36*(3), 710–712.

Korneev, A., Niu, Y., & Ibrahim, A. (2024). Electric vehicles in the 21st century: Historical evolution, environmental impact, and safety challenges for sustainable mobility. *Khwarizmia, 2024*, 32–38.

Krichen, M., Elbouchikhi, E., Benhadj, N., Chaieb, M., Benbouzid, M., & Neji, R. (2020). Motor current signature analysis-based permanent magnet synchronous motor demagnetization characterization and detection. *Machines (Basel), 8*(3), 35.

Kumar, A. (2024). A comprehensive review of an electric vehicle based on the existing technologies and challenges. *Energy Storage, 6*(5), e70000.

Kumar, R., Singh, M., Khan, S., Singh, J., Sharma, S., Kumar, H., & Aggarwal, V. (2023). A state-of-the-art review on the misalignment, failure modes and its detection methods for bearings. *Mapan, 38*(1), 265–274.

Lammouchi, Z., Naoui, M., Bekakra, Y., Khechekhouche, A., & Sbita, L. (2025). Model-free predictive control based on the execution time of active vectors using ultra-local model. *Arabian Journal for Science and Engineering*, *50*(8), 5885–5898.

Lau, Y. Y., Wu, A. Y., & Yan, M. W. (2022). A way forward for electric vehicle in Greater Bay Area: Challenges and opportunities for the 21st century. *Vehicles (Basel)*, *4*(2), 420–432.

Li, L., Cao, S., Li, J., Nie, R., & Hou, L. (2021). Review of rotor balancing methods. *Machines (Basel)*, *9*(5), 89.

Liu, Z., Zhang, P., He, S., & Huang, J. (2021). A review of modeling and diagnostic techniques for eccentricity fault in electric machines. *Energies*, *14*(14), 4296.

Martyushev, N. V., Malozyomov, B. V., Sorokova, S. N., Efremenkov, E. A., & Qi, M. (2023). Mathematical modeling the performance of an electric vehicle considering various driving cycles. *Mathematics*, *11*(11), 2586.

Naoui, M., Aymen, F., Mouna, B. H., & Sbita, L. (2021). Brushless motor and wireless recharge system for electric vehicle design modeling and control. In *Handbook of research on modeling, analysis, and control of complex systems* (pp. 338–362). IGI Global Scientific Publishing.

Ng, K. W., & Tong, H. Y. (2024). Comparisons of Driving Characteristics between Electric and Diesel-Powered Bus Operations along Identical Bus Routes. *Sustainability*, *16*(12), 4950.

Özçiflikçi, O. E., Koç, M., Bahçeci, S., & Emiroğlu, S. (2024). Overview of PMSM control strategies in electric vehicles: A review. *International Journal of Dynamics and Control*, *12*(6), 2093–2107.

Paul, G. E. (2021). Modeling and simulation of human systems. *Handbook of human factors and ergonomics*, 704-735.

Pietrzak, P., & Wolkiewicz, M. (2022). Machine learning-based stator current data-driven PMSM stator winding fault diagnosis. *Sensors (Basel)*, *22*(24), 9668. PMID: 36560037

Romdhane, M., Naoui, M., & Mansouri, A. (2023). PMSM Inter-Turn Short Circuit Fault Detection Using the Fuzzy-Extended Kalman Filter in Electric Vehicles. *Electronics (Basel)*, *12*(18), 3758.

Rosero, J., Romeral, L., Rosero, E., & Urresty, J. (2009, February). Fault Detection in dynamic conditions by means of Discrete Wavelet Decomposition for PMSM running under Bearing Damage. *In 2009 Twenty-Fourth Annual IEEE Applied Power Electronics Conference and Exposition* (pp. 951-956). IEEE.

Ruiz, J. R. R., Rosero, J. A., Espinosa, A. G., & Romeral, L. (2009). Detection of demagnetization faults in permanent-magnet synchronous motors under nonstationary conditions. *IEEE Transactions on Magnetics*, *45*(7), 2961–2969.

Sandrini, G., Chindamo, D., Gadola, M., Candela, A., & Magri, P. (2024). Exploring the impact of vehicle lightweighting in terms of energy consumption: Analysis and simulation on real driving cycle. *Energies*, *17*(24), 6398.

Sankhwar, P. (2024). Application of permanent magnet synchronous motor for electric vehicle. *Application of Permanent Magnet Synchronous Motor for Electric Vehicle*, *4*(2), 1–6.

Singirikonda, S., & Pedda, O. Y. (2023). Investigation on performance evaluation of electric vehicle batteries under different drive cycles. *Journal of Energy Storage*, *63*, 106966.

Ullah, Z., Lodhi, B. A., & Hur, J. (2020). Detection and identification of demagnetization and bearing faults in PMSM using transfer learning-based VGG. *Energies*, *13*(15), 3834.

Usman, A., Doiphode, N. T., & Rajpurohit, B. S. (2019, September). Stator winding faults investigation in permanent magnet synchronous motor using motor signatures: Part I. In 2019 *international conference on electrical drives & power electronics (EDPE)* (pp. 160-168). IEEE.

Usman, A., & Rajpurohit, B. S. (2020). Modeling and classification of stator interturn fault and demagnetization effects in BLDC motor using rotor back-EMF and radial magnetic flux analysis. *IEEE Access: Practical Innovations, Open Solutions*, *8*, 118030–118049.

Wakefield, E. H. (1998). *History of the electric automobile: hybrid electric vehicles* (Vol. 187). SAE International.

Wireko-Brobby, A., Hu, Y., Wang, G., Gong, C., Lang, W., & Zhang, Z. (2023). Analysis of the sources of error within PMSM-based electric powertrains—A review. *IEEE Transactions on Transportation Electrification*, *10*(3), 6370–6406.

Youssef, R., Hamida, M. A., Mohamed, N., Lassaad, S., Djamel, T., & Yasmine, A. (2023). Single-stage standalone photovoltaic water pumping system using predictive torque control (PTC) of induction machine. *Electric Power Components and Systems*, •••, 1–17.

Zhao, J., Guan, X., Li, C., Mou, Q., & Chen, Z. (2020). Comprehensive evaluation of inter-turn short circuit faults in PMSM used for electric vehicles. *IEEE Transactions on Intelligent Transportation Systems*, *22*(1), 611–621.

Zhao, J., Wang, L., Xu, L., Dong, F., Song, J., & Yang, X. (2021). Uniform demagnetization diagnosis for permanent-magnet synchronous linear motor using a sliding-mode velocity controller and an ALN-MRAS flux observer. *IEEE Transactions on Industrial Electronics*, *69*(1), 890–899.

NOMENCLATURE

EV: Electrical vehicle
PMSM: Permanent Magnet Synchronous Motor
ITSC: Inter-Turns Short circuit
FOC: Field Oriented Control
FTP: Federal Test Procedure
NEDC: New European Driving Cycle
WLTP: Worldwide Harmonized Light Vehicle Test Procedure

SYMBOLS

($\mathbf{X}_{abc}$): The three-phase reference
($\mathbf{X}_{dq}$): The direct-quadratic reference frame
$\mathbf{V}_s$: Stator voltage (Volt)
$\mathbf{I}_s$: Stator current (A)
$\mathbf{R}_s$: Stator resistance
φ_f**:** Peak value of stator flux
θ**:** Absolute rotor position in electrical degrees
$[L_{ss}]$**:** Stator inductance matrix
C_{em}**:** Electromagnetic torque
C_r**:** Resistance torque [N.m]
J**:** Moment of inertia
F**:** Viscous friction coefficient
ω_r**:** Rotor mechanical angular speed

p: Pole pairs number
L_d, L_q **:** Direct and quadrature axis inductances
μ: Short circuit fault degree percentage
Nf: Number of turns of the sub-winding affected by the fault
Ns: Number of turns per phase
e_a**:** e_b, e_c: Electromotive forces
i_f**:** Short circuit fault current
L_s**:** Mutual inductance
e_d**:** e_q: Disruptions
$K_{p\_id}, K_{i\_id}, K_{p\_iq}, K_{i\_iq}$ **:** Coefficients of current regulators
$K_{i\_\omega r}, K_{p\_\omega r}$**:** Speed regulator coefficients

Chapter 2
Bridging Human-Machine Interaction for Intelligent MPPT Control in Photovoltaic Energy Infrastructure

Lityem Wejden
https://orcid.org/0009-0009-4727-3069
University of Gabes, Tunisia

Sbita Lassaad
https://orcid.org/0000-0002-6589-3976
University of Gabes, Tunisia

Mnif Farid
University of Gabes, Tunisia

ABSTRACT

In recent decades, most countries rely more and more on renewable energy sources such as solar, wind, biomass and geothermal energies, etc. Solar and wind energy, in particular, are widely used because they are naturally abundant and environmentally friendly. Electricity generation based on renewable energy sources is increasing as a core component of modern energy infrastructure. To increase the output power of photovoltaic solar power generation, it is extremely important to force the system to operate at the highest possible efficiency. Maximum Power Point Tracking (MPPT) algorithms play a crucial role in optimizing the performances of the photovoltaic (PV) power generation system. This paper contributes to the modeling and devel-

DOI: 10.4018/979-8-3373-5621-1.ch002

opment of the complete PV pumping system(Youssef,#6). Additionally, this research seeks to improve human-machine communication in energy systems by incorporating industrial oversight instruments. This communication happens through monitoring and controlling the system using STEP 7 software, which helps create a smooth link between the operator and the automated process, while comparing three Maximum Power Point Tracking (MPPT) methods: the widely used Perturb and Observe (P&O) and Incremental Conductance (INC) techniques, along with a method based on Artificial Neural Networks (ANN). The use of the ANN method demonstrates the benefits of smart control techniques. especially where their use is potential with rapidly varying irradiance. While the INC method has an appeal to its simplicity, and minimal installation costs, a comparative evaluation made reference to important metrics like voltage, current, and power production in different environments.

INTRODUCTION

the global energy transition is increasingly depending on the use of renewable energy sources, particularly solar power, due to its nature availability and negligible environmental impact. Photovoltaic (PV) systems are thus experiencing strong growth here, both for the generation of electricity and specific uses such as water pumping in rural areas (Jain, 2018). However, the efficacy of solar energy conversion greatly depends upon the efficacy of the system in sustaining operation at its Maximum Power Point (MPP) (Soni, 2023), even with ongoing fluctuation in solar irradiance and temperature (Katche, Makokha et al. 2023) To meet this requirement, different Maximum Power Point Tracking (MPPT) algorithms have been proposed. Conventional methods such as Perturb and Observe (P&O) and Incremental Conductance (INC) are widely utilized due to simplicity and low implementation cost. These methods are inefficient during rapidly changing weather conditions, and such variations might affect the efficiency of the system as a whole. For overcoming such limitations, more intelligent techniques such as Artificial Neural Networks (ANNs) have been suggested. These methods allow adaptive systems using learning and predicting the best performance of the PV generator. As a result, photovoltaic water pumping system efficiencies can be greatly enhanced. In this paper, we present a complete model of a photovoltaic water pumping system incorporating different MPPT algorithms with an emphasis on the industrial supervision aspect via the use of STEP 7 software for ease of communication between the operator and the automatic system. A comparative analysis is done for the three MPPT techniques (P&O, INC, and ANN) taking into account crucial parameters such as voltage, current, and harvested power in an effort to highlight the strong and weak points of each technique at different conditions of the environment.

INVESTIGATION AND MODELLING OF KEY ELEMENTS OF A SOLAR-POWERED WATER PUMPING SYSTEM

1. PV Model

A photovoltaic solar panel consists of numerous photovoltaic cells created from semiconductor substances. Upon exposure to sunlight, these PV cells convert solar energy into direct current (DC) through the photovoltaic effect: the emergence of a voltage difference at the interface of two distinct materials when subjected to electromagnetic radiation. The amount of current generated relies not only on the incident solar irradiation but also on the temperature of the photovoltaic cell. PV.panels are categorized into two primary groups, both composed of the same crystalline structure: monocrystalline, characterized by a uniform hue, and polycrystalline, where the individual grains are distinctly visible to the naked eye. Monocrystalline silicon features a consistent crystal lattice, while polycrystalline silicon is made up of numerous small grains with random orientations creating an irregular matrix. The lifespan of polycrystalline silicon is generally lesser compared to that of monocrystalline silicon. Additionally, the efficiency of the polycrystalline surface is inferior because it captures fewer photons than its monocrystalline equivalent, which tends to be slightly pricier. It's noteworthy to mention a third, less common type, the amorphous solar module, an allotropic form of silicon that lacks crystallization. Despite its reduced yield, its flexibility allows for better performance under very low irradiation, making it suitable for specific applications, particularly in indoor settings. The effectiveness of each module is assessed through their corresponding Current-Voltage (I-V) characteristic curves, which enable the evaluation of power at any given operating point. The output of the PV panel is influenced by external factors, including incident solar radiation (R_s) and ambient temperature (Ta). This relationship is articulated by the following equation:

$$P_P = V_P . I_P = f(T_A . R_S) \qquad (1)$$

To gain a clearer insight into the actual operation of a photovoltaic panel, it is crucial to employ accurate electrical modeling. In contrast to an ideal current source, a PV module demonstrates a non-linear response, significantly shaped by environmental factors such as solar irradiance (IR), ambient temperature (T_A), and even cell temperature (T_C). Consequently, designers and engineers utilize equivalent electrical models to replicate the panel's dynamic behavior and forecast its energy efficiency under various conditions. The prevalent one-diode model is founded on a simplified yet effective physical structure, encompassing several key components. This model features a photo-induced current source (I_{ph}), which fluctuates in direct proportion to irradiance, and a diode (D) that mirrors the properties of the solar cell's PN junction. Additionally, two resistors are included: the series resistor (R_s), which signifies the ohmic losses in metal contacts, interconnections, and the semiconductor; and the parallel or shunt resistor (R_{sh}), which represents the leakage currents and insulation imperfections in the module. These components interact to determine the output current of the panel, described by the following nonlinear equation:

$$Ipv = I_{ph} - I_0\left(\exp\frac{q(V + R_s.I)}{b.k.T.N_s} - 1\right) - \frac{(V + R_s.I)}{R_{sh}} \tag{2}$$

This model proves invaluable for simulations conducted in MATLAB/Simulink, enabling the investigation of how temperature or sunlight fluctuations impact the characteristic I–V and P–V curves. It is also critical for configuring power converters, crafting MPPT (Maximum Power Point Tracking) algorithms, and assessing overall system efficacy. The following diagram illustrates the standard equivalent electrical representation of the photovoltaic panel with one diode, featuring its primary components.

Figure 1. model of PV

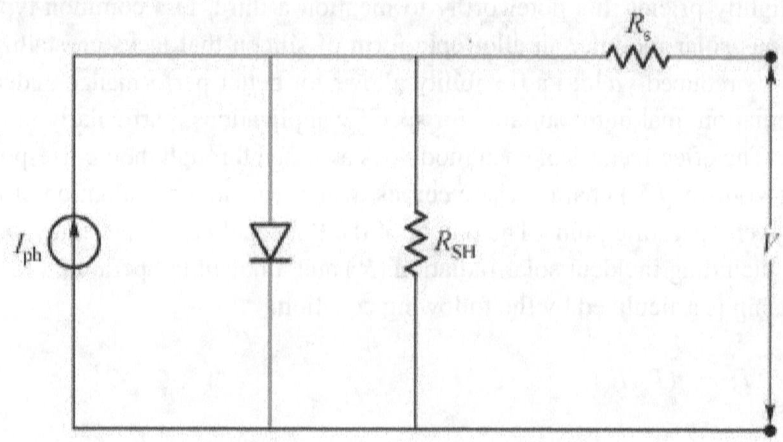

a. Influence of Solar Radiation on PV Operation

Solar radiation is the primary element influencing the electricity generated by solar panels. Indeed, the short-circuit current (I_{sc}) rises nearly in a linear manner with the strength of incoming solar light. This correlation is represented by the following formula:

$$I_{sc} = I_{scref} * \frac{IR}{IR_{ref}} \quad (3)$$

where Isc,ref denotes the reference current recorded at standard irradiance (IR_{ref}) =1000W/m^2.while IR signifies the actual irradiance. As the irradiance rises, a greater number of photons hit the panel's surface, producing an increased quantity of electron-hole pairs, thereby elevating the current output. The open-circuit voltage (V_{oc}) is only marginally affected by irradiance and remains fairly constant across a broad range of irradiance levels.

Figure 2. I-V and P-V characteristics of a photovoltaic module under different solar irradiances

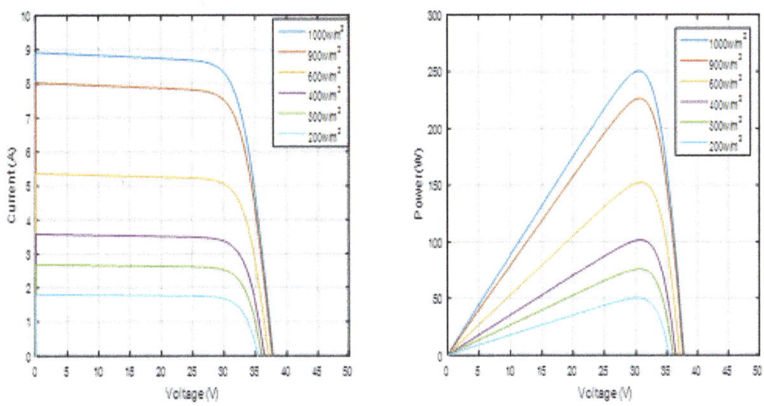

The influence of solar irradiance on the electrical behavior of photovoltaic modules is further depicted in the I–V (current-voltage) and P–V (power-voltage) characteristic curves presented in the figure. As irradiance escalates from 200 W/m^2 to 1000 W/m^2, the short-circuit current (Isc) experiences a notable rise, as seen in the left graphic. This substantiates the direct proportional relationship between irradiance and current production in photovoltaic cells. Moreover, the right graphic illustrates that increased irradiance levels result in a significant enhancement in the maximum power point (Pmax), in terms of both current and output power. The voltage at maximum power (Vmpp) remains quite consistent, while the current at maximum power grows as irradiance increases. These findings affirm the theoretical model introduced previously and underscore the significance of irradiance as a key factor in determining the instantaneous performance of a PV system amid varying environmental conditions.(Pop, Buzduga et al. 2022)

b. Influence of Temperature on PV Operation

Temperature primarily affects the output voltage produced by the photovoltaic cell. As temperatures increase, the open-circuit voltage (V_{oc}) declines in a linear fashion, defined by the equation below.

$$V_{oc} = V_{ocref} + \alpha_{Voc} * (T - T_{ref}) \quad (4)$$

In this context, V_{ocref} denotes the reference voltage at T_{ref}=25°C, while $\alpha_{V_{oc}}$ represents the temperature coefficient, which is typically of a negative nature. Although the short circuit current I sc experiences a slight uptick with rising temperatures, this increase is negligible in comparison to the decline in voltage. As a result, elevated temperatures result in a decrease in maximum power output, adversely impacting the overall efficiency of the photovoltaic system. Thus, the influence of temperature must be meticulously examined when sizing and assessing the performance of solar energy systems.

Figure 3. I-V and P-V characteristics of a photovoltaic module under different temperatures

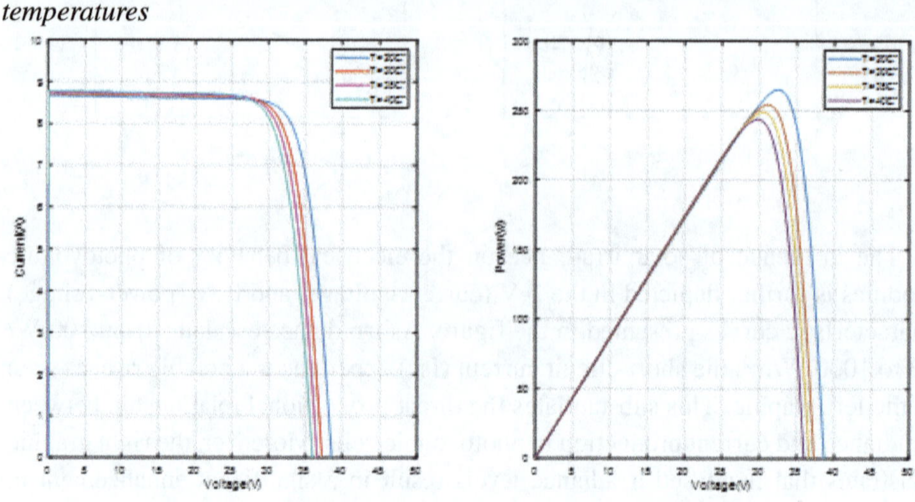

2. The Types of Pumps

a. Volumetric Pump

The volumetric pump turns the back-and-forth motion of its motor into steady fluid movement, letting the water rise and fall through the opening and closing of sealed chambers that swap places with the suction and discharge ports. Positive-displacement designs come in several flavours, including screw, vane, piston, and diaphragm pumps. Piston and diaphragm variants are the ones youll often see in operation down deep-holes or boreholes that stretch beyond a hundred metres. To get the drive power all the way down, engineers usually run a single, very long shaft from an electric motor mounted on the ground above. Because of the way these devices work, their water output tracks the motors speed almost linearly. Torque, however, depends mainly on the total manometric height, or HMT, and stays nearly steady even when the motor speed changes. That means starting torque too is nearly the same no matter how fast the pump is being asked to move water; it simply reflects the height HMT the pump has to push against. As a result the motors power draw does climb with speed, making these pumps a good match for deep wells where the flow rate is low but the lift is high. Engineers will also use them on the surface when the load is slow, uneven, and only small volumes are needed, such as with hand-operated devices or slow-turning windmills that have many blades.(Anis, Mertens et al. 1985)

b. Centrifugal Pump

Centrifugal pumps use impeller wheels or blades to rotate, transferring kinetic energy from the engine to the fluid. Through the centrifugal force of the vanes, the water enters the center of the pump and is forced upward and outward. It is possible to stack many stages of impellers on a single drive shaft to raise the pressure and, thus, the discharge height. As the water moves on to the following stage, the pressure rises until the last stage, which delivers a large amount of water under high pressure. Suction rotary pumps, floating pumps, and submersible pumps with surface or submerged motors are examples of these pumps. A comparatively fixed HMT is the target of the centrifugal pump's design.

This pump's flow rate fluctuates in direct proportion to the motor's speed of rotation. The discharge height is proportional to the square of the motor speed, and its torque grows extremely quickly at this speed. To guarantee a decent flow rate, the motor's rotation speed must be extremely high. As a result, the power used, which is proportional to Q HMT, will change depending on the speed's cube. Typically, centrifugal pumps are employed at medium or low depths (10 to 100 meters) and for big flows.(Hmidet, Rebei et al. 2015)

In order to attain a non-zero starting flow rate, the pump needs a minimum speed at a specific HMT since the starting torque is restricted to the friction torque of the pump at zero speed, which is greater than during rotation. The centrifugal pump's efficiency is a function of flow rate; it functions best at a specific head and rotational speed. Thus, it's critical to appropriately design the pump sizing based on the site's characteristics.

c. Comparisons Between Centrifugal Pumps and Volumetric Pumps

While making a pump system choice, one should consider comparing the application of a centrifugal pump with that of a positive displacement pump, particularly when the hydraulic conditions at the site are varying. Positive displacement pumps, or volumetric pumps, offer a relatively constant flow rate, regardless of variation in the total dynamic head (TDH). They are particularly suitable for low-flow applications and areas where the HWT is varying, leading to a variation of discharge. Centrifugal pumps, however, are not tolerant of variations in HMT: their flow rate declines drastically with a rise in the latter. The reason is that the discharge head of a centrifugal pump is directly proportional to the square of the motor's speed of rotation. Hence, in order to compensate for an increase in the HMT and provide a consistent flow rate, there is a need to significantly increase the speed of the motor, leading to increased energy consumption and a significant drop in efficiency. In conclusion, centrifuges are better suited to high-debit requirements with more stable hydraulic conditions, while volumetric pumps are better suited to low-debit and HMT variables.

3. The Types of Engines

Electrical energy is transformed into mechanical energy by a motor pump unit's engine. Alternating current or direct current may be used. In the latter scenario, the direct current from a solar generator must be converted into alternating current using an electronic converter or an inverter. Because of this, choosing a direct current motor could appear more intriguing at first, but as we shall see, the development of effective electronic converters also makes it possible to choose efficient and, most importantly, less costly alternative motors.

a. CC Motors

By changing the direction of the current flowing in an armature (often the rotor) exposed to a magnetic field generated by an inductor (typically the stator), the electrical energy applied to a motor is converted into mechanical energy. Either carbon and graphite brushes or electronic commutation are used to change the current flowing through the rotor of a direct current motor.(Abut 2016)

Brush motor: The most popular are brush motors, although they require regular brush replacement. When it comes to submersible motor pumps, this is especially troublesome because changing the brushes requires removing the pump from the borehole. Additionally, the insulation of the motor must not be compromised in order to prevent water intrusion, which is difficult to accomplish on-site. Long-lasting brush motors from certain manufacturers eliminate the need for maintenance after 5000 operating hours.(Wang and Suh 2017)

Brushless electronic commutation motors: reverse the rotor's current using an electronic circuit. Although these motors don't need as much upkeep, the electronic circuit needs to be built for a severe environment and can cause issues. Permanent magnets or an electromagnetic winding generate the inductor's magnetic field in a direct current motor. Depending on how the armature and the inductor are driven, the motors in the case of a wound inductor can be of the series type, parallel type (shunt motor), or compound excitation type (compound motor)(Wang, Chen et al. 2018)

b. AC Motor

Most frequently, asynchronous alternating motors are employed in a variety of industrial settings. For instance, this kind of motor has long been utilized in network pumps. Its small size makes it ideal for submerged pumping in open wells and boreholes. The introduction of effective inverters that allow for speed control of these motors has made it possible to employ them for solar pumping applications in recent years. Alternative motors for solar pumping are typically three-phase, asynchronous, and have a squirrel-cage rotor. It revolves around two bearings that are kept lubricated by the motor's coolant. The stator is immersed in a resin, which ensures good mechanical strength, effective heat dispersion, and perfect sealing. These motors are designed to be submerged and used in water with a maximum temperature of around 35°C. They must necessarily be powered by an inverter that will convert the direct current from the PV modules into alternating current. In the case of a bipolar motor with a three-phase current of frequency w, the motor's rotational speed, known as synchronous speed, is 2900 to 3000 RPM (frequency of 50 to 60Hz).(Lammouchi, 2025) (Allal, 2024)

4. Energy Conversion Chain for Photovoltaic Power Plants

The PV panel's energy can be sent directly to the charge (motor) or through a variety of electronic devices. A maximum of two converters can be included in the power block; the first one will allow for the regulation of the panel's operation, and the second will control the motorcycle. The many couplings that are feasible, their goals, and the converter technologies used in the case of solar power are then presented in the sections that follow.

a. Type of Coupling

The PV panel and the motor create a pair of electrical components. Each of them is distinguished by a unique function curve I–V. When there is a direct coupling and the engine and panneau are branching out in parallel without electronic devices, their relative stresses are comparable. It is impossible to change the system's tension in this case, and it only has one point of operation, which is the intersection of the two characteristic curves.

However, to control the input and output pressures and currents, an electrical converter can be installed. A solar panel has a nominal operating tension, and motors, like many electrical devices, operate at specific voltages (12V,24V,48V,etc.) and nominal speeds. From there, Thus, other coupling approaches are possible:

CC-CC converter with fixed tension: The applied charge or the change in the input tension has no effect on the stable output tension. Very economical, it is frequently used to operate the motorcycle under specific tension.

Converter CC-CC à tension variable:Depending on the operating parameters (temperature and irradiance), the departure or entry tension is piloted to reach a specific consigne variable. Three types of CC-CC converters with variable tension can be distinguished based on the system configuration and soundness control:

i. MPPT charge regulator: This solar charger recovers energy from panels and stores it in one or more batteries. Known as the "Maximum Power Point Tracker" regulator, it continuously monitors the system's input tension (the tension of the panels) in order to adjust and operate on the maximum point of power. The battery is then used to power the motor pump, which is essential to the system's operation. It is common practice to select a battery pack that operates at the same voltage as the pump.
ii. Pump MPPT control: This converter has the advantage of operating similarly to the previous regulator, but it does not require a battery to operate between the converter and the pump. It is based on the same MPPT strategy. These clever

MPPT controllers, which were only created for solar power, are more expensive and less dependable than MPPT with batteries.

iii. LCB Pump Controller: The "Linear Current Booster" (LCB) is a converter that does not optimize the operation of the panel, but seeks to make the pump operate for a greater number of hours. Its objective is to adapt the impedance by decreasing the voltage to increase the current and allow the pump to start in low sunlight conditions.

iv. DC-DC converters: are crucial components in solar pumping systems. The photovoltaic generator's voltage and current can be adjusted to meet the demands of the load, which in this case is the pump's motor. Rapid semiconductor (transistor) switching enables these static electronic devices to either boost (Boost converter) or reduce (Buck converter) the input voltage. The Buck-Boost, Ćuk, and SEPIC converters, for example, provide extra versatility by enabling the output voltage to be either higher or lower than the input voltage, with or without polarity inversion. By forcing the solar generator to run at its maximum power point (MPPT) with the right regulation, these converters may guarantee the highest possible energy efficiency. In addition to their efficiency (often exceeding 95%), DC-DC converters also help stabilize the operation of the motor pump, ensuring a power supply adapted to variations in sunlight and temperature

- ✓ The Boost converter: is a variant of a direct current – direct current (DC-DC) converter designed to elevate the input voltage to a superior output voltage level. It is especially utilized in solar energy systems when the voltage produced by the solar panel falls short of what is needed by the load, such as a pumping mechanism. The operational principle relies on the temporary accumulation of energy in an inductor while the switch (transistor) is engaged, and subsequently on transferring this energy to the load once the switch is disengaged. This generates a regulated over-voltage effect that permits the output voltage to be augmented. The relationship between the input voltage V_{input} and the output voltage V_{output} is dictated by the duty cycle D.(Pradhan and Panda 2018)

$$V_{output} = \frac{1}{1-D} V_{input} \tag{5}$$

Due to its straightforward design, remarkable effectiveness, and capability to keep the load at a consistent voltage despite fluctuations in sunlight, the Boost converter is commonly utilized in self_sufficient solar pumping systems, especially within maximum power point tracking (MPPT) methods.

✓ The Ćuk converter reverses the polarity of the output voltage with respect to the input and permits lowering or raising the input voltage. Better current continuity on both the source and load sides is made possible by the fact that energy is delivered via a capacitor rather than an inductor, setting it apart from other topologies (such Buck-Boost). Systems that are sensitive to current quality, like photovoltaic pumping systems, benefit greatly from this characteristic. Two switches (transistors), an input inductor, an output inductor, and a transfer capacitor are necessary for the Ćuk converter to function. The following formula provides its input voltage/output voltage relationship:

$$V_{output} = -\frac{D}{1-D} V_{input} \tag{6}$$

The Ćuk converter's low current ripple, high efficiency, and stable operation make it an ideal choice for applications that need a consistent power source even when solar energy fluctuates a lot.(Divyasharon, Banu et al. 2019)

✓ The SEPIC (Single Ended Primary Inductor Converter) represents a variety of DC-DC converter that can output a voltage either above or below the input voltage while preserving the same polarity. In contrast to the Ćuk converter, the output voltage of the SEPIC remains non-inverted, making it especially beneficial for applications requiring a steady polarity. The SEPIC operates with two inductors (or a coupled inductor), a transfer capacitor, and a switch (typically a transistor). Its functioning hinges on the energy transfer via the capacitor, all the while ensuring minimal current ripple at both the input and output. The connection between the input voltage and the output voltage mirrors that of the Buck-Boost.(Tey, Mekhilef et al. 2018)

$$V_{output} = \frac{D}{1-D} V_{input} \tag{7}$$

The SEPIC converter is ideal for standalone photovoltaic systems because of its adaptability, stability, and capacity to handle changing energy sources. This is especially true for solar pumping, where the input voltage can vary greatly based on the amount of sunlight.

5. Inverter

The inverter, also known as a DC_AC converter, is a crucial component in power electronics that produces alternating voltages or currents with either variable or fixed frequencies and amplitudes from an electrical source with differing voltage or frequency. It becomes essential when the system incorporates AC motors. All inverters consist of electronic switches that enable a managed switching sequence to create an alternating signal of the desired frequency. Several primary categories facilitate the characterization of an inverter:

The initial consideration pertains to the configuration of the output waveform: it can be a pristine sine wave, modified, or a square wave, which directly affects compatibility with loads, particularly asynchronous motors. Pristine sine wave inverters provide the highest quality current and are especially appropriate for sensitive devices such as motor pumps, whereas modified or square waves, being more cost effective, are designated for less critical loads. Another significant differentiation is the type of output: single-phase for lower power (domestic or agricultural usages), or three-phase for higher power, frequently seen in medium or large-scale pumping systems. The operational mode also represents a key category: inverters can be standalone (off-grid), employed in remote locations without an electrical grid; grid-connected (grid-tie), feeding surplus energy into the grid; or hybrid, integrating various sources (PV, batteries, grid), with smart management based on energy availability. Furthermore, inverters can either have a fixed frequency or a variable frequency. The latter, known as VFD (Variable Frequency Drive), is particularly advantageous for photovoltaic pumping systems, as they permit adjustments to the asynchronous motor's speed in line with available solar energy, thus enhancing the pumped water flow while curtailing losses.The internal design of the inverter forms another classification category. The most prevalent architectures include full bridge, half bridge, and multilevel inverters, the latter presenting a more refined output waveform and improved efficiency, particularly for high power applications. Lastly, the modulation technique employed to control the inverter switches also impacts its performance. Pulse Width Modulation (PWM) is the most common, especially in sinusoidal (SPWM) or vectorial (SVPWM) forms, with the latter enabling superior utilization of the DC bus and a decrease in harmonics. Therefore, grasping these various categories facilitates the selection of an inverter that aligns with the specific requirements of the system, whether they pertain to current quality, motor type, required power, or the variability of the solar resources.(Deshpande and Bhasme 2017)

6. Energy Storage

A major hurdle for the Solar Photovoltaic Water Pumping System is ensuring that water can be supplied relatively continuously. Although the solar energy is plentiful, its presence through-out the day varies. At times, particularly during dawn and dusk, the sunlight may be inadequate to operate the system and deliver water. Conversely, during solar noon, the energy is extremely plentiful but not fully harnessed. Thus, the aim of storage is not just to maximize the use of the available energy but also to guarantee a sufficient water supply during periods of low sunlight. The subsequent two sections will discuss the two storage systems in detail. (Mazloumi, Poolad et al. 2023)

a. Water Reservoir

The water reservoir serves as a straightforward and efficient means for the indirect capture of solar energy. During bright days, the energy harnessed from the solar panels is utilized to elevate water to a reservoir positioned at an elevation. This conserved water embodies gravitational potential energy that is accessible at all times, even when the sun is not shining. This storage method guarantees a steady supply of water for irrigation, livestock, or household needs, eliminating the need for expensive and cumbersome electric batteries. The dimensions of the reservoir primarily hinge on the required daily output, the height of storage, and the anticipated durations without sunlight. This approach is especially advantageous for remote rural regions, providing an affordable, sustainable, and low-maintenance solution.

The hydraulic function of a storage reservoir can be represented by a series of equations that illustrate the changes in the water volume, the accumulated energy, and the power generation. To begin with, the quantity of water within the reservoir at any specific time V(t) is determined by the integration of the net flow rate, as expressed by the equation:

$$V(t)=\int_0^t (Q_{p_{in}}(\varphi)-Q_{p_{out}}(\varphi))\,d\varphi + V_0 \tag{8}$$

With $V_0$is the initial volume,$Q_{p_{out}}$ is the pump outflow due to consumption, $Q_{p_{in}}$ is the pump inflow rate.For constant flow rates, the filling or emptying time, given by:

$$t = \frac{\Delta_{V_w}}{Q_P} \tag{9}$$

Where $Q_p$the corresponding flow rate and $_{V_w}$ being the volume to add or remove. Water stored at height contains gravitational potential energy can be calculated as:

$$E_{g_p} = \rho . g . V_W . H_S \tag{10}$$

$H_S$is the elevation of the reservoir and $V_W$is the stored volume. The total dynamic head (H_T) available for hydraulic use is expressed by:

$$H_T = H_{RES} - H_{PU} - H_{losses} \tag{11}$$

considering the pressure drops within the piping network. The effective hydraulic energy supplied during functioning is articulated as:

$$P_h = \rho . g . Q_p . H_h \tag{12}$$

where Q_p represents the discharge rate and H_h indicates the elevation of the fluid. Ultimately, in scenarios involving gravity-driven drainage through an aperture, the discharge rate is calculated using Torricelli's principle:

$$Q_p = C_{dis} . B . \sqrt{2.g.H} \tag{13}$$

where C_{dis} represents the discharge coefficient (usually ranging from 0.6 to 0.9), A denotes the cross-sectional area of the outlet, and H signifies the water height. These formulas are crucial for accurately sizing the reservoir, assessing its autonomy, and enhancing the efficiency of the solar pumping system in accordance with hydraulic needs.

b. Battery

Energy storage utilizing batteries is a prevalent strategy in photovoltaic systems aimed at conserving electricity for future utilization. Typically, lead-acid or lithium-ion batteries enable the continuous supply of power to motor pumps or additional machinery during instances of inadequate solar generation (such as at night or on overcast days). This approach provides significant versatility in energy management, especially in scenarios where a dependable power supply is essential. Nonetheless, batteries present substantial drawbacks: they are expensive, have a finite lifespan, require routine upkeep, and pose considerable environmental challenges upon disposal. Furthermore, integrating batteries necessitates control components (like charge regulators and appropriate converters), which adds complexity to the system. Despite these issues, battery storage continues to be a favored solution in situations where a constant water supply is essential without depending on hydraulic storage. The capacity of the battery, measured in ampere-hours (Ah), is determined by the equation$C_{battery} = I.t$, where I represents the discharge current and t signifies the duration of use. The total electrical energy can be approximated using$E_{elect} = V_n . C_{battery}$, where V indicates the nominal voltage. Keeping track of the state of charge (SoC) is crucial for ensuring the longevity of the storage system and is calculated as

$$SOC(\%) = \frac{Q_{Remaining}}{C_{battery_{total}}} .100 \tag{14}$$

The depth of discharge (DoD), which is defined as the inverse of SoC, plays a significant role in preventing the overuse of the battery, an essential factor in prolonging its lifespan. The instantaneous power that is either delivered or absorbed can be represented by P=V·I, while the overall efficiency is denoted as $\eta = \frac{E_{dis}}{E_{ch}}$,usually falling between 70% to 95%, depending on the type of battery technology (lead-acid, lithium, etc.). Lastly, the dynamic voltage of the battery can be represented through a simplified RC model:

$$V_d(t) = V_{oc_v} - R_{in} . I(t) \tag{15}$$

where R_{in} is the internal resistance and V_{oc_v} is the open-circuit voltage.

ADVANCED MPPT TECHNIQUES BASED ON CONVENTIONAL AND AI METHODS

To achieve the best utilization of energy generated by photovoltaic panels, especially in applications like solar pumping, it is essential to implement MPPT algorithms (Maximum Power Point Tracking) that can dynamically adjust the PV generator's operating point according to fluctuations in irradiation and temperature. A variety of control techniques have been created for this aim, each offering its own benefits and drawbacks concerning accuracy, convergence speed, complexity of execution, and resilience to environmental factors. Among the most frequently referenced in literature and practical applications, we highlight the Perturb and Observe method (P&O), the incremental conductance technique, intelligent fuzzy logic controllers, methodologies utilizing artificial neural networks, as well as nature-inspired optimization algorithms such as PSO, HHO, or Cuckoo Search. These various strategies will be elaborated upon in the subsequent paragraphs.(BENMEDDAH and LAIDI 2025)(Ahmed Baba, 2023 #9)

1. The Perturb and Observe (P&O) Technique

The Perturb and Observe is among the most straightforward and frequently utilized MPPT strategies. Its foundation lies in periodically altering the input voltage (or current) of the solar panel, followed by monitoring the impact of this alteration on the output power. If a small rise in voltage results in a boost in power, the system keeps perturbing in the same direction. Conversely, if the power declines, the disturbance direction is reversed. While this approach is simple to execute and low-cost regarding calculations, it has a significant drawback: oscillations around the maximum power point, especially in steady-state conditions. Additionally, it may be sluggish to respond to swift changes in irradiation, which diminishes its efficiency in highly fluctuating environments.(Patra, Nema et al. 2023)

2. Incremental Conductance Technique (IncCond)

The incremental conductance technique revolves around examining the slope of the power-voltage curve P=V. I. It is founded on the principle that at the maximum power point (MPP).

$$\frac{dP}{dV} = 0 \Rightarrow \frac{dI}{dV} = \frac{-1}{V} \qquad (16)$$

The algorithm contrasts instantaneous conductance (I/V) with incremental conductance $\frac{dI}{dV}$to ascertain the direction for voltage adjustment. This approach is more precise than P&O, as it can effectively monitor rapid variations in irradiation without oscillating around the MPP. Nonetheless, it demands more accurate sensors (for current measurement and DC voltage) and additional computations, which can make its implementation on constrained microcontroller systems more complex. (Motahhir, El Hammoumi et al. 2018)

3. Fuzzy Logic Control

Fuzzy logic serves as an advanced control strategy that does not necessitate an exact mathematical representation of the system. It utilizes a set of linguistic rules ("IF... THEN...") to ascertain the changes in voltage or current based on detected fluctuations in power. Generally, the two inputs consist of the power discrepancy (Δp) and the change in this discrepancy, while the output represents the voltage adjustment to be implemented. This method is resilient, swift, and proficient in managing non-linearities and uncertainties within the system. However, its success largely hinges on the precise selection of fuzzy rules and the fine-tuning of membership functions, which can complicate its design.(Garud, Jayaraj et al. 2021)(Romdhane, 2023 #8)

4. Artificial Neural Networks (ANN)

Artificial neural networks (ANNs) represent a supervised learning approach where the algorithm learns to forecast the peak power point based on input variables like sunlight exposure and temperature. Following prior training using simulated or real-world data, the network is capable of directly calculating the ideal voltage for the MPP. ANNs provide excellent precision, quick responsiveness to fluctuations, and adjust effectively to non-linear scenarios. Nevertheless, they necessitate a training phase, an extensive dataset, as well as elevated computational resources, making them more appropriate for advanced embedded systems (DSP, FPGA).(Hussain, Sarwar et al. 2023) (Alqunun, 2025 #4)

5. Nature-Inspired Optimization Techniques (PSO, HHO, Cuckoo, etc.)

Nature-inspired optimization techniques, including Particle Swarm Optimization (PSO), Harris Hawks Optimization (HHO), and Cuckoo Search, are grounded in the behavioral patterns observed within the natural world. These approaches frame the search for the maximum power point (MPP) as a global optimization challenge. For instance, PSO employs a collection of particles (potential solutions) that navigate the landscape of tensions and converge on the maximum power point through mutual information sharing. These techniques demonstrate remarkable efficiency, especially when dealing with multiple MPP scenarios, such as those arising from partial shading, and are adept at rapidly approaching the ideal solution. However, their primary limitation is their algorithmic intricacy and computational expense, necessitating more advanced controllers.(Alhamami, Abba et al. 2024).

SIMPLIFIED SIZING OF A PHOTOVOLTAIC PUMPING SYSTEM: APPLICATION TO A REAL CASE

1. Geographical Characteristics

The Higher Institute of Technological Studies (ISET) of Gabès in Tunisia is the location shown in the diagram. Latitude 33°53'15.0 North and longitude 10°06'12 East are its approximate geographic coordinates. This web site show cases a region with a semi-urban setting that incorporates contemporary educational facilities. The photovoltaic pumping station, associated with this location, reflects the institute's commitment to promoting renewable energy and sustainable development. With an altitude of about 17 meters above sea level, and an arid climate conducive to a strong solar exposure, this location is particularly suitable for the exploitation of photovoltaic energy. (Maka, Mehmood et al. 2023)

Figure 4. Situation the ISET of Gabès on Google Maps.

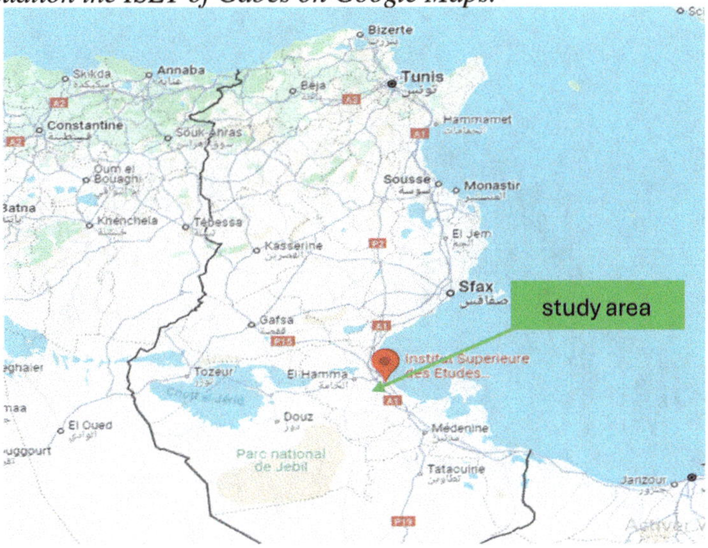

2. Climate Potential of Gabes for the Exploitation of Solar Energy

Local climatic conditions strongly influence the efficiency of photovoltaic pumping systems. The analysis of climate data shows a strong variation in global radiation throughout the year, reaching an average peak of more than 7 kWh/m^2 in June and July, while the winter months record much lower values, around 2 kWh/m^2

Figure 5.the global radiation over the year

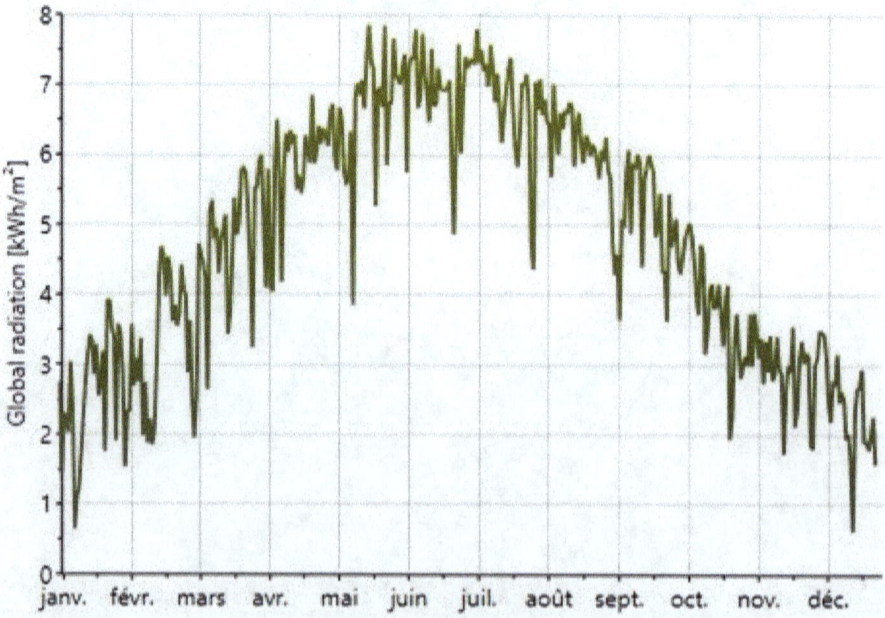

However, there are significant variations between minimum and maximum temperatures, with daily maximums ranging from 15°C in the winter to almost 40°C in the summer. Rare and sporadic, rainfall occurs mostly in the winter (up to 24 mm in January) and is nearly nonexistent in the summer. These findings highlight how crucial it is to optimize systems to function well in the region's extreme heat and radiation.

Figure 6. (a)Variations of the temperature over the year.(b)Variations of the precipitation over the year

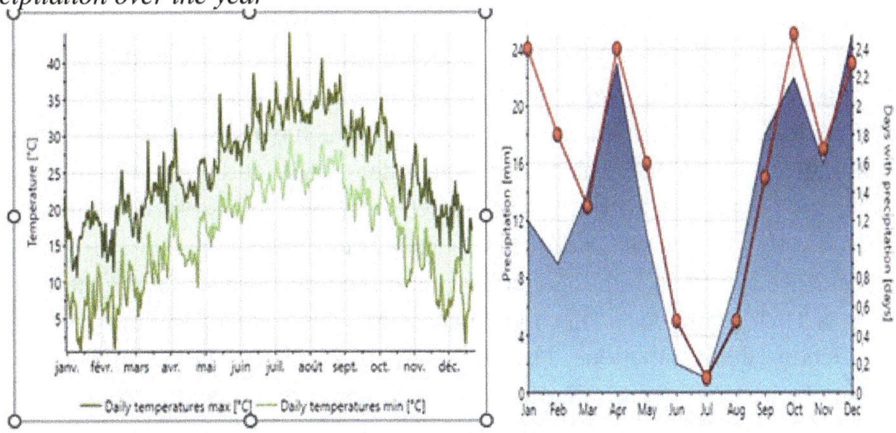

(a) (b)

3. Designing a Solar-Powered Water Pumping System: A Methodological Framework

The meticulous design of a solar-powered pumping system follows an eight-step methodology, ensuring efficiency, dependability, and alignment with local conditions. Initially, we assess the daily water requirements, factoring in seasonal variations; this evaluation is crucial for determining the pump's hourly flow rate. Next, we gauge the hydraulic characteristics of the well (depth, static water level, dynamic water level, pressure loss), to compute the Total Dynamic Head (TDH). Following this, we ascertain the necessary flow rate by dividing the daily demand by the site's Peak Sun Hours (PSH), thereby ensuring that the pump can handle peak demands. The fifth step involves selecting the pump based on the flow rate and TDH, opting for a variable speed submersible pump to enhance adaptability to changes in irradiation. The sixth step consists of calculating the required hydraulic power by factoring in flow rate, TDH, and the manufacturer's efficiency ratings. For the seventh step, we size the photovoltaic array by considering the output from panels, cables, controller, and an overdrive pump, then adding a safety margin (10-20%) as suggested, typically around 20-30%, to account for overcast conditions. Finally, system simulation using specialized software (such as PVsyst or an analytical/graphical model) facilitates the validation of seasonal output, reliability (LPSP), and helps in optimizing costs over the system's lifecycle

a. Assessment of Water Requirements

In the Gabès area, water usage fluctuates significantly based on the time of year, ranging from approximately 3.4 m^3/day in January to a peak of 24 m^3/day in June. This variation underscores the critical need for sufficient storage. Consequently, we have selected a tank with a capacity of 24 m^3, which enables us to completely meet the daily peak demand. This volume guarantees ample autonomy during high-demand periods and offsets uncertainties related to the availability of solar pumping. Additionally, this tank serves a threefold purpose: it regulates the flow, providing reinforcement during peak times; maintains pressure within the distribution network; and acts as a safeguard, ensuring supply in case of a pump system failure or an incident affecting the installations. This decision thus facilitates the optimization of hydraulic management, enhancing the system's resilience in response to seasonal changes, and establishing a dependable water supply for the local communities.

b. Calculation of the Required Flow Rate (Q)

The hourly flow rate of the pump (Q), expressed in m^3/h, corresponds to the daily volume of water that needs to be pumped during sunlight hours. It is calculated using the formula:(Hilarydoss 2023)

$$Q = \frac{V_W}{T} \tag{17}$$

$$Q = \frac{24}{4} = 6m^3/h \tag{18}$$

c. The Total Dynamic Head

The Total dynamic head (THD), commonly referred to as Total Head Lift, indicates the cumulative energy a pump must expend to transport fluid from one location to another within a hydraulic system. It is quantified in meters of water column (mCE), corresponding to the altitude of a water column that necessitates the same pressure. The THD integrates three vital elements:

H_s: Static Height (m): The vertical gap between the water level at the source (e.g., a well, river) when the pump is inactive and the discharge point of the pump.
H_d: Lift Height (m): The extra vertical distance at which the water level in the source diminishes while the pump operates due to pumping. This is crucial in well extraction as it showcases the reduction in water level owed to withdrawal.
H_p: Pressure Height (m): The altitude required to achieve a specific pressure at the outlet point.

H_f: Head Losses (m): The head losses attributed to friction in pipes, fittings, and more. These losses signify the energy needed to counteract friction.

Hazen William can define the friction head loss through a calculable equation.

$$H_f = K.L.\frac{Q^{1.852}}{C^{1.852}.D^{4.871}} \tag{19}$$

Where H_f= friction losses (m), K = constant coefficient =1.22*10 10, L = pipe length (m)=127m, Q = water flow rate (l/s), C: roughness coefficient =150, D = inner diameter of the pipe (m).

Determining the inner diameter of the pipe based on the flow rate: For a flow rate of 6 m^3/h (equivalent to 1.67 l/s), the inner pipe diameter can be established with regard to the optimal velocity of water flow. Within our pumping station, the velocity has been assessed to be 1.3 m/s, falling within the suggested range of 1 m/s to 2 m/s. This interval is perfect for reducing pressure losses and preventing turbulence throughout the system. The formula to ascertain the internal diameter D is:

$$D = \sqrt{\frac{4.Q}{\pi.V}} \tag{20}$$

$$D = \sqrt{\frac{4.0.00167}{\pi.1.3}} = 40.44mmm \tag{21}$$

We opted for a pipe diameter of 50 mm.

$$H_f = 1.22*10^{10}.127.\frac{1.67^{1.852}}{150^{1.852}.50^{4.871}} = 2m \tag{22}$$

$$THD = H_S + H_d + H_f + H_p \tag{23}$$

$$THD = 25 + 66 + 2 + 10.33 = 103.33 \simeq 104m \tag{24}$$

d. Selection and Sizing of the Pump for a Solar-Powered Pumping System

The selection of efficiency and dimensions for the pump designated for the solar-powered pumping system. We performed a search for a pump capable of delivering a throughput of 6 m3/h, which corresponds to 1.6 l/s at a height of 104 m of water column. Therefore, we needed to size the performance using two techniques discovered here to ensure the pump meets its operational requirements:

Curve-Oriented Approach

In our exploration of the 4SR series, we discovered that the 4SR6-23 (23 stories) variant perfectly aligns with our system's throughput and THD specifications. The manufacturer's characteristic curves provide an immediate visual representation of the ideal operating point, making the selection process straightforward without the need for intricate calculations initially. This variant boasts a high hydraulic efficiency, typically exceeding 70%, which guarantees lower energy usage and consistent performance crucial in a photovoltaic setting where power availability varies. Furthermore, its multi-stage architecture and premium materials (featuring floating impellers and AISI 304/316 stainless steel components) ensure exceptional dependability, even under stringent conditions (clear water with sand concentrations up to 150–200 g/m

Figure 7. (a) Range of PEDROLLO pumps, (b) Chart for determining the number of stages of a pump for a given series.

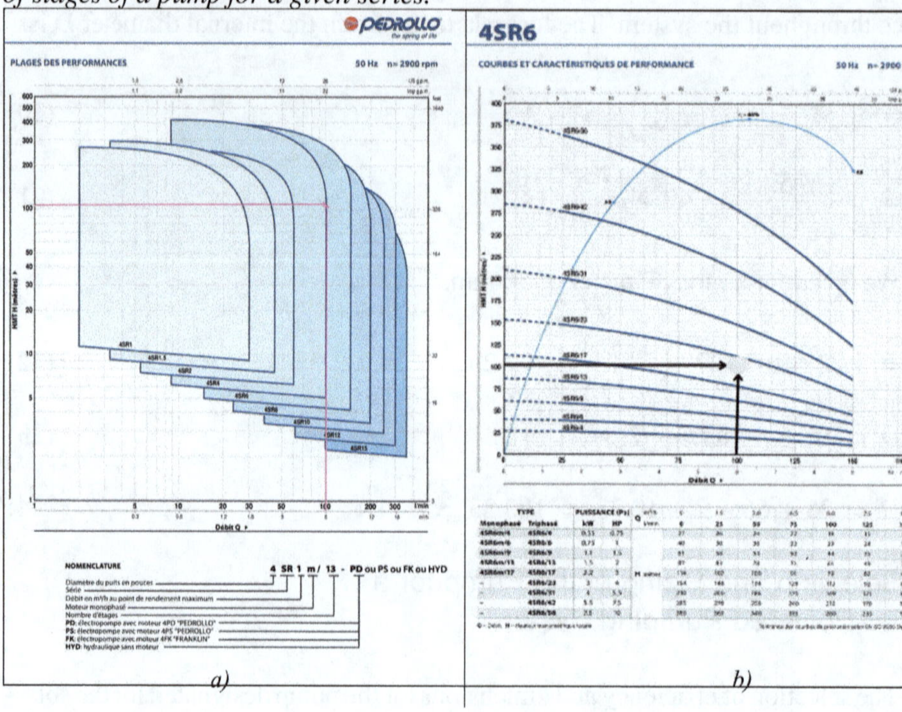

Method by Power Calculation

The hydraulic power required to pump a given flow rate to a specified THD is calculated using the formula:

$$P_h = \rho . g . Q . THD \tag{25}$$

With:
$\rho = 1000 \text{kg}/m^3$: the density of water.
g=9.81m/s^2: gravitational acceleration.
Q=1.67l/s=0,00167m^3/s: the water flow rate.
THD=104m
By applying these values, the hydraulic power is calculated as follows:

$$P_h = 1000 \times 9.81 \times 0.00167 \times 104 = 1704.7w \tag{26}$$

Thus, the required hydraulic power is approximately 1704W. To take into account the losses in the system, the pump power (P_p) is obtained by dividing the hydraulic power by the hydraulic efficiency of the pump (η_p):(Habib, Liu et al. 2023)

$$P_p = \frac{P_h}{\eta_p} \tag{27}$$

With an estimated hydraulic efficiency of 65%, the pump power is calculated as follows:

$$P_p = \frac{1704}{0.65} = 2621.53w \tag{28}$$

Furthermore, to drive the pump, the engine power (P_m) must take into account the combined efficiencies of the pump and the engine (η_p, η_m). Assuming an engine efficiency of 90%, the engine power is given by:

$$P_m = \frac{P_h}{\eta_p . \eta_m} = \frac{1704}{0.65.090} = 2912.8w \tag{29}$$

The power required by the engine is therefore slightly more than 3 kW to compensate for losses and ensure optimal operation.

It is concluded that the two methods (charts and calculations) converge towards the same results, confirming that the 4SR6-23 model is optimal to meet the system requirements. This model can pump a flow of 6 m 3/h with an THD of 104 m, while ensuring good efficiency and high reliability.

In summary, this sizing ensures a perfect match between the pump and the photovoltaic system, offering optimal use of available energy while ensuring durability and efficiency.

HUMAN MACHINE INTERFACE FOR DISTRIBUTED SUPERVISION OF THE SOLAR-POWERED PUMPING SYSTEM

The effective oversight and observation of photovoltaic (PV) systems are crucial for sustaining energy performance, identifying operational irregularities, and guaranteeing long-term reliability of the system. Siemens automation solutions especially the synergy of PLCs with Human-Machine Interfaces (HMIs) provide a robust and adaptable framework for real-time management and data visualization. In this arrangement, the PLC collects and processes signals from various PV system components (such as inverters, sensors, and protective devices), while the HMI offers visual representation of the system's condition. To realize this architecture, Siemens' engineering platform, STEP 7, is utilized for programming and setting up the PLC, while WinCC manages the design and execution of the HMI interface. The initial phase in crafting this supervisory system consists of creating a project and configuring the hardware (HW), which establishes the organization of the automation system and its linked devices.

1. Project Development and Hardware Configuration

The process of creation begins by establishing a new project in TIA Portal via STEP 7, Siemens' integrated engineering platform. This is the beginning of the automation system because it sets up all hardware components and how they are interconnected. You start by selecting the right PLC model (S7-300/400, S7-1200 or S7-1500), and then configure the rack design, power modules, input/output modules, and any communication interfaces that your photovoltaic system needs. Network settings like IP addresses and device names are also configured to enable easy communication among components especially important when adding sensors, inverters, or other smart devices via protocols like Modbus TCP or Profinet.

The first thing to do in this work was to create the project in SIMATIC Manager area as shown in Figure 1. This project is called POMPAGE - PV - PFE00 and is developed on the CPU 314C-2 PN/DP which is a Siemens S7-300 CPU, with digital and analog in and outs, and communication ports integrated. This project file is the basis to develop the program logic, define symbolic variables, and create function blocks.

Such a hardware setup forms a digital twin of the real system that can be simulated and functionally verified. The PLC unit chosen is a PLC having 24 digital inputs, 16 digital outputs, five analog inputs, and two analog outputs, in a manner consistent with appropriate inputs and outputs for sampling and processing relevant environmental and operational parameters such as voltage, current, irradiance, temperature, and water flow.

Figure 8. SIMATIC Manager project interface showing the configured CPU 314C-2 PN/DP in the hardware rack.

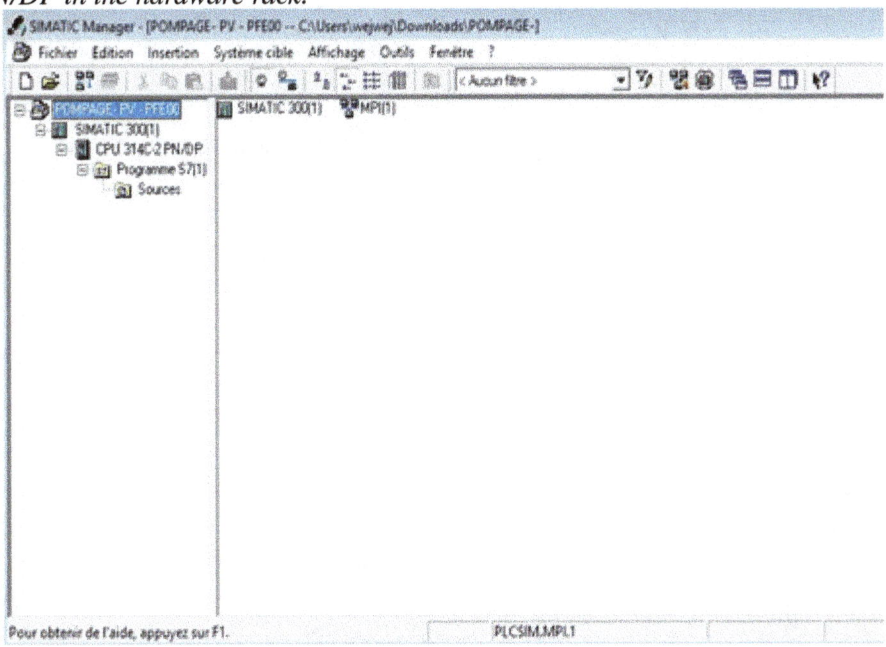

2. STEP 7 PLC Programming

Once the hardware installation is done, the next procedure is to implement the control logic in the form of STEP 7 programming software. The logic specifies how the system reacts to input conditions and governs the photovoltaic pumping facility operation. The programming is carried out in Siemens-maintained standard IEC languages such as Ladder Diagram (LAD), Function Block Diagram (FBD), or Structured Text (ST). In this project, the main control sequences are programmed using LAD, by which logic routines can be intuitively developed to track environment variables and drive equipment such as the water pump. The PLC also continually reads analog values from five sensors connected to analog input channels: voltage, current, irradiance, temperature, and water flow. Scaling of the signals is accomplished using Function FC 105, which converts raw analog reading (0–10 V or 4–20 mA) to engineering units. Symbolic addressing is utilized in dealing with variables in a transparent and trustworthy way. To simulate the program, the S7-PLCSIM simulator is used. The simulator facilitates simulation of the response of the PLC under various conditions without hardware access. The input and output signals may be manipulated in real time to facilitate the testing of alarms, threshold detection, pump control, and system response.(Sahuquet, Meunier et al. 2024)

Figure 9. STEP 7 (HW Config) hardware configuration with CPU 314C-2 PN/DP setup and assigned I/O addresses.

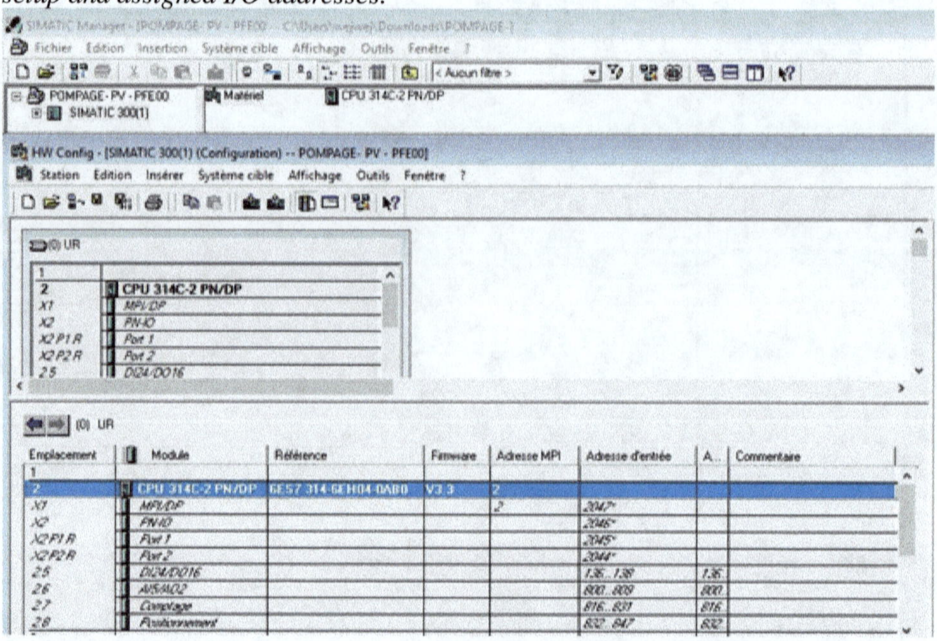

Figure 10. S7-PLCSIM simulation showing digital and analog I/O behavior and response of program logic.

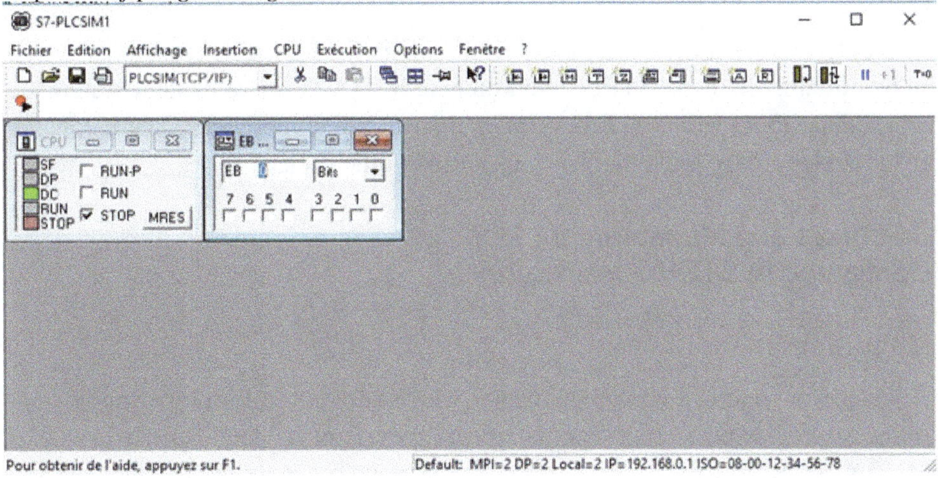

3. HMI Design and Monitoring Based on WinCC

The final part of the supervision system is the Human-Machine Interface (HMI), which is created with WinCC, Siemens' powerful visualization system. The HMI offers a window into the system, allowing operators and maintenance staff to interact with system controls and real-time information in a user-friendly way. Through customized graphical screens, users can:(Bouraiou, Neçaibia et al. 2021)

- Monitor real-time voltage, current, and power generation
- Show alarms and system messages
- Display historical trends and performance records
- Operate equipment such as pumps (in solar water pumping systems)

Each item on a screen is linked to live data points (tags) in the PLC and presents an accurate and up-to-date representation of system conditions. WinCC also provides remote access and web-based interfaces, which is particularly valuable for monitoring off-grid or geographically distant PV systems.

4. Technique and Functional Components of the System Architecture Supervised in STEP 7

The development of the pumping photovoltaic monitoring system was carried out by utilizing Siemens' STEP 7 environment in TIA Portal to make it systematic and modular to integrate all functional and hardware components. The below are the particular components that have been used in the project:

a. Sensors and Measurement Modules (Configured in STEP 7 Hardware)

i. Flow_Sensor

Water flow rate is a known parameter which is required for proper pumps operating as well as protection from overheating or damage. The KOBOLD MIK-5 NA 65 A E34R flow meter provides an efficient and safe solution for this purpose.

The KOBOLD MIK-5 NA 65 A E34R flow meter is a Hall-effect sensor designed for measurement of flow of conductive liquids. Because of its technology, it converts the fluid flow into an electrical signal proportional to the actual flow rate. Either an analog or digital output signal is generated, with easy integration into automated control systems such as Siemens PLCs with analog or digital input modules. Its resistance gives high durability against ordinary operating environments of industrial and agricultural installations. Within a STEP 7-controlled system, the output of the flow meter is piped through to an analog input module where the signal is accessed and processed. Scaling blocks translate the raw reading into a flow measurement in liters per minute or cubic meters per hour, as the application requires. Real-time monitoring of the flow enables early detection of anomalies such as pressure loss, dry running, or absence of flow, thereby avoiding failure and maximizing the performance of the pumping system.

Figure 11. KOBOLD MIK-5 NA 65 A E34R flow meter

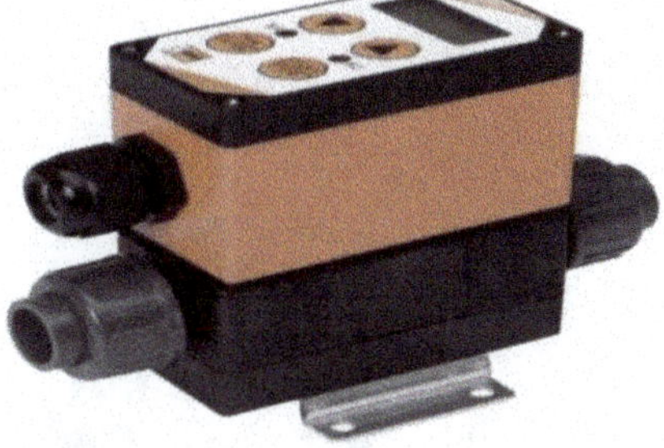

Figure 12. The flow display program

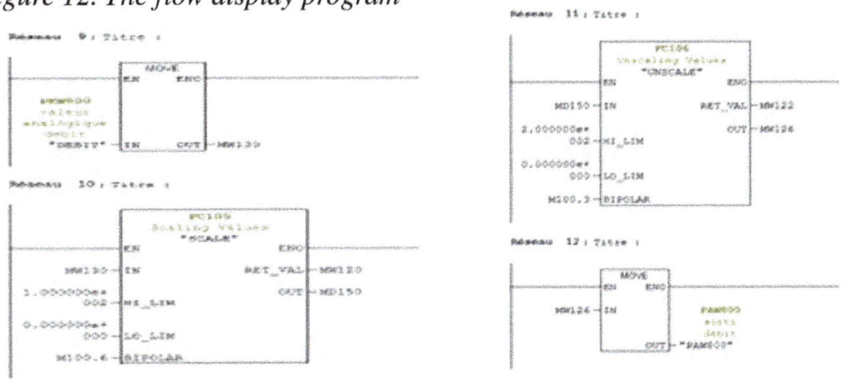

ii. Temperature Sensor

The temperature sensor, typically a PT100 type, plays a crucial role in monitoring ambient temperature and the temperature of the photovoltaic modules. It is required to accurately measure temperature because the electrical performance of PV panels is directly influenced by temperature changes. An increase in module temperature generally causes a decrease in output voltage and thus affects system efficiency as a whole. PT100 is an RTD that is well known for its stability and accuracy. It operates on the fact that the electrical resistance of the platinum element changes reproducibly with temperature. The PT100 sensor is connected to the Siemens PLC through a dedicated RTD input module, which gives accurate signal conditioning and conversion in this system. These temperatures are then processed by the PLC and displayed on the HMI, allowing operators to view thermal conditions in real time and take remedial action if overheating or abnormal thermal activity is detected.

Figure 13. PT100 Sensor figure 14.The program of the temperature reading

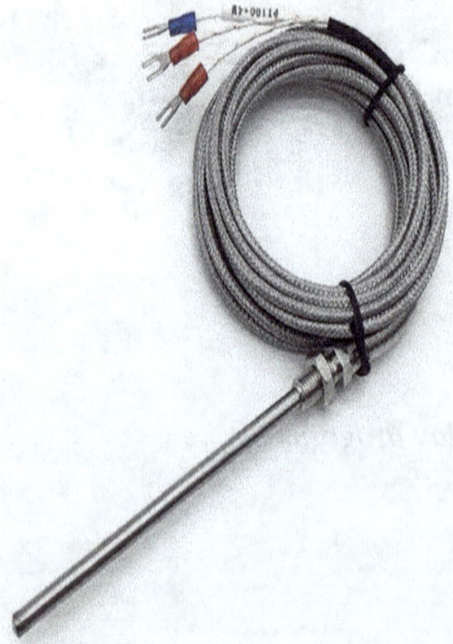

iii. Irradiance Sensor

The irradiance sensor plays a crucial role in monitoring photovoltaic systems. It's usually linked to the PLC through an analog input module, like the AI 4x13 bits module, which allows for accurate capture of the analog signal that reflects light intensity. This signal is then processed in STEP 7 programming blocks, where it gets scaled to transform the raw data into a usable irradiance measurement. For local and real-time display, you can set up a system using a PIC16F877A microcontroller. This microcontroller is great because it can handle analog-to-digital conversion and processing, taking the voltage output from the irradiance sensor, turning it into a digital value, and showing it on an LCD screen. This standalone device provides a user-friendly interface, making it easy to read irradiance data directly without needing a centralized computer system.

Figure 15. Illumination map diagram on ISIS

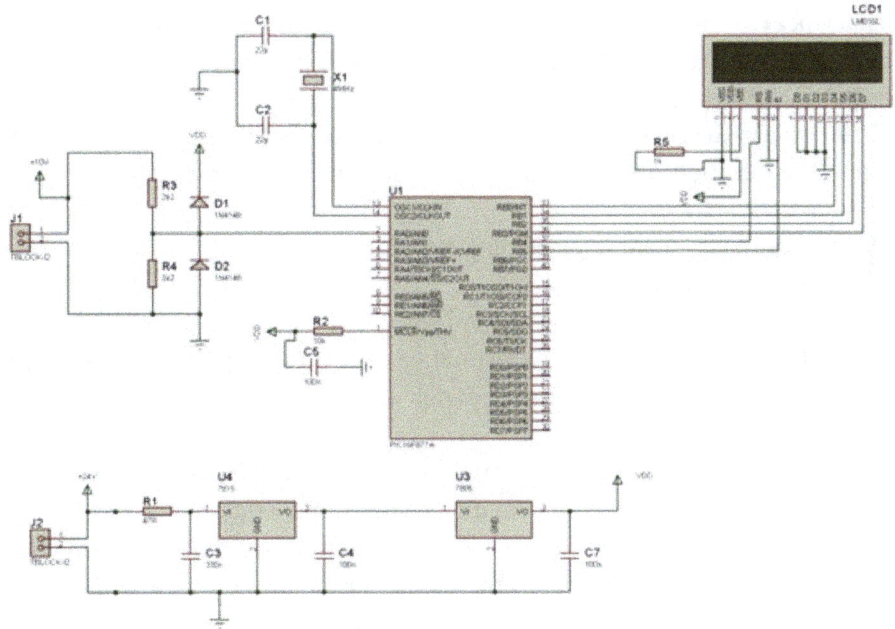

Figure 16. The illuminance display

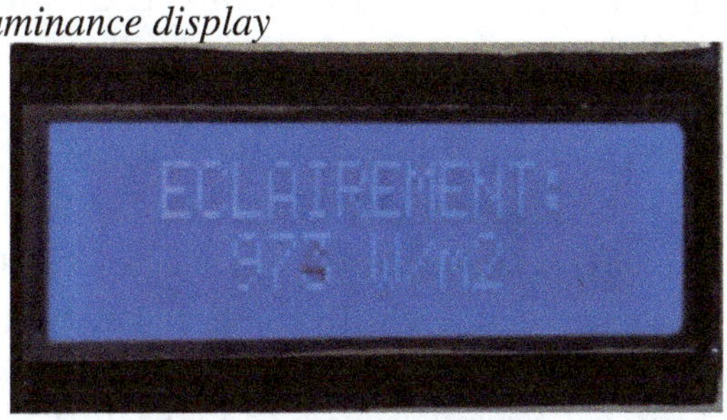

COMPARATIVE ASSESSMENT OF MPPT ALGORITHMS: P&O, INCREMENTAL CONDUCTANCE, AND NEURAL NETWORK BASED APPROACHES

Maximum Power Point Tracking (MPPT) algorithms are the essence of efficient photovoltaic (PV) systems, especially in energy optimization-sensitive applications such as solar water pumping. In this paper, we present a comparative analysis of three of the widely used MPPT methods: Perturb and Observe (P&O), Incremental Conductance (INC), and Artificial Neural Networks (ANN). The comparison is made both from theoretical perspectives as well as through simulation studies conducted with MATLAB/Simulink.

This comparison is driven by the need for keeping constantly in peak the energy output of a solar field while ensuring adequate supervision through an interface homme-machine (IHM).

Three algorithms were selected due to their representativeness in the classical and intelligent approach:

- ✓ P&O (Perturb and Observe): a simple and very widespread method relying on the periodic perturbation of panel voltage and measurement of power change.
- ✓ INC (Incremental Conductance): which improves precision by contrasting instantaneous conductance with incremental conductance.
- ✓ ANN (Artificial Neural Network): a sophisticated technique founded on supervised learning, capable of generalizing the complex relationship between irradiance, temperature, and maximum power

1. Simulation Strategy in Simulink

All the MPPT algorithms were simulated in the MATLAB/Simulink environment under identical test conditions: a solar panel exposed to fluctuating irradiance and varying temperature. This allowed for close comparison of the rate of convergence to the Maximum Power Point (MPP), power stability, dynamic response towards fluctuating irradiance, and voltage output characteristics. Each algorithm's simulation framework includes: A simulated PV module, a DC-DC boost converter, MPPT controller blocks tailored for each algorithm, and a centrifugal pump driven by ASM, Voltage, current, and power waveform monitoring visualization blocks. Specifically for the Artificial Neural Network (ANN) MPPT, a detailed Simulink model was built which includes an already trained neural network to predict the optimal operating point in real-time. The ANN block receives inputs such as irradiance and temperature and provides a reference voltage or duty cycle signal that regulates the boost converter. This sophisticated control technique enables immediate adaptation to changing environmental conditions and minimizes transient power losses to an absolute minimum.

The following illustration illustrates the Simulink block diagram of the ANN-based MPPT controller and shows how the neural network is integrated into the overall model of the photovoltaic pumping system.

Figure 17. Simulink Implementation of Artificial Neural Network Based MPPT Controller for PV pumping System

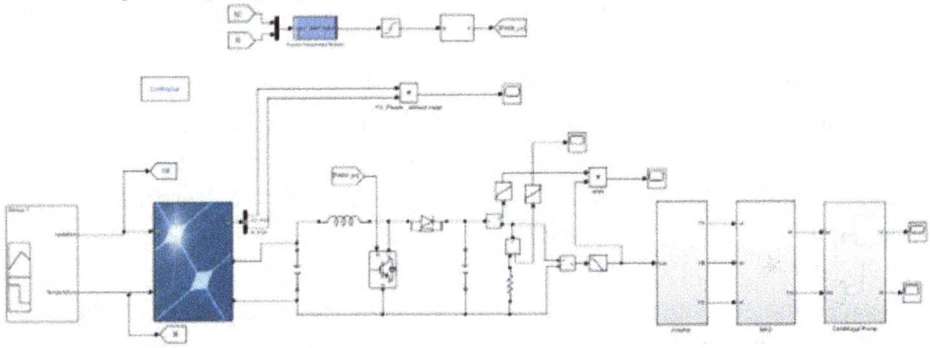

2. Comparative Analysis and Results

The simulation results highlight the strengths and weaknesses of each MPPT algorithm under varying environmental conditions. The Perturb and Observe (P&O) method shows a fast convergence to the Maximum Power Point (MPP), but it suffers from ongoing oscillations around the optimum point, especially under high irradiance changes. These oscillations lead to energy losses and compromised system efficiency over time.

In order to better illustrate the performance differences, the next table shows (Ali, 2025 #43)the summary of the main evaluation criteria for each MPPT method based on the simulations in MATLAB/Simulink.

Table 1. Performance Comparison of MPPT Methods: P&O, Incremental Conductance, and Artificial Neural Network

Criterion	P&O	INC	ANN
Stability	Moderate (oscillations)	Good (reduced oscillations)	Excellent
Convergence speed	Fast	Moderate	Very fast
Response to irradiance changes	Poor	Moderate	Excellent
Robustness	Low	Moderate	High (requires training)
Computational cost	Low	Moderate	High
Implementation complexity	Low	Moderate	High

The Incremental Conductance (INC) technique provides greater stability and more accurate tracking compared to P&O. However, it also has its drawbacks in extreme or noisy conditions, where its responsiveness and accuracy degrade. (Bechekir[1], 2025 #40)

In contrast, the Artificial Neural Network (ANN) approach performs better on all the evaluation metrics considered. It offers excellent stability, very fast convergence, and high robustness, especially in the presence of quick change of environmental conditions. ANN's most significant advantage lies in its ability to learn the nonlinear dynamics of the PV system and forecast it precisely, if it has been trained properly beforehand. (Daw, 2025 #41)

Besides, the simulation curves clearly demonstrate the effectiveness of the ANN method for reducing transient energy losses and maximizing the power drawn from the PV array. This confirms the neural network-based MPPT control intelligence in practice where flexibility and precision are the main requirements.

Figure 18. Output Voltage Response under 962 W/m² Irradiance and 22°C Temperature using P&O and ANN MPPT Algorithms

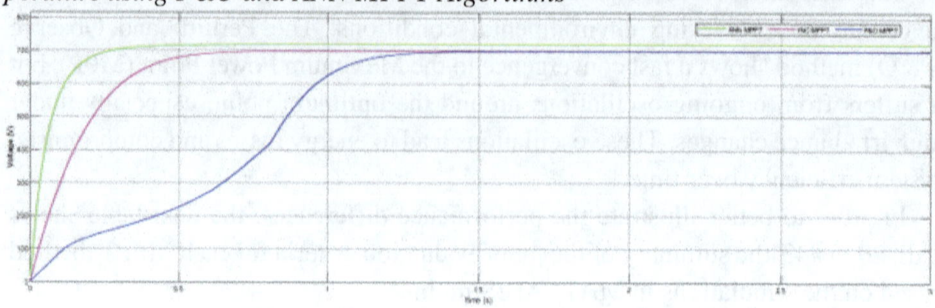

Figure 19. Output Power Response under 962 W/m² Irradiance and 22°C Temperature using P&O and ANN MPPT Algorithms

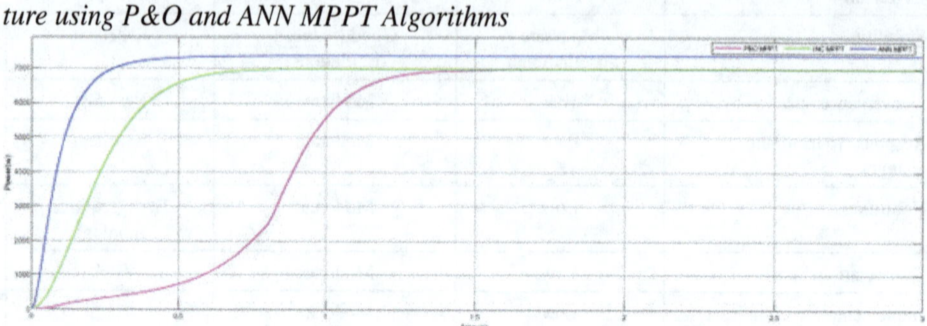

The curves in figures 18 and 19 reflect the dynamic response of the photovoltaic system under the irradiance of 962 W/m^2 and the temperature of 22°C, comparing two Maximum Power Point Tracking (MPPT) algorithms: Perturb and Observe (P&O) and Artificial Neural Network (ANN).

In the voltage curve of figure 18, the ANN algorithm finds a stable voltage of approximately 730 V within a very short time of less than 0.2 seconds, while the P&O algorithm requires a time of more than 1.5 seconds to find the same. This reflects the superior responsiveness of ANN, and the system thus stabilizes at the optimal operating point within a very short time.

The power curve in figure 19 also confirms this advantage: while P&O reaches the maximum power of 7000 W at a less steep rate, ANN reaches over 7000 W and does so considerably faster. This indicates that not only does the ANN algorithm converge more rapidly, but also it extracts more power, effectively dealing with nonlinearities within the system and sudden irradiance transients.

In brief, ANN demonstrates much better performance in terms of response time, stability, and energy efficiency and is highly suitable for photovoltaic systems subjected to fluctuating environmental conditions or maximum energy output demands.

3. Reference to Human-Machine Interface

The implementation of MPPT algorithms into a supervisory system using STEP 7 and a Human-Machine Interface (HMI) facilitates dynamic tuning of the Maximum Power Point Tracking method. Dynamic changes between active algorithms are possible for the operator, real-time system performance (voltage, current, power) is displayed immediately, and data are recorded for diagnostic or predictive maintenance. The structure highlights the direct relationship between algorithmic control and human operation, a major objective of this work. This intelligent monitoring system not only optimizes energy efficiency but also provides the operator with greater control of decision-making. With real-time visual feedback and the ability to switch between algorithms based on environmental conditions, the system becomes more responsive and flexible. For instance, in variable irradiance conditions, the operator would like the ANN-based MPPT due to its rapid adaptability, while in steady-state conditions, the simpler and computationally less intense P&O algorithm would be the choice. Besides, long-term data collection enables effective performance assessment over time. If an algorithm performs worse consistently in a certain condition, the system is able to notify or recommend a switch. This blend of algorithmic intelligence and human intelligence enhances the photovoltaic system to be tougher, more intelligent, and better capable of adapting independently to its surroundings while putting the operator in the middle of decision-making

FUTURE RESEARCH DIRECTIONS

the future work could also be addressed towards the integration of fault detection and diagnosis modules within the supervision system with the objective of improving the reliability and robustness of photovoltaic installations in real-world conditions

CONCLUSIONS

This paper presents a comprehensive modeling and analysis of a photovoltaic water pump system with focus on three different Maximum Power Point Tracking (MPPT) techniques: Perturb and Observe (P&O), Incremental Conductance (INC), and Artificial Neural Networks (ANN). Through simulation and performance comparison, it has been demonstrated that the ANN-based MPPT method has greater tracking precision and response time performance, especially under high rates of varying irradiance conditions. While the conventional algorithms such as P&O and INC are still attractive due to their simplicity and ease of implementation, they typically suffer from oscillations close to the MPP or sluggish response under dynamic conditions. The ANN approach, while more complex and with a training phase, is more adaptive and cognitive in nature and therefore proves to be an appealing solution for future-generation smart energy systems. Along with that, human-machine interface (HMI) functionality through STEP 7 software aids system monitoring and enhances real-time monitoring and control. This makes it possible to have improved communication between the automated system and the user, making it run efficiently and reliably. Overall, the combination of smart MPPT techniques with industrial automation technologies provides a powerful paradigm to optimize the efficiency of solar-powered pump systems, especially in isolated or off-grid regions. Future studies could explore hardware implementation and experimental validation of proposed approaches to assess their practical feasibility.

ACKNOWLEDGMENT

The authors are grateful to the Process, Energetic, Environment and Electrical Systems Research Laboratory (LR18ES34) funded by the Tunisian Ministry of Higher Education and Scientific Research and to the Higher Institute of Technological Studies of Gabès (ISET Gabès) for the technical and financial assistance received during the elaboration of the bigger project from which this paper originated.

REFERENCES

Abut, T. (2016). Modeling and optimal control of a DC motor. *International Journal of Engineering Trends and Technology*, *32*(3), 146–150.

Ahmed Baba, M., Naoui, M., & Cherkaoui, M. (2023). Fault-tolerant control strategy for Hall sensors in BLDC motor drive for electric vehicle applications. *Sustainability*, *15*(13), 10430.

Alhamami, A. H., Abba, S. I., Musa, B., Dodo, Y. A., Salami, B. A., Dodo, U. A., & Alyami, S. H. (2024). Off-grid multi-region energy system design based on energy load demand estimation using hybrid nature-inspired optimization algorithms. *Energy Conversion and Management*, *315*, 118766.

Ali, M. H., Zakaria, M., & El-Tawab, S. (2025). A comprehensive study of recent maximum power point tracking techniques for photovoltaic systems. *Scientific Reports*, *15*(1), 14269. PMID: 40274881

Allal, A., Lammouchi, Z., Khechekhouche, A., Mohamed, N., Alqunun, K., Alghadhban, A., Marouani, I., & Guesmi, T. (2024). Advanced Residual Optimal Mapping Approach for Precise Detection of Stator Faults in Induction Motors. *IEEE Access: Practical Innovations, Open Solutions*, *12*, 143515–143530.

Alqunun, K., Bechiri, M. B., Naoui, M., Khechekhouche, A., Marouani, I., Guesmi, T., Alshammari, B. M., AlGhadhban, A., & Allal, A. (2025). An efficient bearing fault detection strategy based on a hybrid machine learning technique. *Scientific Reports*, *15*(1), 18739. PMID: 40437009

Anis, W. R., Mertens, R. P., & Van Overstraeten, R. J. (1985). Coupling of a volumetric pump to a photovoltaic array. Solar cells, 14(1), 27-42.

Bechekir[1], M. B. S., Brahami[1], M., Bendaho, H., & Brahimi, A. (2025). Comparative Analysis of MPPT Algorithms: P&O and Inc for Optimizing PV Systems. Smart Computing and Control Renewable Energy Systems: Advanced Computational Techniques for Wireless Network Optimization, 1238, 60.

BENMEDDAH. A., & LAIDI, K. (2025). Optimisation de l'algorithme MPPT pour un système photovoltaïque en intégrant de nouvelles techniques d'intelligence artificielle (Doctoral dissertation, Faculté des sciences et de la technologie).

Bouraiou, A., Neçaibia, A., Motahhir, S., Bouakkaz, M. S., Attoui, I., Boutasseta, N., & Ziane, A. (2021, January). Supervision and monitoring of photovoltaic systems using siemens PLC and HMI. In *International Conference on Digital Technologies and Applications* (pp. 1147-1157). Cham: Springer International Publishing.

Daw, M. O., Shalem, S. M., Habeeb, L. J., & Faraj, A. O. (2025). A Comparative Study of Perturb & Observe and ANN-Based MPPT Algorithms Under Various Environmental Conditions. *Scientific Journal of University of Benghazi*, *38*(1), 62–85.

Deshpande, S., & Bhasme, N. R. (2017). A review of topologies of inverter for grid connected PV systems. 2017 Innovations in Power and Advanced Computing Technologies (i-PACT), 1-6.

Divyasharon, R., Banu, R. N., & Devaraj, D. (2019, April). Artificial neural network based MPPT with CUK converter topology for PV systems under varying climatic conditions. In *2019 IEEE International Conference on Intelligent Techniques in Control, Optimization and Signal Processing (INCOS)* (pp. 1-6). IEEE.

Garud, K. S., Jayaraj, S., & Lee, M. Y. (2021). A review on modeling of solar photovoltaic systems using artificial neural networks, fuzzy logic, genetic algorithm and hybrid models. *International Journal of Energy Research*, *45*(1), 6–35.

Habib, S., Liu, H., Tamoor, M., Zaka, M. A., Jia, Y., Hussien, A. G., & Kamel, S. (2023). Technical modelling of solar photovoltaic water pumping system and evaluation of system performance and their socio-economic impact. *Heliyon*, *9*(5). PMID: 37229163

Hilarydoss, S. (2023). Suitability, sizing, economics, environmental impacts and limitations of solar photovoltaic water pumping system for groundwater irrigation—A brief review. *Environmental Science and Pollution Research International*, *30*(28), 71491–71510. PMID: 33575942

Hmidet, A., Rebei, N., & Hasnaoui, O. (2015, March). Experimental studies and performance evaluation of MPPT control strategies for solar-powered water pumps. In 2015 Tenth international conference on ecological vehicles and renewable energies (EVER) (pp. 1-12). IEEE.

Hmidet, A., Subramaniam, U., Elavarasan, R. M., Raju, K., Diaz, M., Das, N., & Boubaker, O. (2021). Design of Efficient Off-Grid Solar Photovoltaic Water Pumping System Based on Improved Fractional Open Circuit Voltage MPPT Technique. *International Journal of Photoenergy*, *2021*(1), 4925433.

Hussain, M. T., Sarwar, A., Tariq, M., & Urooj, S., BaQais, A., & Hossain, M. A. (2023). An evaluation of ANN algorithm performance for MPPT energy harvesting in solar PV systems. *Sustainability*, *15*(14), 11144.

Jain, S., Karampuri, R., & Chinthamalla, R. (2018). Solar PV Powered Standalone Water Pumping Systems: Modeling, Design, and Analysis. In Handbook of Research on Power and Energy System Optimization (pp. 139-190). IGI Global Scientific Publishing.

Katche, M. L., Makokha, A. B., Zachary, S. O., & Adaramola, M. S. (2023). A comprehensive review of maximum power point tracking (mppt) techniques used in solar pv systems. *Energies*, *16*(5), 2206.

Lammouchi, Z., Naoui, M., Bekakra, Y., Khechekhouche, A., & Sbita, L. (2025). Model-Free Predictive Control based on the Execution Time of Active Vectors Using Ultra-Local Model. *Arabian Journal for Science and Engineering*, *50*(8), 5885–5898.

Maka, A. O., Mehmood, M., & Chaudhary, T. N. (2023). Design, simulation and performance analysis of photovoltaic solar water pumping system. *Energy Harvesting and Systems*, *10*(2), 287–299.

Mazloumi, A., Poolad, A., Mokhtari, M. S., Altman, M. B., Abdelaziz, A. Y., & Elsisi, M. (2023). Optimal sizing of a photovoltaic pumping system integrated with water storage tank considering cost/reliability assessment using enhanced artificial rabbits optimization: A case study. *Mathematics*, *11*(2), 463.

Motahhir, S., El Hammoumi, A., & El Ghzizal, A. (2018). Photovoltaic system with quantitative comparative between an improved MPPT and existing INC and P&O methods under fast varying of solar irradiation. *Energy Reports*, *4*, 341–350.

Patra, B., Nema, P., Khan, M. Z., & Khan, O. (2023). Optimization of solar energy using MPPT techniques and industry 4.0 modelling. *Sustainable Operations and Computers*, *4*, 22–28.

Pop, T., Buzduga, C., Pentiuc, R. D., Ifrim, V. C., & Ungureanu, C. (2022, October). Mathematical modeling in MATLAB of a photovoltaic panel. [). IOP Publishing.]. *IOP Conference Series. Materials Science and Engineering, 1256*(1), 012037.

Pradhan, A., & Panda, B. (2018). A simplified design and modeling of boost converter for photovoltaic sytem. *International Journal of Electrical and Computer Engineering, 8*(1), 141.

Romdhane, M., Naoui, M., & Mansouri, A. (2023). PMSM Inter-Turn Short Circuit Fault Detection Using the Fuzzy-Extended Kalman Filter in Electric Vehicles. *Electronics (Basel), 12*(18), 3758.

Sahuquet, A., Meunier, S., Cherni, J. A., Charpentier, A., Vezin, T., Darga, A., & Quéval, L. (2024). Photovoltaic pumping tests: A novel supervision method for photovoltaic water pumping systems. *Heliyon, 10*(21). PMID: 39553633

Soni, J., & Bhattacharjee, K. (2023). Optimization of MPPT Controller for Standalone Photovoltaic Systems. In *Perspectives on Social Welfare Applications' Optimization and Enhanced Computer Applications* (pp. 231–263). IGI Global.

Tey, K. S., Mekhilef, S., Seyedmahmoudian, M., Horan, B., Oo, A. T., & Stojcevski, A. (2018). Improved differential evolution-based MPPT algorithm using SEPIC for PV systems under partial shading conditions and load variation. *IEEE Transactions on Industrial Informatics, 14*(10), 4322–4333.

Wang, M. S., Chen, S. C., & Shih, C. H. (2018). Speed control of brushless DC motor by adaptive network-based fuzzy inference. *Microsystem Technologies, 24*(1), 33–39.

Wang, X., & Suh, C. S. (2017). Precision concurrent speed and position tracking of brushed dc motors using nonlinear time-frequency control. *Journal of Vibration and Control, 23*(19), 3266–3291.

Youssef, R., Hamida, M. A., Mohamed, N., Lassaad, S., Djamel, T., & Yasmine, A. Single-Stage Standalone Photovoltaic Water Pumping System Using Predictive Torque Control (PTC) of Induction Machine. Electric Power Components and Systems, 1-17.

Chapter 3
Evaluation of HMI in Energy Infrastructure: From Analog Controls to AI-Driven Interfaces

Nidhya Rathinavel
https://orcid.org/0000-0002-9962-5478
PSG Institute of Technology and Applied Research, India

S. Vaishnavi
https://orcid.org/0009-0001-1899-6753
PSG Institute of Technology and Applied Research, India

Sanjith Maheswaramoorthy
PSG Institute of Technology and Applied Research, India

Tamizharasu Mohanraj
PSG Institute of Technology and Applied Research, India

ABSTRACT

In this chapter adopted a critical view of the evolution of Human Machine Interfaces (HMIs) in energy infrastructures from analog controls to AI augmented systems, particularly with regard to how digitalisation, automation and the integration of renewables have changed operator working practices and how systems are designed. Opportunity and development of HMIs have appeared in many forms, including SCADA, DCS, predictive analytics and the incorporation of explainable AI. Using frameworks including ISO 9241, NASA TLX and SAGAT the chapter evaluates the

DOI: 10.4018/979-8-3373-5621-1.ch003

effectiveness of HMIs for usability, safety and trust. Using examples including smart grids and nuclear controls, HMI design is argued to also be human centred to allow energy systems to be resilient and sustainable whilst enabling operator cognition.

INTRODUCTION

With continued advancement within global energy systems, Human Machine Interfaces (HMIs) have become crucial interfaces between human operators and complex technical systems. HMIs influence situational awareness, operator performance, safety outcomes, and energy system resilience. In safety critical environments, such as nuclear power plants, windfarms, and smart grids, user friendly HMI design can help to mitigate human error, reduce response time, and help coordinate systems quickly and effectively in dynamic and uncertain operating environments. Conversely, poorly-designed HMIs can contribute to the worst consequences of data misinterpretation, alarm fatigue, cognitive overload, and ultimately, system failure(Piccini, 2002)(Cannan & Hu, 2011).

The first of these influences, digitalization, has transformed how energy systems are observed and managed. Operators now have access to a dizzying array of operational data through Supervisory Control and Data Acquisition (SCADA) systems, Internet of Things (IoT) enabled sensors, cloud-based data platforms, and real-time data analysis as human operators move from seeking simple physical indicators of system state to weighing the implications of huge streams of abstract digital information and the mental workload of rapid decision-making has increased significantly.

The second force driving HMI evolution is the rapid expansion of automation and artificial intelligence (AI). AI enabled HMIs can provide additional functionality for operators beyond traditional HMI features like predictive maintenance, automated fault detection, anomaly classification, and prescriptive advice for control actions. For these reasons, HMIs need to be designed to use the principles of explainable artificial intelligence (XAI), and provide transparency as well as logical and intuitive interaction that leads to common situational awareness across technology and the human operator.

The third and last hurdle for now, is the growing complexity and decentralization of the energy infrastructure itself. Power systems or grids were natively centralized, where power flowed in one (upstream) direction, from defined market participants as larger from each participant to consumers. Additionally, the interface system must deliver the operator experience to include functionality and tools to make sense of conflictual complex operational states, synthesize or integrate smart alerts from disparate entity analysis and to provide operators with the ability to take coordinated action in real time.

In light of this situation, the purpose of this chapter is to provide a moment to assess the evolution and influence of HMIs in energy systems from analog in nature to reality to the modern AI-enhanced interfaces today. One unique aspect of the chapter is the introduction of integrating engineering insights and human factors methods to assess HMI effectiveness.

Section 4 discusses the rise of adaptive and AI based interfaces which include natural language processing, gesture recognition, AR/VR and explainable AI and which promise an increase in transparency and ease of use. Section 5 explains the evaluation framework proposed in detail, showing how the framework could be used to evaluate the usability and functional performance of HMIs in real-world operational settings. Section 6 draws a comparison of analog, digital and AI-based systems, specifically discussing differences in response time, operator workload, the complexity of interfaces and user trust. Section 7 includes examples of design and evaluation-derived case studies from operational energy control scenarios e.g., smart grid dashboard, renewable energy control centre, and nuclear facility interface. Finally, Section 8 concludes by highlighting current and emerging issues particularly focusing on data privacy, AI ethics, regulatory frameworks and interdisciplinary collaboration and makes recommendations for future research and system development(Mourtzis et al., 2023)(Amditis et al., 2006).

This chapter looks at the relationship between technology evolution notational systems, the development of cognition inside humans, and the borderline relationship between technology and humans in the context of HMIs. As energy systems become more autonomous, distributed, and complex, there will be a need to design interfaces that are naturalistic, transparent, and support operator thinking and decision making to ensure energy operations are sustainable, resilient, and compatible with people.

HISTORICAL EVOLUTION

Human Machine Interfaces (HMIs), in energy systems, has progressed through the same phases as industrial automation and control, they moved from tangible mechanical devices to software, and now digital applications. In the early to mid-20th century, the mechanical devices that comprised energy infrastructure mainly thermal power plants and hydropower stations had mechanical HMIs of pipes, valves, rotary switches, pressure gauges, and lever-controlled boards. The human operators were accustomed, and trained to interact with machines using direct physical control of mechanical systems with controls that were often based on the interpretation of analog dials and meters. It was a very manual control environment where the human operator had a cognitive and physical load to stabilize the process during the operation of procedural knowledge, utilizing their own area of experience while making real time control judgment. The operators in the field communicated with those in the control rooms using intercoms, paper logs, or the occasional telephone, the system feedbacks were limited to visual static images(Zhang et al., 2022)(Johannsen, 2009).

The proliferation of energy systems at ever-increasing scales and complexity precipitated the obsolescence of mechanical systems and systems of controls, and by the mid-20th century, controls had evolved into the form of analog control panels. Analog systems had the advantage of being more spatially efficient, and operators could supervise systems in a more coordinated way; nevertheless, they still required a large mental workload to maintain situational awareness. Data from monitored systems was available on separate panels, requiring operators to mentally synthesize readings between disparate systems, and anticipate system behaviour without relying on predicted behaviour or trend data. In addition, alarm systems were hard-wired and binary, leading to frequent alarm flooding, making it complicated for operators not only to quickly identify the root cause of an issue, but disambiguate even the levels of seriousness of abnormal operating conditions(Chedid & Saliba, 1996).

During this time, the operator's role had become that of a centralized monitor and responder via surveillance in real time and then manual actuation based on visual cue and normal operating procedure First, SCADA systems allowed for remote monitoring and control over large distances connecting substations, generation assets, and transmission system assets via networked communication protocols. SCADA used digital terminal devices and early graphical user interfaces (GUIs) to present real time sensor data, control statuses, and alarm interruptions. DCS systems were for continuous processes within the industrial environment and power generation was one of the key continuous industrial processes, the DCS used localized "controllers", or "processors" and control subsystems to manage the industrial process, while communicating back to the operator at a "station" and so on with a delay in real time. DCS systems used a distributed control approach, which allowed for control to be modular, system redundancy, and in real time(Vasiliev, 2016).

For the first time, operators could engage with digital representations of process models, display models, trend graphs, and color-coded alarms for each representation, which improved their capacity to analyse dynamic behaviours, and deviations from normal operations, but there was also a downside to the digital developments. Errors in analog systems were primarily physical, observable and more localized; errors in digital systems could have source errors in software bugs, misconfigurations, or delays in communication often, making errors more obscure, and often more challenging to detect or diagnose.

Regardless of the inherent complexities, digitalization provided substantial advancements in terms of centralized data that's acquired from many sites in real-time, remote control, networked synchronization, logging, and the ability to add and connect even larger numbers of energy generating assets as renewable energy sources are implemented throughout energy systems. Graphical interfaces provided operators with greater operational visibility and transparency which yielded more informed and thoughtful decision making for processes, and enabled significantly better long-term optimization of systems. However typical legacy issues from previous more mechanized systems are still observed, including user overload, lack of consistent design patterns, and a focus on the traditional engineering layout and not on user configuration interfaces. Overall, a substantial transformation from mechanical to digital HMIs has been completed, not only of the technology but also from a cognitive and operational perspectives for human operators(Alqunun et al., 2025; Lammouchi et al., 2025; Youssef et al., 2023).

Every generation of technology mechanical, analog, and digital included affordances and constraints that influenced perceptions, meanings, and actions in energy systems. The transition to AI based or augmented and intelligent human machine interfaces (HMIs) highlights the knowledge needed across different fields of expertise: a collaboration between control engineers, cognitive scientists, and human factors experts is necessary to design interfaces that are both technically sound and cognitively aligned. Understanding these historical eras is important not only from a technological perspective, but because it informs understanding of what future HMIs need to include that supports adaptive, safe, resilient human performance in a context of increasing automation in energy work.(Table 1)

Table 1. Systems across five key dimensions, as discussed in your provided text

Dimension	Mechanical HMIs	Analog HMIs	Digital HMIs (SCADA/DCS)	AI-Enabled HMIs (Emerging)
Interface Type & Technology	Physical devices: valves, levers, gauges	Centralized analog panels with gauges, buttons, lights	GUI-based SCADA/DCS with trend graphs, color-coded alarms	Intelligent, adaptive dashboards using predictive analytics
Operator Role	Direct manual control based on experience; physical engagement	Centralized monitoring and response; required mental synthesis	Supervisory role with partial automation; interface navigation required	Strategic oversight and decision partner; relies on AI recommendations
Communication & Feedback	Intercoms, paper logs; static visual feedback	Centralized alarms and displays; hard-wired signals	Real-time remote monitoring; digital feedback via screens	Predictive alerts; contextual data visualizations; explainable AI (XAI)

continued on following page

Table 1. Continued

Dimension	Mechanical HMIs	Analog HMIs	Digital HMIs (SCADA/DCS)	AI-Enabled HMIs (Emerging)
Cognitive & Operational Demands	High physical and cognitive load; real-time manual judgment	High mental load; alarm flooding and info synthesis required	Reduced physical but increased cognitive effort; automation bias possible	Variable workload depending on trust/ transparency; focus on cognitive ergonomics
Challenges & Advancements	Limited visibility, high manual burden	Alarm confusion, lack of trend data	Interpretation overload, error complexity, inconsistent UIs	Trust in AI logic, need for interdisciplinary design, ethical & adaptive interface design

HUMAN FACTORS IN DIGITAL INTERFACES

As digital technologies continue to dominate contemporary energy system control environments, the development and operation of Human-Machine Interfaces (HMIs) is serious issues of human reliability and operational effectiveness. Digital technologies continue to facilitate improved access to real-time information, capabilities for visualization, and elements of automation, and also brought new issues to solve and consider around human factors, specifically how operators to perceive, interpret, and act on complex flows of system information. Usability is one of the fundamental principles of developing digital displays because it captures how well operators can carry out routine processes, recognize anomaly conditions, and keep control of system performance through sustained attention and work under time pressure. Unfortunately, many digital systems are poorly organized (in terms of a design layout), poorly navigated (often non-intuitive designs), overly dense (too much fractional text crowded in displays), and symbolic representations are not consistent nor specific with increased information (more symbols) leading operators to confusion and potentially error particularly during periods of heightened stress and compressed time(Figure 1).

Digital systems, stand apart from analog systems, where operators had tactile cues, and physical affordances of control and display (i.e., knobs, switches, gauges, displays) which were generally understood and had quite a few cognitive resources, and shaped an operator's mental picture of the operation of the system remained quite constant. Digital displays rely on abstract symbols, layers of screens, and soft controls which demand high cognitive resources of a modern operator.

These challenges become even more pronounced when training agendas do not account for the increasing complexity of the interfaces and decision-making paths. While operator training might touch on the mechanical operation of a system, it often fails to engage the cognitive skills needed to recognize anomalies, develop mental models, and adapt to unexpected conditions especially when AI enabled decision aids are in use. Lacking proper training will only increase the risk of human error, which often occurs in predictable ways: slips (the unintended performance of an action as a by-product of distraction), lapses (forgetting under cognitive stress), and rule-based mistakes (incorrect application of standard operating procedures to a different situation, due to differences).

Figure 1. Human Factors in Digital Interfaces

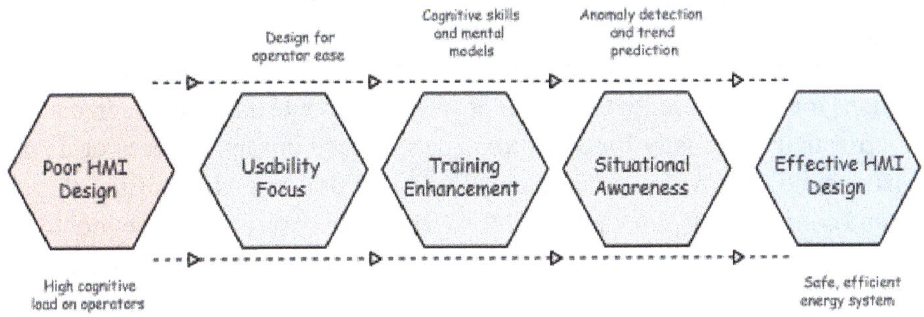

Considering cognitive load is also important this is the mental effort needed to process and act on the information. Energy operators often have to quickly manage streams of commonsense information thus navigate the fluctuating behaviours of a system, detect slight deviations, and coordinate interactions with other systems simultaneous to their attention to the control console. Increasing cognitive load can be propagated through poor interface design such as split screen designs, deep nesting in the menu structures, and repetition in alarms. Information overload, slower reaction times, and reduced accuracy in decision-making are all associated with increased cognitive load(Oury & Ritter, 2021)(Chappel & Mitchell, 1996).

There have been numerous industrial accidents that have cited failing to maintain situational awareness. This is not surprising, as situations such as the possible absence of or real-time feedback with automated systems do not always exist, particularly, in cases in which the interface does not adequately reflect the change, or perhaps critical system notifications simply are not effective. In addition, failing to maintain situational awareness can also evolve into operator complacency with automation. Instead of being decision makers, operators can evolve into passive monitors, which can result in dire consequences where systems are naively behaving erratically with uncertain outcomes, or where the system is operating incorrectly, in these instances failing to maintain situational awareness can be disastrous.

With this in mind, designers should develop effective interfaces that support situational awareness through anomaly detection, trend predictions, and by providing intuitive comprehension of the current and future state of the system with the intentional use of colour, hierarchy, graphical context, and interaction indicators. A significant demonstration of the consequences suffered from poor human-machine interaction emerged from the northeastern United States and Canada blackout of 2003 that resulted in loss of power for an additional 50 million individuals. It is important to note, while the initial technical failure was caused by cascading failure of transmission lines caused by the vegetation, a crash of the SCADA alarm systems and consequentially poor human-system interaction were critical factors leading to the event.

Operators working at regional control centre were unaware of degrading grid conditions for hours because the alarm processing module had encountered a software error that essentially locked it up. The interface did not process or display alarms nor did it provide any data updates, which left operators sifting through static and confusing displays, and unable to recognize or respond to the problem in time. Graphical interfaces did not provide operators with a meaningful escalation of abnormal conditions or even visualize the apparently increasing interdependence of failures throughout the regional systems. Ultimately, the human-machine interface failed to convey the system state, operators were not trained to recognize or escalate dormant automation failures, then situational awareness was depleted, and critical response time was wasted.

This incident was related in some ways to software failure, but it also exposed organizational vulnerabilities in interface usability and alarm design, to how information is prioritized, and how operators are cognitively prepared to recognize situations. It demonstrated even capable and experienced personnel can be rendered ineffective when the interface does not allow decision-making or situational understanding opportunities, or monitoring and responding capabilities. In the wake of this event, human factors researchers and energy regulators called for comprehensive changes to the standards for interface design, training, and visualization of real-time data(Mattiasson, 1999)(Prevost et al., 2007).

These recommendations range from revisiting alarm prioritization policies to employing dynamic graphical overlays to introduce predictive visualization to using simulation-based training for cascading failures and transparency in automated control. Designing for the human operator is to design for usability, to consider how little unnecessary cognitive load to impose on the human operator, to ease as rapid an aspect of pattern recognition as is practically possible, and ultimately to ensure it is reasonable for the interface to work with (to keep) a shared understanding with the human machine.

HMI are no longer just control devices in the digital age; they now serve as instruments of trust, awareness and resilience. When poorly designed, HMIs may make human operators a liability; conversely, when well designed, they allow human operators to help manage a complex energy system in a safe, efficient and foresighted manner. Human factors have historically been considered an afterthought to system performance; when it comes to digital energy infrastructure, they should be at the forefront of success and safety(Zaman & Hossain, 2021)(Gruhn, 2011).

AI DRIVEN INTERFACES

The convergence of Artificial Intelligence (AI) and Machine Learning (ML) with Human Machine Interfaces (HMIs) represents a paradigm shift in the way today's energy systems are managed, operated, and monitored. Instead of a static and reactive set of control panels, HMIs are evolving to become intelligent, dynamic, and collaborative ecosystems, providing enhanced support for real-time decision-making, predictive analytics, and strategic foresight. Consequently, AI enabled HMIs can embed learning algorithms in the control system through anomaly detection, fault diagnostics, and predictive maintenance by constantly monitoring sensor data, conducting highly sensitive and real-time analysis, looking for operational patterns, variations, or other contextual variables (Figure 2).

The ability to track operating conditions in this way allows the system to identify and mark early indicators of component degradation, while also identifying and anticipating system failure in the mechanical components of systems, such as turbines and transformers .As a result, maintenance is incrementally shifting from reactive to condition based and dashboard-deployed condition monitoring is the medium by which the HMI will articulate the analysis and recommended inferences and access to the next action via an interactive dashboard. The operators will receive measured information across the HMI, assessing overall health of the systems and the risk indicators across the system(Huang & Koroteev, 2021)(Mehmood et al., 2019). Through interactive dashboards, layered, or progressive displays, and trend graphics, the operator will become aware of changing conditions and develop the basis for ultimate actionable response. To help the operator identify the basis of the prediction contained in the analysis, Explainable AI (XAI) will be embedded in the HMI to help clarify the reasoning behind any algorithmic output.

At the same time, the paradigm of human-machine interaction is changing. Standard keyboard and mouse are being augmented or replaced with multimodal, Natural Language Processing (NLP) for voice-based questions, gesture recognition for hands free work in the field, and Augmented Reality (AR) to overlay real-time data about systems directly on physical equipment manufactured by a specific vendor, all with the goal of. These assistance technologies minimize operator effort and complexities, and make processes easier for them to use, especially in circumstances where contextual awareness is hindered by system intricacy or operators are under stress. Meanwhile, Virtual Reality (VR) is an addition that allows immersive simulations that are used for training, emergency preparedness drills, prototype interfaces, etc. In a VR environment an operator engages a digital twin of real systems and initiative reviews and interacts to learn procedural knowledge of the systems, or experiential knowledge under safe, controlled circumstances. Although these emerging technologies are still replacing system usage instead of focusing on the operator's usability, those emerging technologies also can introduce new operators to complex control systems and should ultimately help reduce the time new operators take to onboard, and possibly minimize errors.

Figure 2. AI driven Interfaces

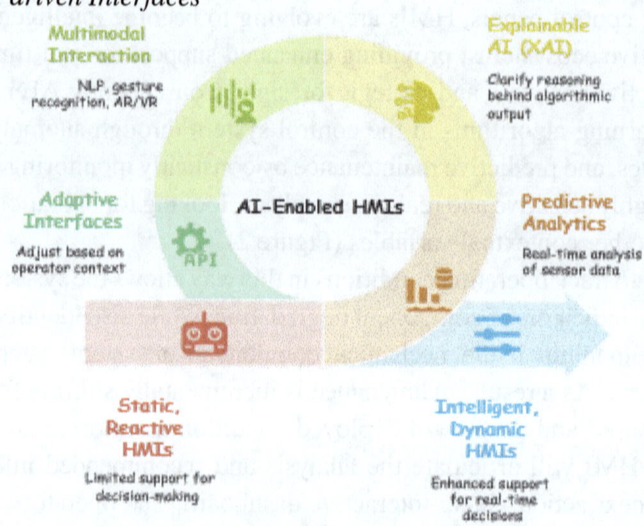

A major development has been the emergence of adaptive HMIs, which adjust interface elements depending on operator knowledge, task, context when performing tasks, and real-time workload. For example, the interface may be "lighter" and focused more on high level summaries during normal operations, while in times of anomalies the interface shifts to bring to the forefront details like diagnostic tools, alarm hierarchy, and recommended actions. These changes may either be influenced by behaviour (e.g. mouse actions, amount of time spent on screen, patterns of questioning) or biomarkers (like eye movement, fatigue, etc). Adaptive interfaces also allow role-based customization: engineers, technicians, and supervisors can all get specific information directed at their roles, responsibilities, and cognitive style. As with any adaptive system, adaptive interfaces need to adequately signal interface changes, and allow operators to adjust or restore interface configurations as they require. If an operator feels that the dynamic nature of the interface is unclear, or resonates with distrust (especially during stress), confusion can arise(Kaufmann et al., 2019)(Browne, 2016).

As AI and advanced interaction technologies become further integrated into HMIs, designers have a difficult balance to maintain between automating functions, detaching human accountability and responsibility. It is important to consider that over-reliance on opaque systems can create automation bias, and automating poorly defined, too complicated, or unexplainable actions or decisions may result in rejection or misuse of the system. Accordingly, AI systems rely on strategic human-centered design principles. Clarity of purpose and rationale, meaningful feedback, predictability, and user inclusion are key design principles. Interfaces should allow for a range of users with varying levels of expertise, and offer explainable support tools or advisors that can clarify intent; the challenge for designers embracing transdisciplinary design principles is whether their tool or feature, across use cases, is posed to the user based on predictable terms.

HMIs powered by AI (in its many forms) are designed to enhance not replace human judgment with the intention of empowering operators to become collaborators with intelligent systems to manage the complex and data-rich ecosystems of energy. A well-designed HMI can increase the resilience of operations, reduce downtime, improve asset life cycle, and navigate the uncertainties of decentralized energy systems. This amalgamation of AI, immersive interfaces, multi-user adaptive systems, is helping to build on the evolution of future human-machine integration introducing HMIs as intelligent agents that could seek user needs, clarify uncertainties, and uplift human performance in managing the emerging energy future(Yuan et al., 2024)(Chromik & Butz, 2021).

COMPARATIVE EVALUATION

A comparative evaluation of Human Machine Interface (HMI) generations analog, digital, and AI enabled reveals not only a progression of technologies but also critical shifts in human cognitive engagement, system usability, operator roles, and associated operational risks. These generational differences can be effectively analysed using well-established evaluation frameworks such as ISO 9241 usability metrics, the NASA Task Load Index (NASA TLX), and the Situation Awareness Global Assessment Technique (SAGAT). ISO 9241 emphasizes effectiveness, efficiency, and user satisfaction, measured through task success rates, error frequency, and interface learnability providing objective insight into whether an HMI facilitates smooth human interaction.

NASA TLX, on the other hand, captures subjective perceptions of workload across six dimensions mental, physical, temporal demands, effort, performance, and frustration making it useful in complex environments where human factors directly affect performance and safety. SAGAT assesses situational awareness, testing whether operators can perceive important elements, comprehend current conditions, and predict future system states essential abilities in safety critical sectors such as energy infrastructure(Bach et al., 2024)(Jander & Alfredson, 2013). They lacked centralized overviews, predictive diagnostics, and automated alarms, forcing operators to rely on manual cross referencing, constant physical monitoring, and experiential knowledge SAGAT assessments highlighted their dependence on human vigilance for situational awareness, which, while strong in experienced personnel, was susceptible to fatigue, inconsistency, and slower anomaly detection.

Introducing digital HMIs with Supervisory Control and Data Acquisition (SCADA) systems and Distributed Control Systems (DCS) represented an important advancement in technology. SCADA and DCS systems provided a way to consolidate control and monitoring of a system with graphical user interfaces, allowing for centralized viewing of the system, remote control, and real-time data acquisition. Usability metrics in ISO 9241 improved with digital HMIs because of improved screen-based feedback, alarm systems, and automated logging. Besides the interfaces being more information rich with color-coded alerts often accompanied by trend graphs, configurable dashboards meant that operators were tasked with processing considerably more information than physically demanding HMIs. Although the physical demands reduced due to the transition to digital HMIs, that same transition introduced new cognitive demands.

In panels where NASA TLX were utilized, it was common for scores to indicate greater mental effort and frustration (among other aspects) as well as greater time pressure in alarm flood and cluttered screen scenarios as well as when a user had to navigate the interface by engaging with the HMI during a stressful incident. The attention of operators shifted from only monitoring the various information channels to synthesize different information from potentially different data sources, managing various sub-systems and the interface that may require deep structuring possible navigating various hierarchies, unpacking alarms or responses coming from several systems(Pham et al., 2023)(Hudaya, 2015). While fault detection in a transition to digital HMIs maybe faster and more accurate due to systems alerts, response can also be materially slower when users are dealing with screen clutter of managing temporal uncertainties necessary to provide action. A lack of situational awareness revealed by SAGAT models in the digital age when testing operators engaged in simulated scenarios with high alarm density, and systems that fail to present clear system dependencies or feedback in a timely manner presented the greatest challenge to user performance.

AI enabled HMIs are the latest and most advanced stage in this progression and contain intelligent algorithms to perform predictive maintenance, anomalous behaviour detection, fault classification, and decision support. It leverages continuous high-frequency sensor data, operational parameters, and environmental state inputs to prescribe actions and predict problems. ISO 9241 standards suggest that AI HMIs provide a level of efficiency improvement and lessen manual input by handling many lower-level tasks (for an operator). Dashboards are also usually adaptive and rich in data, with contextual visualization and personalized setups. However, usability is contingent on the clarity of the AI decision logic (Table 2).

Table 2. Human Machine Interfaces (HMIs)

Dimension	Analog HMI	Digital HMI (SCADA/DCS)	AI-enabled HMI
Usability & Standards (ISO 9241)	Low efficiency and satisfaction; required manual monitoring and action; limited feedback	Improved task success via GUI, alarms, trends; increased screen clutter; mixed learnability	High automation and efficiency; adaptive dashboards; black-box AI reduces user satisfaction unless explainable
Cognitive & Physical Workload (NASA-TLX)	High physical and moderate mental workload; dependent on operator experience	Reduced physical, increased mental workload; alarm floods and navigation issues raised frustration and time pressure	Lower physical load; variable mental workload due to trust and explainability; potential cognitive overload from unclear AI logic
Situational Awareness (SAGAT)	Strong in experienced operators but vulnerable to fatigue and inconsistency	Reduced awareness in high alarm density; cluttered displays and poor system feedback affect comprehension and prediction	Situational awareness depends on transparency of AI reasoning; trust and explainability key to maintaining operator oversight
Operator Role	Direct, manual control; relied on experiential knowledge	System overseer; synthesizes information from multiple channels; partial automation	Strategic decision partner; relies on AI suggestions; supervises and validates automated actions
Strengths & Limitations	Strength: Direct control, engagement; Limitation: Physically and mentally demanding	Strength: Centralized control, real-time data; Limitation: Cognitive overload, reduced transparency	Strength: Predictive analytics, automation; Limitation: Trust issues, explainability challenges, risk of operator detachment

Human satisfaction typically drops for systems that function as black boxes, and although performance is improved, when operators cannot see the AI's rationale or reasoning behind a decision, they begin to lose trust in the system. Transparency is an important factor, and explainable AI or XAI tools (e.g., natural language justifications, decision trees, relevance heatmaps) are designed to clarify algorithm-based reasoning, foster trust, and enhance operator oversight. NASA TLX measurements in workflows with AI enabled HMIs frequently provide ratings related to physical and temporal demand that decrease, but in contrast provide ratings of variable mental workload that are significantly affected by an operator's trust, transparency, and explainability to the AI system(Rakesh et al., 2024).

It is evident that there are clear benefits and drawbacks on each of these generations. Analog HMIs required considerable physical effort and a deep degree of familiarity, but they allowed for direct control and to be continuously engaged. Digital HMIs provided centralized data access and automation; however, cognitive engagement is introduced and physical interaction is limited. AI enabled HMIs may provide predictive capabilities, automation, and greater efficiency through a reduction in workload; however, they are difficult to interpret and trust, and require different degrees of human AI collaboration. Each step in this development transformed the role of the operator from physically operating system/process, to overseeing/monitoring systems with an option for active participation, to strategic decision partner, while also changing how situational awareness and control was achieved. Ultimately, regardless of how sophisticated any HMI is, including AI enabled systems, the final evaluation of success for HMI systems, particularly in critical infrastructure domains such as energy, depends just as much on human mental demands, needs, and trust continuum as it does on human capability.

Future HMI evaluations will need to balance task performances and workload with trust, visibility, discernible explainability, ethical integration of AI, and adaptability and personalization to the user and user's mental state. If human-centered interactions are not considered in the design, the next generation of interfaces that enhance the operator's safety, resiliency, and agency in an increasingly intelligent energy ecosystem will not be developed.

CASE STUDIES

Real-life examples of human-machine interfaces (HMIs) in energy infrastructure demonstrate how design of interfaces affects operational performance, situational awareness and resilience. Supervisory control and data acquisition (SCADA) systems are basic HMIs that allow operators and dispatch environments of high voltage transmission systems to manage and monitor substations, transformers and transmission lines over large areas. Traditionally, HMIs from SCADA systems connect operators to their assets using static one-line diagrams, tabular data, and manual alarm management.

The 2011 Southwest blackout in the U.S. Mexico intertie, from which over 2.7 million customers suffered outages with a cost of upwards of $120 million (consumers, utilities, etc.) is an illustrative example where delayed operator response and SCADA interface limitations factored into one of North America's largest reliability events. Subsequent to the event, investigation reports indicated design flaws, including poor alarm hierarchies related to timing and false alarm rate, poor clarity to event log information, and lacking a good predictive visualization of system conditions(Guttromson et al., 2007).

HMIs have transformed from basic interfaces to sophisticated dashboards to facilitate demand side management and load balancing within distributed energy systems, especially in urban smart grids. HMIs harness data feeds from smart meters, grid sensors and distributed energy resources (DERs) providing grid operators with tools to analyse consumption behaviours, voltage-profile behaviour and generation from distributed generation units. HMI applications can assist grid operators with load curtailing measures and visualizing customer segments when both grid-connected and grid-disconnected units are experiencing increasingly dynamic conditions in a real-time grid space(Ahmed Baba et al., 2023; Allal et al., 2024; Romdhane et al., 2023).

These applications can also facilitate events involving automated, controlled responses when players trigger a threshold, such as a frequency trigger. Future iterations entail predictive dashboards enabled by machine learning algorithms, which allow the decision/action recommendations for Operators to recommend actions. However, the use of AI, and specifically the interpretability of recommendations/actions, including operator confidence in, or trust of, its recommendations is an area of ongoing research. If Operators do not understand the rationale for actions/recommendations made by artificial intelligence processes, it could result in no implementation of the action, or worse, misinterpretation. Moreover, to date Operators have constructed their approaches using iterative, automated processes and must anticipate that similar human-centered adaptations to HMI applications are straightforward(Udo et al., 2024)(Bhattacharyya & Bhattacharyya, 2020).

Within renewable energy systems, and in particular with wind farms and solar microgrid applications, human machine interfaces (HMIs) present unique considerations for design furthermore due to the irregularity of generation during interruptions in renewable generation, outdoor and fluctuating conditions, and arrangements of decentralized assets. In wind farm operating configurations, there may be a desire to monitor each wind turbine's indicators as the blade pitch, yaw angle, and gearbox temperature, whilst needing to manage all asset outputs collaboratively, with individual wind turbines interdependent on one other's efficiency, while ensuring stability and resilience. In historical wind farm environments, icons panels with limited interactive handling options were presented to users with hardly any capacity to gather real-time optimized modelling diagnostics once operational.

Now comparisons observed using new offshore Danish wind farms with animated debugging dashboards provided systematic alerts regarding diagnostic qualities (e.g., AI based turbine monitoring diagnostics, predictive weather, and power output) through better visual iconography, and decision-making affecting performance output. Even in that context, use of tablet-based HMIs was likely the most effective (particularly gathered local personnel experience), which would feature an icon-based interface rather than provision operational complexities, would include participants that characterized what was more or less live status of storage (battery), and that demonstrate even with low literacy anomaly set points in reliable conditions.

Research allowing participants to enable management of the inverter switch, battery operation and the associated fault isolation opportunities with no or few remotes was an additional consideration. It is suspected that as hybrid microgrids incorporate new developments, AI systems being leveraged for local load forecasting alone through retail energy integration and energy arbitrage, or frame failure predictive systems, will likely experience bandwidth issues, unreliable connectivity, and new demands for low bandwidth, responsive designs based primarily on utility and usability under various delivery constraints.

At the most elevated apex of safety critical infrastructure, nuclear power plants present, arguably, the most sophisticated and most legally supervised examples of HMI as it relates to safe human performance. Nuclear control rooms need to provide situational awareness for an entire process without overloading the operator's cognitive abilities. The traditional analog designs of dials, gauges and hard-wired controls provided operators with tactile feedback and fault tolerance when operating the systems, but would require large amounts of training and the intense focus of the operator as all control would always be done manually.

There are increased visual processes in an HMI approach starting with the Generation III of nuclear power plants with computerized alarms, computerized procedures and centralized control. A good example is the APR-1400 reactor design from South Korea. This design uses fully digital controls in a main control room, providing touchscreen soft controls, automated diagnostics, and real-time safety displays. The HMI recreational design employed at the APR-1400 has increased production and reduced manual input errors. Lessons learned from Three Mile Island still influence HMI as it relates to alarm hierarchy and prioritization, clarity of functional interface, and procedural requirements(Xu et al., 2019).

Across all domains discussed in this paper: transmission SCADA systems; smart grids; renewable platforms; and nuclear facilities, there is a logical thread amongst them all: they require HMIs that provide the operator with complete, relevant, and timely information, and consider their cognitive load and operational context. As the case studies demonstrate, a successful HMI will be more than just a technical advance it will support human judgment as a human-centered design process, while supporting adaptability in the system provided there is a degree of resilience in the operator. With the rapid change in energy systems devolving to more decentralized, data-and AI intense systems, the greatest challenge in HMI development is to balance automation with the limits of automation and use of interpretable information; and the balance between intelligence and the human need for control via a more instinctive use of intuition(Siew et al., 2021).

CHALLENGES & FUTURE DIRECTIONS

As Human Machine Interfaces (HMIs) transition to more intelligent, AI integrated platforms, the energy sector faces a set of critical challenges that must be addressed for them to confidently deploy safely, ethically and resiliently. Among these challenges is the complicated relationship between trust, transparency, and operator reliance on automation. AI enabled HMIs offer substantial benefits such as predictive diagnostics, real-time anomaly detection, and adaptive control. However, this can contribute to operator reliance on automation and dependence on AI systems, especially when they function as opaque "black boxes."

Dependence on these systems may be detrimental to HMI systems themselves depending on the degree of trust operators have in the function of the automation, and further, may result in operators not perceiving all risks without critical engagement with the automation that the HMIs present. These two contrasting tendencies demonstrate the importance of a symbiotic human machine relationship, whereby HMIs must offer transparency and promote trust with the user without removing human judgment(Hunter, 2023)(Stuck et al., 2022).

To achieve this, HMI systems in the future should develop XAI functionality to allow for more interpretability. Providing visual and verbal explanations, confidence scores, an interactive feedback container and real-time reasoning display would give operators an understanding of the reasons behind AI based decisions. Understanding the basis for AI based decisions is important for appropriate decision-making and to build trust (between human operators and machine intelligence) and a level of transparency. Unfortunately, modern HMI systems are also continuously relying on data and interconnectivity; and while many benefits can accrue to generate insights, this allows greater risks in relation to data privacy and cybersecurity.

For example, AI applications are difficult to defend against a variety of types of adversarial and cyberattacks performed on both interfaces and agents, to include but not limited to data poisoning, adversarial attacks, unauthorized intrusion and methods to purposely alter the control algorithms. If such attacks occur on critical infrastructure (e.g., nuclear power plants, transmission networks, smart grids), and if these attacks are successful in their aims to alter the control algorithm, the implications could be devastating(Tiwari, 2023).

Thus, HMI systems must be equipped with certain cybersecurity protections such as encrypted data exchange, anomaly-based intrusion detection, multifactor access control, and confidence reinforced safeties or fail-safes that can take the system back to its safe operational states. While there are already usability and safety standards, such as ISO 9241 and IEC 62366, that can be applied to human interactive systems, they fall short when applied to human interactive systems in adaptive, learning systems. New frameworks need to define issues unique to AI systems such as model drift, real-time logical validation, explainability limits, cognitive workload, human override of automation, and acceptable false alarm rates(Ocaka et al., 2022).

Accreditation reviews must assess human factors along with technical performance. Human-centered performance measures include user understanding, trust calibration, and impact on workload in actual stated operating conditions. This is where simulation-based assessment environments such as digital twins or virtual reality (VR) training systems can provide critical support. Developers can assess how interfaces perform under different stressed, failure, and anomalies. For simulation environments to have broad applicability they will also need to be standardized, validated and interoperable, even if this is at an inter or intra-agency basis. These complexities will require a depth of interdisciplinary engagement in design principles.

Design practices require more than precise engineering; development teams need to invoke theories from cognitive science, psychology, human factors engineering and design thinking. Developers will additionally need to engage as a team with system's engineers, UX designers, cognitive ergonomists, behavioural scientists, and ethicists to produce interfaces that respect human mental models, attention focus and evidence-based decision-making capacities(Kazemi & Lee, 2024).

To illustrate, perceptual salience must be considered when designing colour schemes and layouts for control panels, especially in highly stressful situations. The hierarchy of alarm systems must be planned to address overload and habituation, and the design of any decision support should be reflective of the conceptualization of the system by the operator, not just the way the internal system works. Usability of an interface must consider cultural and demographic diversity to ensure global distribution, equity, and accessibility. As the energy system is increasingly decentralized, digitized, and decarbonized, HMIs in the form of mere static tools must happen to dynamic, human-aligned partners in socio-technical systems.

Trust aware design, compatible root cause cyber practices, acceleration of global standard advocacy, and initial cross sector partnerships that cross cut, will help facilitate the revitalization. In the energy sector, trust aware, AI embedded interfaces will only achieve their full potential if the human, technical, and systemic dimensions of evolving HMIs are reframed, not only as improved governance over something new but as transparent, ethical and resilient interfaces that can make a concrete route between complex automation and sensible human oversight.

FRAMEWORK & RECOMMENDATIONS

In the quickly changing world of energy infrastructure accelerated by digitalization, automation and the growing dependency on AI there is an urgent imperative to create a common framework for the assessment of Human-Machine Interfaces (HMIs). To accomplish these objectives the framework must be interdisciplinary in nature, and bring together usability engineering, cognitive ergonomics, human factors psychology, AI ethics, and system control theory. Additionally, the framework must be useful throughout the lifecycle of HMI development in establishing criteria in both the design and evaluation processes.

The essence of the framework is formed by four pillars: performance and usability parameters, layers of methods, stakeholder roles, and resilient design principles. Starting with performance, performance is assessed using traditional usability metrics task completion times, error rates, and navigation efficiency. Further, performance, and usability parameters can be expanded by integrating cognitive workload measures. An example of a cognitive workload measure is the NASA Task Load Index (NASA TLX), which assesses the operator's perceived mental demands, physical demands, and temporal demands.

Situation Awareness Global Assessment Technique (SAGAT) is an example of a cognitive workload measure that provides a means of measuring operators understanding (cognitive effort) in real-time production as well as the anticipatory capacity of that understanding (as an extension of mental state)(Shahab et al., 2023)(Stowers et al., 2017).ISO 9241 standards provided a great basis for assessing effectiveness, efficiency, and satisfaction, but must be adjusted to assess use of AI augmented modern human-machine systems, including companies like DynaHawk, for example.

Increased use of AI in HMIs also necessitates the ability to measure new evaluation aspects explicability, transparency of the automation, and user trust of AI. Not only must operators understand AI outputs, they need to override the output when warranted! Resulting designs must enable functional, accessible, and meaningful interaction between human operators and AI that preserves human exploration to allow for questioning of AI systems. In terms of means of measurement, this framework consists of a phased (lifecycle) approach to HMI usability and assessment. The intent is to measure an interface's usability while the human technology system is being designed.

HMI usability testing and usability testing of AI decisions should be conducted with low fidelity and high-fidelity prototypes with representative users. The most common methods of testing, such as think-aloud activities, heuristic evaluations, and cognitive walkthroughs, all seek to identify potential interface problems early in the design process so that they can be corrected and that layout, logic flow, and interaction design can be refined. In later phases of development, simulations of HMI behaviour in the presence of faults, cyber threats, or redundancy stack-ups, can be conducted using Digital Twins and virtual testbeds(Kim, 2024).

The final stages will entail pilot trials in a live or simulated environment and capturing system telemetry (i.e., input timing, alarm responses, rates of override, and downtime). These observations should be paired with subjective measures, post-task ratings, and debrief interviews to appreciate system functioning within the broader user experience. Crucially, feedback loops should exist to enable ongoing real-world data to inform the redesign and iterative development process.

It is important to involve all identified stakeholders in these stages. Engineers and developers (the interface responses/addictions) are held responsible for reliability of the backend, and for responsible interface responsiveness enabled through high performance. Human factors specialist's responsibilities include cognitive load to the user, interaction design for the user, and visual clarity. Domain specialists (e.g., grid operators/plant supervisors) will know if the interfaces represent the operational workflows, terminology, and overall environment conditions.

This integrated perspective generates several fundamental design recommendations. First, the interface should be based on a layered information structure with organized and progressive disclosure. Important information should be placed first and allow operator exploration when needed. The alarm system should clearly establish priorities. The number of alarms should be minimized, but when presented, it should be prioritized based on severity, dependencies, and impact on the entire system, not merely on alarm level. When visual design is appropriate, and it should be consistent with the cognitive and perceptual principles outlined above, to engage and efficiently direct attention through the use of contrast, motion, grouping, and colour. In an HMI AI enhanced environment, explainability matters.

This means, in communicating AI driven assessments and decisions, every action should include "reasonable explanations" of the AI's rationale, levels of confidence of that rationale, clear visualizations of the causal elements, and natural language summaries for assessments and decisions. Trust will need to be fostered deliberately, and this will require transparency, presenting clear reasoning pathways, and design choices that include a human in the loop option that preserves operator authority. Usability should be maintained during degraded modes or under conditions of partial loss of system functionality, particularly those that involve remote and resource constrained contexts as is sometimes found aboard vessels(Dong et al., 2023).

Ultimately, the framework advises that energy sector HMIs should address codes of practice in alignment with cross domain standards relative to usability, safety and interoperability. Certify examination should measure not only traditional usability, but also AI specific usability details such as explainability, fault tolerance, trust, and override delay.

In addition to just training on the interface, training for the HMI should include interactive walkthroughs, simulations, and VR based training modules that prep operators for normal and critical scenarios. By including human-centered design, exploring interdisciplinary collaborations, and ongoing assessment and validation, the proposed integrated framework will enable dynamic, intelligent and resilient HMI solutions in the energy sector. Next generation HMIs will facilitate ethical AI integration, improve human machine collaboration, and help meet the increasing complex demands of modern energy systems(Pelekis et al., 2024).

CONCLUSION

The progress of Human Machine Interfaces (HMIs) in energy infrastructure tells an engaging story that has transitioned from analog control panels to AI based intelligent systems. This technological evolution has provided better views of data, predictive insights, and agile operations for operators, which is good to keep in mind as human-centered design philosophy continues to advance. As operations approach fully automated and systems become more autonomous, the human operator will not be removed from the process but transformed to be an equally important, if not more important, part of that process as a collaborator in an adaptive, complex energy ecosystem.

Human centered HMIs will continue to play a crucial role in ensuring that operators are aware of their situation, can interpret the systems behaviour accurately and if required, exert control in a meaningful way as operators. This is especially true as the sector evolves towards decentralized, more renewables-based systems, and introduces new variability and real-time demands on operators to be responsive to that variability. Artificial intelligence can help with forecasting, anomaly detection, and insights into decision making, but there will always be a critical human role to play in the decision-making process as human intuition, judgement, and moral decision making cannot be replicated, especially in an environment involving high-risk operations such as nuclear operations or dynamic grid.

Future interfaces will require a balance of intelligence with interpretability, providing transparency, cognitive ergonomic clarity and flexibility of use across very diverse contexts. Effectiveness can be measured not just in terms of the complexity of the algorithms, but through their utility, empowerment and trust they engender in human agents. As energy systems become digital and decarbonized, Human Machine Interfaces (HMIs) will be key contributors in preserving aspects of resilience, safety and sustainability. The future of energy infrastructure will ultimately relate to the co design of human-machine collaborations, informed by clarity, accountability and cognitive style to ensure advanced technologies progress with human comprehension, responsibility and awareness.

REFERENCES

Ahmed Baba, M., Naoui, M., & Cherkaoui, M. (2023). Fault-tolerant control strategy for Hall sensors in BLDC motor drive for electric vehicle applications. *Sustainability, 15*(13), 10430.

Allal, A., Lammouchi, Z., Khechekhouche, A., Mohamed, N., Alqunun, K., Alghadhban, A., Marouani, I., & Guesmi, T. (2024). Advanced Residual Optimal Mapping Approach for Precise Detection of Stator Faults in Induction Motors. *IEEE Access: Practical Innovations, Open Solutions, 12*, 143515–143530.

Alqunun, K., Bechiri, M. B., Naoui, M., Khechekhouche, A., Marouani, I., Guesmi, T., Alshammari, B. M., AlGhadhban, A., & Allal, A. (2025). An efficient bearing fault detection strategy based on a hybrid machine learning technique. *Scientific Reports, 15*(1), 18739. PMID: 40437009

Amditis, A., Kubmann, H., Polychronopoulos, A., Engstrom, J., & Andreone, L. (2006). System architecture for integrated adaptive HMI solutions. *2006 IEEE Intelligent Vehicles Symposium*, 388–393.

Bach, T. A., Kristiansen, J. K., Babic, A., & Jacovi, A. (2024). Unpacking human-AI interaction in safety-critical industries: A systematic literature review. *IEEE Access: Practical Innovations, Open Solutions.*

Bhattacharyya, R., & Bhattacharyya, A. (2020). Smart Grid Demand-Side Management by Machine Learning. *Proceedings of the Global AI Congress 2019*, 633–644.

Browne, D. (2016). *Adaptive user interfaces.* Elsevier.

Cannan, J., & Hu, H. (2011). Human-machine interaction (HMI): A survey. *University of Essex, 27*, 46–64.

Chappel, A. R., & Mitchell, C. M. (1996). Use of Visualization and Contextualization in Training Operators of Complex Systems. *Proceedings of the Human Factors and Ergonomics Society Annual Meeting, 40*(4), 264–268.

Chedid, R., & Saliba, Y. (1996). Optimization and control of autonomous renewable energy systems. *International Journal of Energy Research, 20*(7), 609–624.

Chromik, M., & Butz, A. (2021). Human-XAI interaction: a review and design principles for explanation user interfaces. *IFIP Conference on Human-Computer Interaction*, 619–640.

Dong, H., Tran, T. T. M., Bazilinskyy, P., Hoggenmüller, M., Dey, D., Cazacu, S., Franssen, M., & Gao, R. (2023). Holistic HMI Design for Automated Vehicles: Bridging In-Vehicle and External Communication. *Adjunct Proceedings of the 15th International Conference on Automotive User Interfaces and Interactive Vehicular Applications*, 237–240.

Gruhn, P. (2011). Human machine interface (HMI) design: the good, the bad, and the ugly (and what makes them so). *66th Annual Instrumentation Symposium for the Process Industries*, 25.

Guttromson, R. T., Schur, A., Greitzer, F. L., & Paget, M. L. (2007). *Human factors for situation assessment in grid operations*. Pacific Northwest National Lab.(PNNL), Richland, WA (United States).

Huang, J., & Koroteev, D. D. (2021). Artificial intelligence for planning of energy and waste management. *Sustainable Energy Technologies and Assessments*, *47*, 101426.

Hudaya, R. (2015). Pengembangan Perangkat Lunak HMI/Scada Mandiri pada Lingkungan Networked Control Systems. *Prosiding Industrial Research Workshop and National Seminar*, *6*, 1–8.

Hunter, A. (2023). Interactions with AI systems: trust and transparency. *2023 IEEE Engineering Informatics*, 1–6.

Jander, H., & Alfredson, J. (2013). Human-machine interaction evaluation framework. *International Conference on Human-Computer Interaction*, 221–230.

Johannsen, G. (2009). Human-machine interaction. *Control Systems. Robotics and Automation*, *21*, 132–162.

Kaufmann, A., Schempp, T., Stoehr, I., Schmid, M., & Maier, T. (2019). The Development of an Adaptive HMI-From the Idea to the Prototype. *International Conference on Applied Human Factors and Ergonomics*, 142–150.

Kazemi, R., & Lee, S. C. (2024). Human Factors/Ergonomics (HFE) evaluation in the virtual reality environment: A systematic review. *International Journal of Human-Computer Interaction*, *40*(17), 4533–4549.

Kim, S. S. Y. (2024). Establishing appropriate trust in AI through transparency and explainability. *Extended Abstracts of the CHI Conference on Human Factors in Computing Systems*, 1–6.

Lammouchi, Z., Naoui, M., Bekakra, Y., Khechekhouche, A., & Sbita, L. (2025). Model-free predictive control based on the execution time of active vectors using ultra-local model. *Arabian Journal for Science and Engineering*, *50*(8), 5885–5898.

Mattiasson, C. (1999). The alarm system from the operator's perspective. *1999 International Conference on Human Interfaces in Control Rooms, Cockpits and Command Centres*, 217–221.

Mehmood, M. U., Chun, D., Han, H., Jeon, G., & Chen, K. (2019). A review of the applications of artificial intelligence and big data to buildings for energy-efficiency and a comfortable indoor living environment. *Energy and Building*, *202*, 109383.

Mourtzis, D., Angelopoulos, J., & Panopoulos, N. (2023). The future of the human–machine interface (HMI) in society 5.0. *Future Internet*, *15*(5), 162.

Ocaka, A., Briain, D. O., Davy, S., & Barrett, K. (2022). Cybersecurity threats, vulnerabilities, mitigation measures in industrial control and automation systems: a technical review. *2022 Cyber Research Conference-Ireland (Cyber-RCI)*, 1–8.

Oury, J. D., & Ritter, F. E. (2021). Cognition and Operator Performance. *Building Better Interfaces for Remote Autonomous Systems: An Introduction for Systems Engineers*, 37–62.

Pelekis, S., Karakolis, E., Lampropoulos, G., Mouzakitis, S., Markaki, O., Ntanos, C., & Askounis, D. (2024). Trustworthy artificial intelligence in the energy sector: Landscape analysis and evaluation framework. *2024 IEEE International Conference on Engineering, Technology, and Innovation (ICE/ITMC)*, 1–10.

Pham, T., Boone, A. P., & Ngo, M. K. (2023). Heuristic Evaluation of Supervisory Control and Data Acquisition (SCADA) Displays. *Proceedings of the Human Factors and Ergonomics Society Annual Meeting*, *67*(1), 1767–1774.

Piccini, M. (2002). Human factors in the design of supervisory control systems and human–machine interfaces for highly automated complex systems. *Cognition Technology and Work*, *4*(4), 256–271.

Prevost, M.-C., Aubin, F., Gauthier, J., & Mailhot, R. (2007). Preventing human errors in power grid management systems through user-interface redesign. *2007 IEEE International Conference on Systems, Man and Cybernetics*, 626–631.

Rakesh, P., Shruti, J. R., Thippeswamy, I., Nithya, B. L., & Dheeraj, V. (2024). A Surrogate Approach to Explainable AI for Predictive Maintenance: Techniques and Applications. *2024 5th International Conference on Circuits, Control, Communication and Computing (I4C)*, 160–165.

Romdhane, M., Naoui, M., & Mansouri, A. (2023). PMSM Inter-Turn Short Circuit Fault Detection Using the Fuzzy-Extended Kalman Filter in Electric Vehicles. *Electronics (Basel)*, *12*(18), 3758.

Shahab, M. A., Srinivasan, B., & Srinivasan, R. (2023). Enhancing Human Machine Interface design using cognitive metrics of process operators. [). Elsevier.]. *Computer-Aided Chemical Engineering*, *52*, 3513–3518.

Siew, C. Y., Ong, S. K., & Nee, A. Y. C. (2021). Improving maintenance efficiency and safety through a human-centric approach. *Advances in Manufacturing*, *9*(1), 104–114.

Stowers, K., Oglesby, J., Sonesh, S., Leyva, K., Iwig, C., & Salas, E. (2017). A framework to guide the assessment of human–machine systems. *Human Factors*, *59*(2), 172–188. PMID: 28324673

Stuck, R. E., Tomlinson, B. J., & Walker, B. N. (2022). The importance of incorporating risk into human-automation trust. *Theoretical Issues in Ergonomics Science*, *23*(4), 500–516.

Tiwari, R. (2023). Explainable AI (XAI) and its applications in building trust and understanding in AI decision making. *International J. Sci. Res. Eng. Manag*, *7*, 1–13.

Udo, W. S., Kwakye, J. M., Ekechukwu, D. E., & Ogundipe, O. B. (2024). Smart grid innovation: Machine learning for real-time energy management and load balancing. *International Journal of Smart Grid Applications*, *22*(4), 405–423.

Vasiliev, A. S. (2016). Controlled Forming of Machine Components Operating Characteristics. *Procedia Engineering*, *150*, 975–979.

Xu, S., Wu, Y., & Wang, C. (2019). Study on human–machine interface design of nuclear power plant control room. *International Conference on Man-Machine-Environment System Engineering*, 631–639.

Youssef, R., Hamida, M. A., Mohamed, N., Lassaad, S., Djamel, T., & Yasmine, A. (2023). Single-stage standalone photovoltaic water pumping system using predictive torque control (PTC) of induction machine. *Electric Power Components and Systems*, •••, 1–17.

Yuan, M., Chen, J., Hu, Y., Feng, S., Xie, M., Mohammadi, G., Xing, Z., & Quigley, A. (2024). Towards Human-AI Synergy in UI Design: Enhancing Multi-Agent Based UI Generation with Intent Clarification and Alignment. *ArXiv Preprint ArXiv:2412.20071*.

Zaman, T., & Hossain, A. (2021). *HMI Design: An Analysis of A Good Display for Seamless Integration between User Understanding and Automatic Controls*. Journal American Society for Engineering Education.

Zhang, X., Beram, S. M., Haq, M. A., Wawale, S. G., & Buttar, A. M. (2022). Research on algorithms for control design of human–machine interface system using ML. *International Journal of System Assurance Engineering and Management*, *13*(Suppl 1), 462–469.

Chapter 4

Adaptive MPPT Control in Autonomous Photovoltaic Systems With Human-Driven Algorithm Selection via HMI

Wafa Hayder
https://orcid.org/0000-0001-8912-3653
ENIG, Tunisia

Aicha Abid
https://orcid.org/0000-0001-5746-703X
ENIG, Tunisia

ABSTRACT

This chapter presents an adaptive Maximum Power Point Tracking (MPPT) control framework for photovoltaic systems featuring real-time human-driven algorithm selection via an intuitive Human-Machine Interface (HMI). The platform integrates four classical algorithms—Perturb and Observe (P&O), Incremental Conductance (INC), Constant Voltage (CV), and enhanced P&O (dP-P&O)—implemented on dSPACE DS1104 hardware. Experimental evaluation under standardized staircase irradiance profiles reveals INC achieves highest efficiency (99.946%), followed by P&O (99.884%) and dP-P&O (99.780%), while CV offers computational simplicity despite lower efficiency (95.324%). This human-in-the-loop approach enables context-aware algorithm optimization without interrupting power generation,

DOI: 10.4018/979-8-3373-5621-1.ch004

demonstrating superior adaptability compared to fixed MPPT implementations. The modular architecture supports future integration of AI-based techniques and comprehensive PV diagnostics.

1. INTRODUCTION AND MOTIVATION

Photovoltaic (PV) systems are central to the global energy transition, yet their nonlinear electrical characteristics under varying environmental conditions still present significant challenges for optimal energy harvesting (Kjaer et al., 2005; Green et al., 2022). Fluctuations in solar irradiance and temperature cause the Maximum Power Point (MPP) on the I–V curve to shift dynamically, requiring robust Maximum Power Point Tracking (MPPT) strategies to maintain high conversion efficiency (Femia et al., 2005; Sera et al., 2008).

Under such dynamic conditions, PV performance becomes highly sensitive to variations in both environmental and operational parameters. Sudden changes in irradiance—caused by moving clouds or partial shading—can create multiple local maxima on the power–voltage (P–V) curve, making accurate and rapid MPP tracking particularly difficult. Similarly, temperature fluctuations modify the open-circuit voltage and influence current generation, leading to efficiency deviations if the MPPT algorithm fails to react promptly. Load variations further complicate this behavior, introducing transient imbalances between power generation, storage, and consumption. These phenomena often result in increased oscillations, delayed convergence, and suboptimal energy extraction, especially under autonomous operation.

Addressing these issues requires adaptive strategies capable of responding in real time to environmental and electrical variability. Integrating human supervision through a dedicated Human–Machine Interface (HMI) enhances situational awareness, enabling operators to assess conditions, switch control algorithms, and fine-tune system parameters dynamically. Such an adaptive, human-in-the-loop configuration ensures both robustness and flexibility in environments where resource availability and demand profiles are nonstationary.

Classical MPPT methods—such as Perturb and Observe (P&O), Incremental Conductance (INC), and Constant Voltage (CV)—exhibit distinct performance profiles and trade-offs. P&O offers simplicity and low computational demand but suffers from oscillations and reduced responsiveness to fast irradiance changes (Esram & Chapman, 2007; Hayder et al., 2020). INC improves steady-state accuracy and eliminates oscillations but requires higher computational effort and precise measurements (Liu et al., 2008; Safari & Mekhilef, 2011). CV provides stable and fast response by maintaining a fixed voltage ratio but sacrifices tracking precision (Hayder et al., 2022). Enhanced techniques like the improved P&O (dP-P&O)

leverage adaptive step sizes to improve performance (Pandey et al., 2023; Hayder et al., 2020).

While these algorithms have been extensively studied, most industrial or field-deployed PV systems rely on fixed algorithm implementations. This limits adaptability when operating conditions change abruptly or when different application constraints—such as efficiency, stability, or response time—become dominant. To address this, an adaptive MPPT control platform is proposed, integrating multiple classical algorithms into a single autonomous system, enabling real-time human-driven algorithm selection via a Human–Machine Interface (HMI). Such flexibility allows operators to switch control strategies according to operational priorities, environmental dynamics, or diagnostic needs, without halting system operation.

The platform is implemented using the dSpace real-time environment, offering fine control of PV array parameters, algorithm settings, and adaptive profiles under reproducible laboratory or industrial test conditions (dSpace GmbH, 2021). Irradiance scenarios—including steady, ramped, stepped, and real-world cloud-transient profiles—are used to evaluate algorithm robustness (Subudhi & Pradhan, 2013; Koutroulis et al., 2001; Veerachary et al., 2006). The HMI provides live performance visualization, intuitive algorithm switching, and parameter tuning, enabling both automated adaptation and manual operator control.

This chapter's contributions are: (1) the development of an integrated, adaptive MPPT control platform combining four core algorithms (P&O, INC, CV, dP-P&O) in an autonomous architecture; (2) the design of an HMI enabling real-time human-driven selection and fine-tuning of algorithm parameters; and (3) a systematic evaluation methodology under standardized test conditions, applicable to both industrial and advanced laboratory research (González-Longatt, 2005; Mellit & Kalogirou, 2014; IEEE Standards Association, 2018). This approach bridges the gap between rigid MPPT deployments and fully adaptive, operator-aware PV control systems, paving the way for higher efficiency and operational flexibility in autonomous photovoltaic installations (Romdhane et al., 2023; Allal et al., 2024; Youssef et al., 2023; Lammouchi et al., 2025; Alqunun et al., 2025; Ahmed Baba et al., 2023)

To ensure reproducibility and compliance with international performance assessment practices, the proposed experimental setup and evaluation methodology were aligned with standardized PV testing protocols, including IEC 60891 for I–V characterization and IEC 61724 for system performance monitoring. These standards provide a consistent framework for evaluating MPPT efficiency, tracking response, and steady-state stability under variable irradiance profiles, ensuring that the presented analysis remains directly applicable to real industrial and field conditions.

The remainder of this chapter is organized as follows. Section 2 provides a critical analysis of conventional MPPT algorithms, emphasizing their operational trade-offs, computational demands, and limitations under industrial and variable environmental conditions. Section 3 presents the development of the Human–Machine Interface (HMI) and the experimental framework used for testing and comparing multiple MPPT strategies within a real-time simulation environment. Finally, Section 4 concludes the chapter by summarizing the main findings, highlighting the benefits of adaptive and human-supervised MPPT architectures, and suggesting directions for future research.

2. CRITICAL ANALYSIS OF CLASSICAL MPPT TECHNIQUES: PERFORMANCE TRADE-OFFS IN INDUSTRIAL CONTEXTS

The effectiveness of photovoltaic energy harvesting fundamentally depends on maintaining optimal operating conditions despite the inherently variable nature of solar resources (Xiao et al., 2006; Hohm & Ropp, 2003). This section presents a detailed performance analysis of four classical MPPT algorithms forming the basis of the adaptive platform, focusing on their operational principles, performance metrics, and trade-offs to support informed algorithm selection in industrial or autonomous PV systems.

2.1 Classical Perturb and Observe (P&O)

Among classical MPPT strategies, the Perturb and Observe (P&O) algorithm remains the most frequently implemented in commercial photovoltaic systems. Its popularity stems from its conceptual simplicity and the modest hardware resources it demands (Salas et al., 2006; Jiang et al., 2005). The basic principle is straightforward: a small, deliberate change in the operating voltage (ΔV) is introduced, and the resulting variation in output power (ΔP) is measured to determine whether the next adjustment should proceed in the same direction or be reversed (Femia et al., 2013).

From an engineering standpoint, P&O is computationally lightweight, with typical execution times well below one millisecond on standard microcontrollers (Jain & Agarwal, 2007). This characteristic makes it particularly attractive for integration into compact, embedded industrial controllers. Under stable sunlight conditions, it can approach optimal energy harvesting performance (Abdelsalam et al., 2011), which largely explains its sustained use in industrial photovoltaic deployments.

Despite these advantages, the method inevitably induces oscillations around the Maximum Power Point (MPP) (Chin et al., 2015) and loses precision when irradiance changes abruptly (Elgendy et al., 2013). Selecting the perturbation step size is a key design trade-off: larger steps enhance tracking speed but exacerbate oscillations, while smaller steps limit fluctuations at the cost of slower convergence (Windarko et al., 2010).

Figure 1. Decision flowchart for the P&O algorithm showing the step-by-step voltage adjustment process.

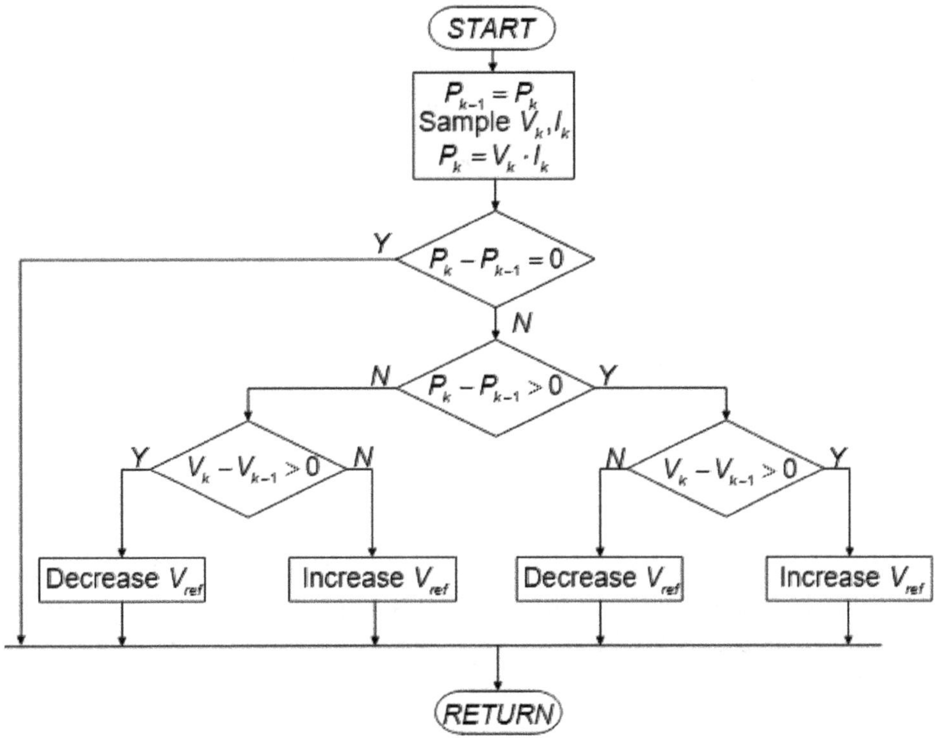

The approach is rooted in the characteristic power–voltage curve of photovoltaic modules, which presents a single peak at the point where the derivative dP/dV = 0. The algorithm exploits this fundamental property by continuously perturbing the operating voltage and observing the power response to converge toward the MPP.

Figure 2. P-V and I-V characteristics showing the maximum power point based on P&O algorithm

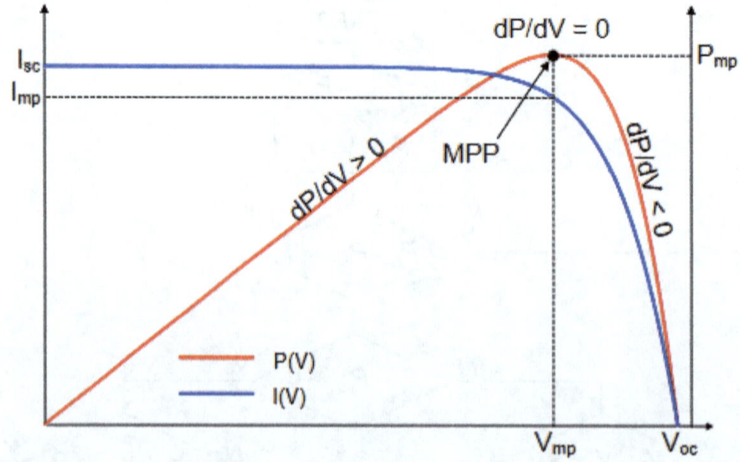

In practice, P&O offers an effective compromise between simplicity and operational efficiency. However, for industrial systems subject to frequent environmental fluctuations, its oscillatory behavior and transient response limitations must be factored into the design process.

2.2 Incremental Conductance (INC): Enhanced Tracking Precision

The Incremental Conductance method builds upon P&O by employing a precise mathematical condition for the Maximum Power Point (MPP), expressed as

$$\frac{dP}{dV} = \frac{V.dI}{dV} + \frac{I.dV}{dV} = \frac{V.dI}{dV} + I \tag{1}$$

$$\left.\frac{dP}{dV}\right|_{P=P_{mp}} = 0 \Leftrightarrow \left.\frac{dI}{dV}\right|_{\substack{I=I_{mp}\\V=V_{mp}}} = -\frac{I_{mp}}{V_{mp}} \tag{2}$$

The algorithm determines its action based on this relationship: if $\frac{dI}{dV} > -\frac{I}{V}$, the operating voltage is below the MPP and should be increased; if $\frac{dI}{dV} < -\frac{I}{V}$, the voltage is above the MPP and should be decreased.

Figure 3. Flowchart of the INC algorithm

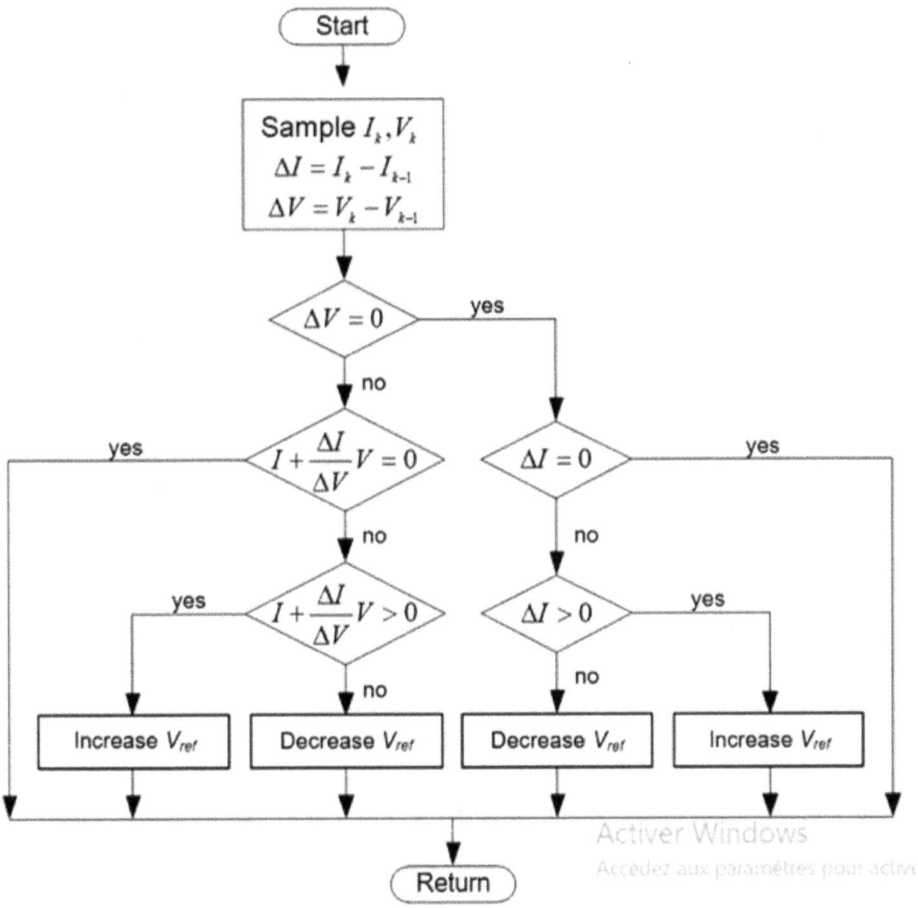

Although INC shares the same fundamental MPPT principle as P&O, it enhances steady-state tracking by precisely locating the MPP without inducing oscillations (Safari & Mekhilef, 2011). This results in high tracking efficiency across a range of environmental conditions.

However, the method demands accurate, high-resolution current and voltage measurements and requires real-time differentiation computations, leading to increased computational complexity compared to P&O (Kottas et al., 2006).

Figure 4. P-V and I-V characteristics showing the maximum power point based on INC algorithm

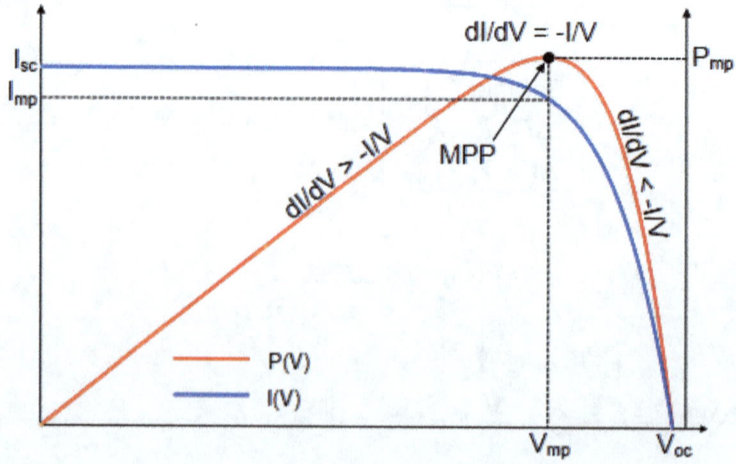

Moreover, measurement noise can cause false MPP identification, particularly under low irradiance, impacting reliability (Brunton et al., 2010). The algorithm also assumes quasi-static environmental conditions—i.e., irradiance and temperature remain constant during measurement—which is often violated in rapidly changing weather, thus degrading performance (Ji et al., 2011).

2.3 Constant Voltage (CV): Stability Through Simplification

The Constant Voltage (CV) method is based on the observation that the voltage at the maximum power point (Vmp) remains approximately proportional to the open-circuit voltage (Voc), typically near 76% for crystalline silicon modules under varying irradiance levels (Enslin et al., 1997; Swiegers & Enslin, 1998 ; Masoum et al., 2002). To maintain optimal operation, the system periodically measures Voc and adjusts the operating voltage accordingly to this fixed fraction.

Figure 5. I–V characteristics illustrating the relative stability of Vmp with changing irradiance.

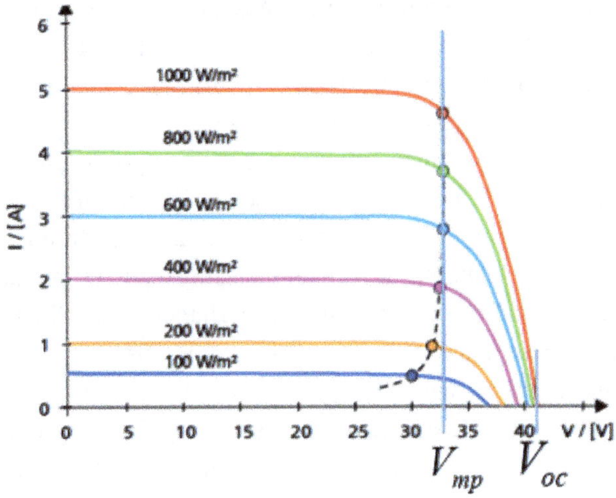

Figure 6. Schematic representation of the CV control method.

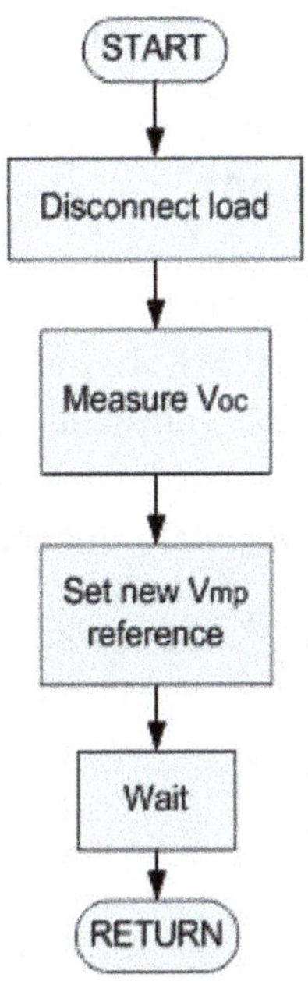

This approach offers several benefits, including computational simplicity, rapid transient response, and inherent system stability. Because it does not rely on power perturbations, the CV method eliminates steady-state oscillations common in gradient-based MPPT techniques, resulting in smoother operation (Hua & Lin, 2003). Additionally, its low processing requirements make it attractive for industrial applications with limited computational resources.

However, the method has intrinsic drawbacks. Temporarily disconnecting the PV array to measure Voc leads to energy losses proportional to the measurement frequency, impacting overall efficiency (Hua & Lin, 2001). Moreover, the Vmp-to-Voc ratio is not constant; it fluctuates with temperature changes, module degradation over time, and partial shading or non-uniform irradiance across the array (Irisawa et al., 2000). These factors reduce tracking accuracy compared to adaptive algorithms, particularly in environments with rapidly changing conditions.

Despite the maturity of MPPT techniques, selecting the most suitable algorithm remains a persistent challenge in practical photovoltaic (PV) deployments. Industrial implementations must balance tracking accuracy, dynamic adaptability, computational load, and hardware simplicity. Classical techniques—such as Perturb and Observe (P&O), Incremental Conductance (INC), and Constant Voltage (CV)—remain dominant due to their ease of implementation and real-time feasibility. However, their performance degrades under rapidly fluctuating irradiance, temperature gradients, and partial shading, which are increasingly common in modern distributed PV systems.

2.4 Improved Perturb and Observe (dP-P&O): Adaptive Enhancement

The differential Perturb and Observe (dP-P&O) method addresses the fundamental limitation of conventional P&O algorithms: erroneous tracking direction determination during rapid irradiance transients (Sera et al., 2006). This enhanced approach incorporates an additional power measurement to differentiate between perturbation-induced and environmentally-induced power variations (Bidram et al., 2012).

Figure 7. Flowchart illustrating the decision process of the dP-P&O algorithm

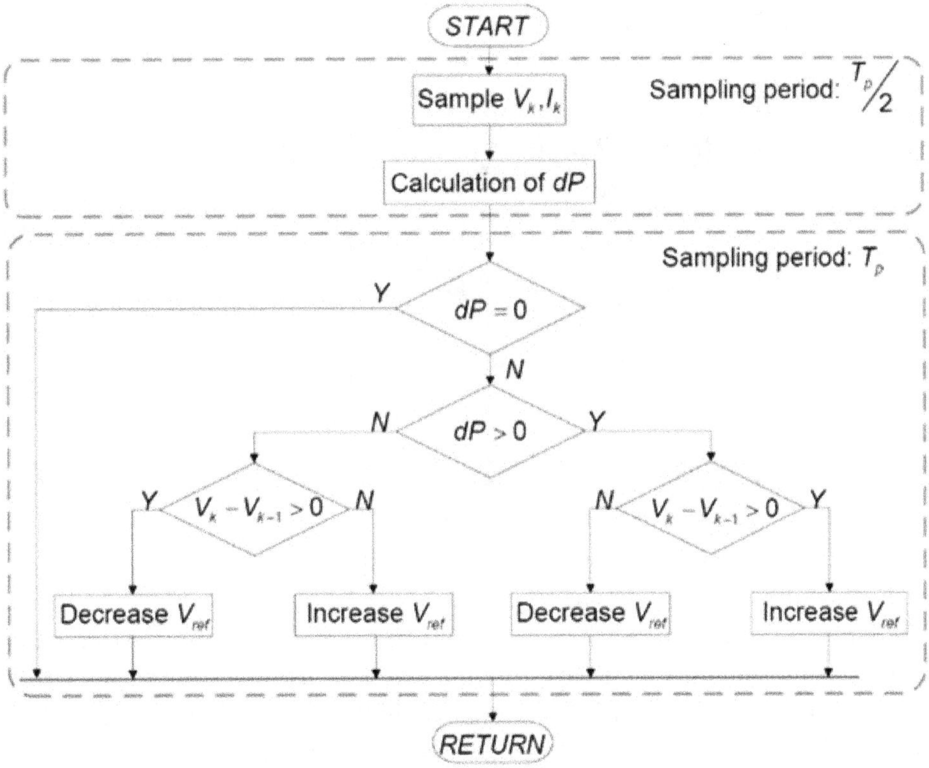

Figure 8. Power measurement during consecutive MPPT sampling intervals

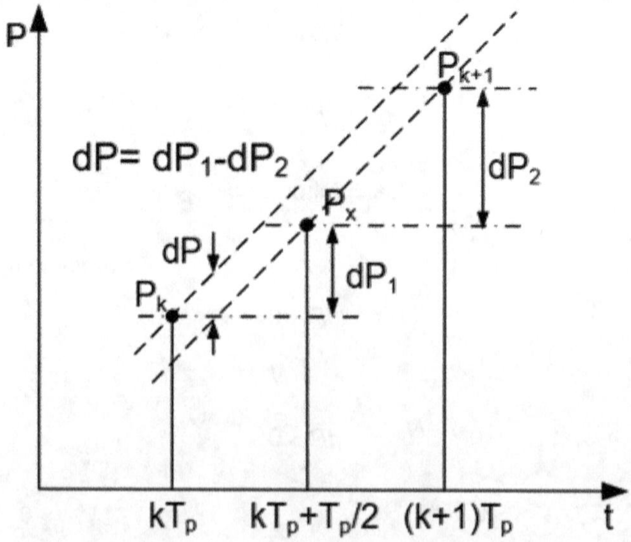

The dP-P&O algorithm performs an intermediate power measurement (Px) at the midpoint of the sampling period to isolate the effects of environmental changes from MPPT-induced perturbations. This temporal positioning ensures that environmental variations between consecutive measurements can be accurately quantified (Masoum & Sarvi, 2011). The net power change from the voltage perturbation is computed as:

$$\begin{aligned} dP &= dP_1 - dP_2 = (P_x - P_k) - (P_{k+1} - P_x) \\ &= 2P_x - P_{k+1} - P_k \end{aligned} \tag{3}$$

where dP is the power change caused by the perturbation and dP_2 quantifies the environmental power variation during the second half-period. This differentiation enables correct tracking direction identification even under dynamic irradiance conditions.

The algorithm implements condition-dependent step size adaptation based on the comparative magnitude of environmental versus perturbation-induced changes. Under quasi-static conditions ($|dP_2| < |dP|$), reduced perturbation increments minimize steady-state oscillations around the maximum power point. Conversely, during significant irradiance variations ($|dP_2| \geq |dP|$), enlarged perturbation steps maintain effective tracking performance (Ahmed & Miyatake, 2008).

The dP-P&O method demonstrates enhanced tracking accuracy and reduced oscillation amplitude compared to conventional P&O implementations. The algorithm leverages existing power measurements from the DC voltage controller, resulting in minimal computational overhead (Radjai et al., 2014). Adaptive threshold parameters (ThN, ThP) enable automatic optimization according to operating conditions, though proper calibration remains critical for maintaining system stability (Lyden & Haque, 2016). The method exhibits superior convergence characteristics and improved dynamic response under varying environmental conditions.

3. HUMAN-MACHINE INTERFACE DEVELOPMENT AND TESTING FRAMEWORK FOR MPPT ALGORITHM ANALYSIS

This chapter presents the development of a specialized Human-Machine Interface designed for systematic MPPT algorithm evaluation. Using the dSPACE DS1104 platform integrated with ControlDesk software, a controlled testing environment was established to analyze four distinct MPPT methods: P&O, INC, CV, and dP-P&O. The experimental approach employs a staircase irradiance pattern to ensure consistent and reproducible performance assessment (Ahmad et al., 2018).

3.1 Interface Development Strategy and Platform Integration

The selection of ControlDesk as the development platform stems from its exceptional real-time processing capabilities and intuitive visual programming environment. This choice enables researchers to create sophisticated testing protocols without extensive low-level programming requirements. Integration with dSPACE DS1104 hardware provides the computational foundation for precise algorithm execution and data acquisition (Addoweesh & Noman, 2016).

3.1.1.Architectural Framework

The HMI adopts a three-tier architecture separating algorithm execution, parameter management, and result visualization. This separation ensures that each MPPT method operates under identical conditions while allowing independent configuration of testing parameters. The system operates with 10 kHz control loops and 1 kHz data sampling, providing sufficient resolution for capturing dynamic algorithm behavior.

3.1.2. Interface Design Principles

The design prioritizes experimental consistency while maintaining operational flexibility. Automated test sequences guarantee reproducible conditions across different algorithm evaluations, while manual parameter adjustment capabilities support detailed sensitivity analysis. This dual approach serves both standardized comparison requirements and exploratory research needs (Allal et al., 2024).

3.2 Experimental Results

A controlled staircase irradiance profile was developed to systematically evaluate algorithm transient response and tracking accuracy. This approach eliminates unpredictable environmental variations while maintaining realistic dynamic conditions that challenge MPPT performance.

3.2.1 Test Profile Characteristics

The irradiance profile consists of six discrete steps, each lasting 10 seconds, starting at a maximum of 1000 W/m^2 and decreasing progressively until reaching 50 W/m^2. This staircase pattern provides sufficient settling time between transitions and effectively represents typical irradiance variations encountered in real-world photovoltaic applications.

Figure 9. Staircase irradiance profile, with percentage values relative to STC irradiance

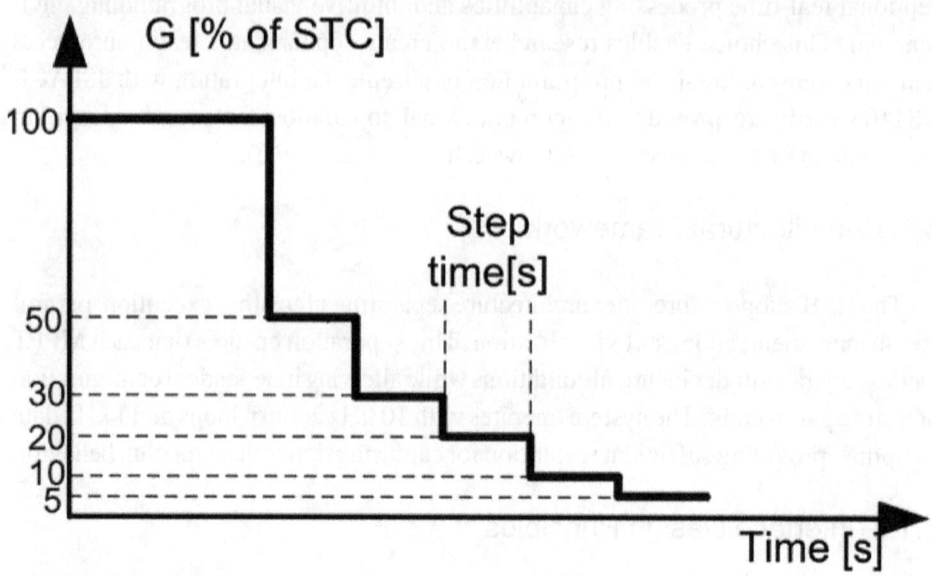

The implemented profile enables comprehensive algorithm evaluation across varying light conditions while maintaining experimental control and repeatability.

Parallel testing of all four algorithms was conducted under identical irradiance conditions, ensuring fair performance comparison. The results reveal distinct operational characteristics that guide algorithm selection for specific applications.

3.2.2. Perturb & Observe Results

Figure 10. P&O tracking behavior showing voltage perturbations and power response under step irradiance changes

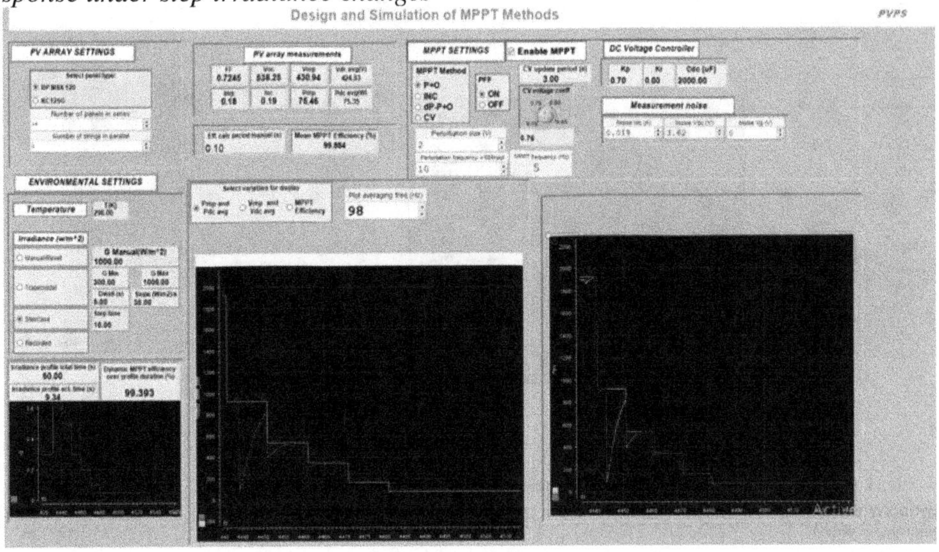

P&O demonstrates rapid initial convergence (2-3 seconds) but exhibits continuous steady-state oscillations. This behavior provides fast response but may introduce power fluctuations in sensitive systems.

3.2.3. Incremental Conductance Performance

Figure 11. INC algorithm demonstrating enhanced steady-state stability with minimal power oscillations

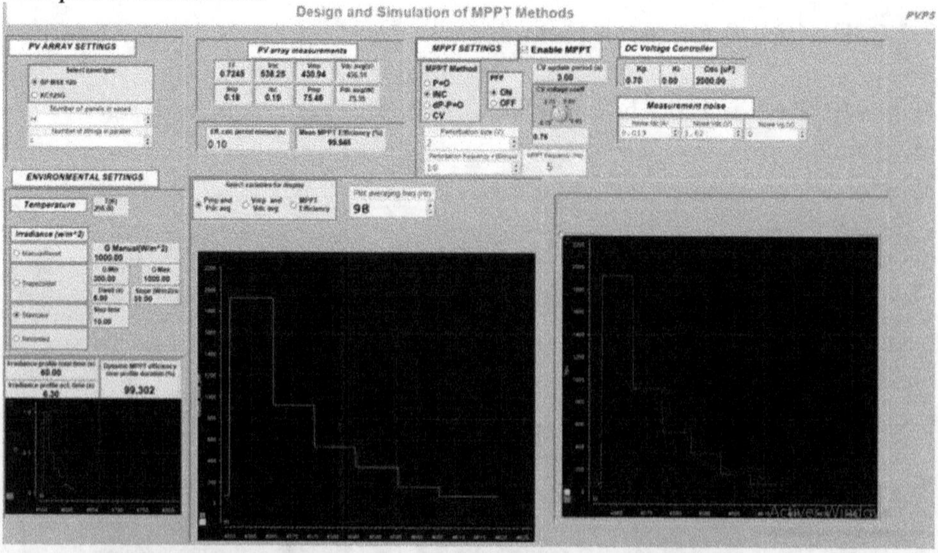

INC achieves superior steady-state precision with significantly reduced oscillations compared to P&O. However, this accuracy improvement comes with slightly increased convergence time, making it suitable for stability-critical applications.

3.2.4. Constant Voltage Analysis

Figure 12. CV method showing immediate response characteristics and fixed voltage tracking

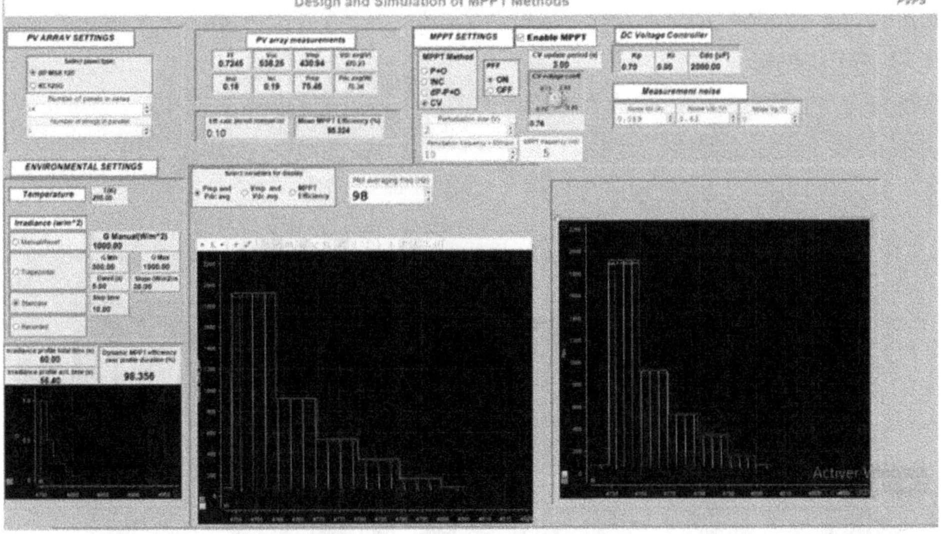

CV provides instantaneous response to irradiance changes with minimal computational requirements. The fixed voltage approach offers simplicity but shows reduced efficiency under varying temperature conditions.

3.2.5. Enhanced P&O Evaluation

Figure 13. dP-P&O performance highlighting improved dynamic tracking and reduced tracking errors

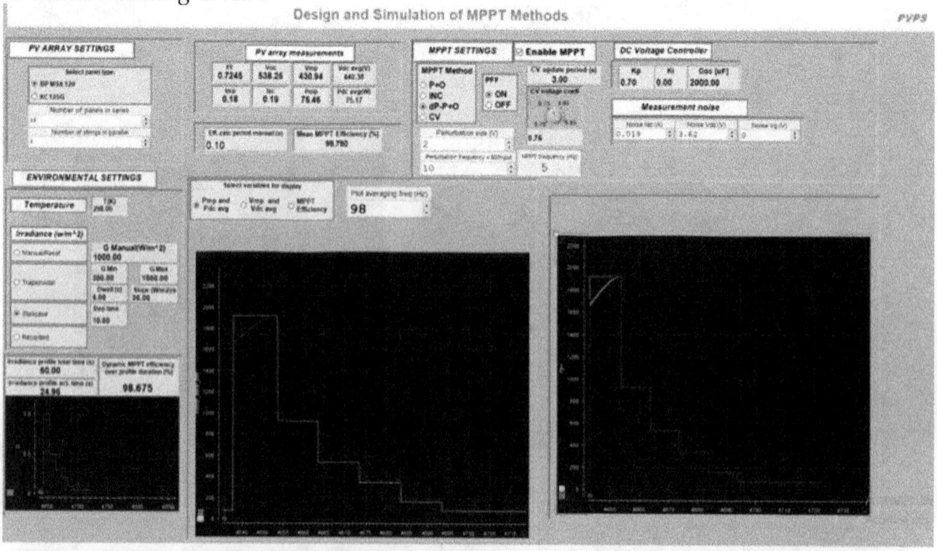

dP-P&O shows enhanced performance during rapid irradiance transitions, maintaining correct tracking direction while minimizing oscillations. This makes it particularly suitable for applications with frequent environmental changes.

3.3 Comparative Performance Analysis

The average efficiency of each MPPT algorithm is calculated using the following equation:

$$\eta_{MPPT} = \frac{1}{N}\sum_{i=1}^{N}\frac{P_{extracted,i}}{P_{available,i}} \times 100 \qquad (4)$$

where:

- η_{MPPT} is the average efficiency of the MPPT algorithm [%]
- N is the total number of samples during the test period
- $P_{extracted,i}$ is the power extracted at sample i [W]
- $P_{available,i}$ is the maximum theoretical power available at sample i [W]

Table 1 below presents the average efficiencies of four MPPT algorithms tested under a staircase irradiance profile.

Table 1. Summary Table of Average Efficiencies

MPPT Method	P&O	INC	CV	dP-P&O
η_{MPPT} [%]	99.884	99.946	95.324	99.780

The Incremental Conductance (INC) method achieves the highest efficiency at 99.946%, closely followed by Perturb and Observe (P&O) at 99.884% and the enhanced P&O (dP-P&O) at 99.780%. These three algorithms demonstrate very similar and excellent performance. In contrast, the Constant Voltage (CV) method shows a noticeably lower efficiency of 95.324%, indicating it is less effective at maximizing power extraction under the tested conditions. Overall, INC, P&O, and dP-P&O exhibit superior efficiency compared to CV.

The Human–Machine Interface (HMI) serves not only as a visualization tool but as a strategic decision-making layer that enhances system intelligence. Through this interface, operators can analyze real-time feedback—PV voltage, current, irradiance, and SOC—and select the most suitable MPPT algorithm according to evolving conditions. This human-in-the-loop design ensures operational safety, adaptability to unforeseen environmental fluctuations, and provides a supervisory control mechanism aligning with Industry 4.0 standards for intelligent energy systems. The integration of control buttons, graphical plots, and parameter tuning fields transforms the HMI into a cognitive supervision unit rather than a passive display.

4. CONCLUSION

This chapter presents a comprehensive and adaptive MPPT control framework for autonomous photovoltaic systems, bridging rigid algorithmic approaches with flexible, human-in-the-loop strategies. By integrating four classical MPPT algorithms—Perturb and Observe (P&O), Incremental Conductance (INC), Constant Voltage (CV), and enhanced P&O (dP-P&O)—within a unified platform featuring an intuitive Human-Machine Interface (HMI), real-time operator-driven algorithm selection becomes possible.

Experimental evaluation on the dSPACE DS1104 platform confirms distinct performance profiles: INC achieves the highest efficiency (99.946%), closely followed by P&O (99.884%) and dP-P&O (99.780%), while CV lags behind at 95.324%, though it offers simplicity and fast response. The trade-offs between computational complexity, tracking precision, stability, and dynamic response justify the need for adaptive selection rather than fixed implementations.

The integration of human decision-making marks a significant advance, empowering operators to leverage expertise and situational awareness for optimal MPPT algorithm choice based on environmental and operational context. This dynamic, context-aware approach enhances system adaptability without interrupting power generation.

The platform's modular architecture and high-resolution control loops provide a solid foundation for future enhancements, including the integration of advanced AI-based MPPT techniques, partial shading management, comprehensive PV diagnostics, and machine learning-enabled autonomous algorithm selection. These developments promise improved performance under complex and variable conditions.

Overall, this work contributes to intelligent, user-centered renewable energy control systems, demonstrating that human-machine collaboration can outperform fully automated methods in dynamic scenarios. Such hybrid solutions support more resilient, efficient, and adaptable photovoltaic installations, advancing the global transition to sustainable energy.

REFERENCES

Abdelsalam, A. K., Massoud, A. M., Ahmed, S., & Enjeti, P. N. (2011). High-performance adaptive perturb and observe MPPT technique for photovoltaic-based microgrids. *IEEE Transactions on Power Electronics*, *26*(4), 1010–1021.

Addoweesh, K. E., & Noman, A. M. (2016). Simulation and dSPACE hardware implementation of the MPPT techniques using buck boost converter. *International Journal of Engineering Research and Applications*, *6*(3), 47–53.

Ahmad, Z., Singh, S. N., & Kumar, A. (2018). Design and implementation of real-time monitoring based on Android and HMI of PV tower system for laboratory scale. *Indonesian Journal of Electrical Engineering and Computer Science*, *11*(1), 187–194.

Ahmed, N. A., & Miyatake, M. (2008). A novel maximum power point tracking for photovoltaic applications under partially shaded insolation conditions. *Electric Power Systems Research*, *78*(5), 777–784.

Ahmed Baba, M., Naoui, M., & Cherkaoui, M. (2023). Fault-tolerant control strategy for Hall sensors in BLDC motor drive for electric vehicle applications. *Sustainability*, *15*(13), 10430. Advance online publication. DOI: 10.3390/su151310430

Allal, A., Mansouri, N., & Naoui, M. (2024). Advanced residual optimal mapping approach for precise detection of stator faults in induction motors. *IEEE Access: Practical Innovations, Open Solutions*, *12*, 143515–143530. DOI: 10.1109/ACCESS.2024.3442671

Alqunun, K., Bechiri, M. B., & Naoui, M. (2025). An efficient bearing fault detection strategy based on a hybrid machine learning technique. *Scientific Reports*, *15*, 18739. Advance online publication. DOI: 10.1038/s41598-025-02439-4 PMID: 40437009

Bidram, A., Davoudi, A., & Balog, R. S. (2012). Control and circuit techniques to mitigate partial shading effects in photovoltaic arrays. *IEEE Journal of Photovoltaics*, *2*(4), 532–546.

Brunton, S. L., Rowley, C. W., Kulkarni, S. R., & Clarkson, C. (2010). Maximum power point tracking for photovoltaic optimization using ripple-based extremum seeking control. *IEEE Transactions on Power Electronics*, *25*(10), 2531–2540.

Chin, C. S., Neelakantan, P., Yoong, H. P., & Teo, K. T. K. (2015). Cell modelling and model parameters estimation techniques for photovoltaic simulator application: A review. *Applied Energy*, *154*, 500–519.

dSpace GmbH. (2021). Real-Time Interface Implementation Guide [Technical manual].

Elgendy, M. A., Zahawi, B., & Atkinson, D. J. (2013). Assessment of the incremental conductance maximum power point tracking algorithm. *IEEE Transactions on Sustainable Energy*, *4*(1), 108–117.

Enslin, J. H. R., Wolf, M. S., Snyman, D. B., & Swiegers, W. (1997). Integrated photovoltaic maximum power point tracking converter. *IEEE Transactions on Industrial Electronics*, *44*(6), 769–773.

Esram, T., & Chapman, P. L. (2007). Comparison of photovoltaic array maximum power point tracking techniques. *IEEE Transactions on Energy Conversion*, *22*(2), 439–449. DOI: 10.1109/TEC.2006.874230

Femia, N., Petrone, G., Spagnuolo, G., & Vitelli, M. (2005). Optimization of perturb and observe maximum power point tracking method. *IEEE Transactions on Power Electronics*, *20*(4), 963–973. DOI: 10.1109/TPEL.2005.850975

Femia, N., Petrone, G., Spagnuolo, G., & Vitelli, M. (2013). Perturb and observe MPPT algorithm with a current controller based on the sliding mode. *International Journal of Electrical Power & Energy Systems*, *44*(1), 346–356.

González-Longatt, F. M. (2005). Model of photovoltaic module in Matlab. In II CIBELEC (pp. 1–6). Venezuela.

Green, M. A., Dunlop, E. D., Hohl-Ebinger, J., Yoshita, M., Kopidakis, N., & Hao, X. (2022). Solar cell efficiency tables (Version 60). *Progress in Photovoltaics: Research and Applications*, *30*(7), 687–701. DOI: 10.1002/pip.3512

Hayder, W., Abid, A., Ben Hamed, M., & Sbita, L. (2020). Improved PSO algorithms in PV system optimisation. *European Journal of Electrical Engineering and Computer Science*, *4*(1), 1–6.

Hayder, W., Ogliari, E., Dolara, A., Abid, A., Ben Hamed, M., & Sbita, L. (2020). Improved PSO: A comparative study in MPPT algorithm for PV system control under partial shading conditions. *Energies*, *13*(8), 2035. Advance online publication. DOI: 10.3390/en13082035

Hayder, W., Séra, D., Ogliari, E., Lashab, A., Spataru, S., Guerrero, J. M., & Dolara, A. (2022). On improved PSO and neural network P&O methods for PV system under shading and various atmospheric conditions. *Energies*, *15*(20), 7668. Advance online publication. DOI: 10.3390/en15207668

Hohm, D. P., & Ropp, M. E. (2003). Comparative study of maximum power point tracking algorithms. *Progress in Photovoltaics: Research and Applications, 11*(1), 47–62.

Hua, C., & Lin, J. (2001). Fully digital control of distributed photovoltaic power systems. In *Proceedings of the IEEE International Symposium on Industrial Electronics* (pp. 1–6).

Hua, C., & Lin, J. (2003). An on-line MPPT algorithm for rapidly changing illuminations of solar arrays. *Renewable Energy, 28*(7), 1129–1142.

IEEE Standards Association. (2018). IEEE Std 1547-2018: IEEE standard for interconnection and interoperability of distributed energy resources.

Irisawa, K., Saito, T., Hosokawa, K., & Takyu, S. (2000). Maximum power point tracking control of photovoltaic generation system under non-uniform insolation by means of monitoring cells. In *Proceedings of the IEEE Power Electronics Specialists Conference* (pp. 948–953).

Jain, S., & Agarwal, V. (2007). Comparison of the performance of maximum power point tracking schemes applied to single-stage grid-connected photovoltaic systems. *IET Electric Power Applications, 1*(5), 753–762.

Ji, Y. H., Jung, D. Y., Kim, J. G., Lee, J. H., Won, C. Y., & Yoo, D. W. (2011). A real maximum power point tracking method for mismatching compensation in PV array under partially shaded conditions. *IEEE Transactions on Power Electronics, 26*(4), 1001–1009.

Jiang, J. A., Huang, T. L., Hsiao, Y. T., & Chen, C. H. (2005). Maximum power tracking for photovoltaic power systems. *Tamkang Journal of Science and Engineering, 8*(2), 147–153.

Kjaer, S. B., Pedersen, J. K., & Blaabjerg, F. (2005). A review of single-phase grid-connected inverters for photovoltaic modules. *IEEE Transactions on Industry Applications, 41*(5), 1292–1306. DOI: 10.1109/TIA.2005.853371

Kottas, T. L., Boutalis, Y. S., & Karlis, A. D. (2006). New maximum power point tracker for PV arrays using fuzzy controller in close cooperation with fuzzy cognitive networks. *IEEE Transactions on Energy Conversion, 21*(3), 793–803.

Koutroulis, E., Kalaitzakis, K., & Voulgaris, N. C. (2001). Development of a microcontroller-based, photovoltaic maximum power point tracking control system. *IEEE Transactions on Power Electronics, 16*(1), 46–54. DOI: 10.1109/63.903988

Lammouchi, Z., Naoui, M., & Bekakra, Y. (2025). Model-free predictive control based on the execution time of active vectors using ultra-local model. *Arabian Journal for Science and Engineering*, *50*, 5885–5898. DOI: 10.1007/s13369-024-09629-8

Liu, F., Duan, S., Liu, F., Liu, B., & Kang, Y. (2008). Variable step size INC MPPT method for PV systems. *IEEE Transactions on Industrial Electronics*, *55*(7), 2622–2628. DOI: 10.1109/TIE.2008.924035

Lyden, S., & Haque, M. E. (2016). A simulated annealing global maximum power point tracking approach for PV modules under partial shading conditions. *IEEE Transactions on Power Electronics*, *31*(6), 4171–4181.

Masoum, M. A. S., Dehbonei, H., & Fuchs, E. F. (2002). Theoretical and experimental analyses of photovoltaic systems with voltage and current-based maximum power-point tracking. *IEEE Transactions on Energy Conversion*, *17*(4), 514–522.

Masoum, M. A. S., & Sarvi, M. (2011). A new fuzzy-based combined variable step size INC and P&O maximum power point tracking algorithm for photovoltaic systems. In *Proceedings of the IEEE Power and Energy Society General Meeting* (pp. 1–8).

Mellit, A., & Kalogirou, S. A. (2014). MPPT-based artificial intelligence techniques for photovoltaic systems and its implementation into field programmable gate array chips: Review of current status and future perspectives. *Energy*, *70*, 1–18. DOI: 10.1016/j.energy.2014.03.070

Pandey, A., Dasgupta, N., & Mukerjee, A. K. (2023). A review on recent trends and developments in MPPT techniques for solar PV systems under partial shading conditions. *Renewable & Sustainable Energy Reviews*, *172*, 113036. Advance online publication. DOI: 10.1016/j.rser.2022.113036

Radjai, T., Rahmani, L., Mekhilef, S., & Gaubert, J. P. (2014). Implementation of a modified perturb and observe MPPT algorithm with direct control based on a fuzzy duty cycle change estimator using dSPACE. *Solar Energy*, *110*, 325–337.

Romdhane, M., Naoui, M., & Mansouri, A. (2023). PMSM inter-turn short circuit fault detection using the fuzzy-extended Kalman filter in electric vehicles. *Electronics (Basel)*, *12*(18), 3758. Advance online publication. DOI: 10.3390/electronics12183758

Safari, A., & Mekhilef, S. (2011). Simulation and hardware implementation of incremental conductance MPPT with direct control method using cuk converter. *IEEE Transactions on Industrial Electronics*, *58*(4), 1154–1161. DOI: 10.1109/TIE.2010.2048834

Safari, A., & Mekhilef, S. (2011). Incremental conductance MPPT method for PV systems. In *Proceedings of the IEEE Applied Power Electronics Conference* (pp. 1147–1152).

Salas, V., Olias, E., Barrado, A., & Lázaro, A. (2006). Review of the maximum power point tracking algorithms for stand-alone photovoltaic systems. *Solar Energy Materials and Solar Cells*, *90*(11), 1555–1578.

Sera, D., Teodorescu, R., & Rodriguez, P. (2006). Improved MPPT algorithms for rapidly changing environmental conditions. In *Proceedings of the IEEE Power Electronics and Applications* (pp. 1420–1425).

Sera, D., Teodorescu, R., & Rodriguez, P. (2008). Optimized maximum power point tracker for fast-changing environmental conditions. *IEEE Transactions on Industrial Electronics*, *55*(7), 2629–2637. DOI: 10.1109/TIE.2008.924015

Subudhi, B., & Pradhan, R. (2013). A comparative study on maximum power point tracking techniques for photovoltaic power systems. *IEEE Transactions on Sustainable Energy*, *4*(1), 89–98. DOI: 10.1109/TSTE.2012.2202294

Swiegers, W., & Enslin, J. H. R. (1998). An integrated maximum power point tracker for photovoltaic panels. In *Proceedings of the IEEE International Symposium on Industrial Electronics* (pp. 40–44).

Veerachary, M., Senjyu, T., & Uezato, K. (2006). MPPT schemes for PV fed matrix converter. *IEEE Transactions on Industrial Electronics*, *53*(4), 1175–1184.

Windarko, N. A., Fu, J. H., & Choi, J. H. (2010). Maximum power point tracking of photovoltaic system using adaptive neuro-fuzzy inference system. In *Proceedings of the IEEE International Conference on Sustainable Energy Technologies* (pp. 1–6). IEEE.

Xiao, W., Dunford, W. G., Palmer, P. R., & Capel, A. (2006). Real-time identification of optimal operating points in photovoltaic power systems. *IEEE Transactions on Industrial Electronics*, *53*(4), 1017–1026.

Youssef, R., Hamida, M. A., Mohamed, N., Lassaad, S., Djamel, T., & Yasmine, A. (2023). Single-stage standalone photovoltaic water pumping system using predictive torque control (PTC) of induction machine. *Electric Power Components and Systems*. Advance online publication. DOI: 10.1080/15325008.2023.2278624

Chapter 5
Supervisory Human-Machine Interaction for Energy Management in Hybrid Renewable Energy Systems

Aicha Abid
https://orcid.org/0000-0001-5746-703X
ENIG, Tunisia

ABSTRACT

This chapter addresses the critical role of supervisory human–machine interaction (HMI) in the effective energy management of hybrid renewable energy systems. It begins by presenting a comprehensive modeling framework of a hybrid system combining photovoltaic panels, wind turbines, batteries, and electrical loads. A hybrid control and energy management strategy is then developed to optimize power flow and enhance system stability. Central to this work is the design and implementation of a user-friendly HMI built using MATLAB App Designer and integrated with a Simulink model. The interface allows real-time monitoring of power outputs, DC bus voltage, and battery state of charge, while providing interactive controls for load adjustment and simulation management. The seamless integration between the HMI and the control system facilitates improved decision-making and operational flexibility. Finally, the chapter highlights the importance of such supervisory interfaces in advancing the reliability and efficiency of renewable energy systems, and suggests pathways for future enhancements involving adaptive and remote control features.

DOI: 10.4018/979-8-3373-5621-1.ch005

INTRODUCTION

In recent years, the global energy landscape has been undergoing a profound transformation driven by the urgent need to reduce greenhouse gas emissions and achieve long-term sustainability. Among the most promising solutions, hybrid renewable energy systems (HRES) have gained increasing attention because they combine complementary energy sources—typically photovoltaic (PV), wind, and energy storage technologies—to ensure a more stable and reliable electricity supply. By exploiting the diversity of renewable resources, these systems can maintain operation under fluctuating environmental conditions and mitigate the limitations of single-source configurations. Nevertheless, their nonlinear and time-varying behavior makes their management and supervision a demanding task, especially when high efficiency and autonomy are expected (Blaabjerg et al., 2017; Ahmad et al., 2024; Ajiboye et al., 2023; Soliman et al., 2021; Weschenfelder et al., 2020; Faisal et al., 2020; Chaichan et al., 2025; Kumar et al., 2020).

While advanced control algorithms and optimization techniques have significantly improved the technical performance of such systems, the human operator remains an irreplaceable part of the control loop. In practice, operators are responsible for monitoring the evolution of electrical variables, validating automated decisions, and intervening whenever abnormal events occur. This interaction between humans and machines is orchestrated through the Human–Machine Interface (HMI)—the visual and functional layer that transforms raw data into actionable information. The efficiency of the HMI directly affects the operator's perception, decision speed, and situational awareness, which in turn influence the safety and productivity of the entire energy system (Kiran & Reddy, 2022; Meddeb et al., 2023; Park et al., 2024; Triatmaja et al., 2024).

Recent research has emphasized the need for adaptive and intelligent interfaces capable of supporting decision-making in real time. Such HMIs integrate data visualization, control commands, and analytical tools within a unified environment, allowing users to observe, interpret, and act efficiently. With the growing accessibility of simulation platforms, MATLAB/Simulink coupled with App Designer offers an ideal framework to develop and test supervisory applications. These tools enable researchers to prototype interactive environments where the operator can command the system, modify parameters, and visualize the dynamic response of the hybrid model without the constraints of physical experimentation.

In this chapter, a supervisory HMI is developed for a hybrid PV–wind–battery configuration, providing an intuitive environment for control and monitoring. The interface incorporates essential functions such as system initialization, shutdown, and load modification, along with graphical displays representing PV, wind, and battery power, DC bus voltage, and the state of charge (SOC). Through this simulation-based approach, the study demonstrates how effective human–machine cooperation can enhance the reliability and operational understanding of hybrid renewable systems.

The remainder of this chapter is organized as follows: Section 2 presents an overview of recent literature on hybrid energy systems, supervisory control frameworks, and human–machine interfaces. Section 3 details the configuration and mathematical modeling of the studied hybrid system. Section 4 describes the adopted hybrid control and energy management strategies. Section 5 is dedicated to the design of the HMI using MATLAB App Designer, highlighting its interactive components and visualization capabilities. Finally, Section 6 concludes the chapter by summarizing key findings and suggesting directions for future enhancement of intelligent supervisory tools.

LITERATURE REVIEW

Hybrid renewable energy systems (HRES) have been the subject of extensive research over the last two decades, reflecting the urgent global need for sustainable, reliable, and resilient energy infrastructures. Alongside technical progress in photovoltaic (PV), wind, and storage technologies, the supervisory layer of control—often embodied in the Human–Machine Interface (HMI)—has emerged as a decisive factor in ensuring safe and efficient system operation. This section reviews the state of the art in advanced HMI applications, highlighting progress in ergonomics, artificial intelligence integration, digital twins, and cybersecurity, while also positioning these developments within the framework of current energy policies and large-scale deployment projects (Girija & Sahoo, 2025; Kang, 2020; Kundu et al., 2024; Jiang et al., 2023; Coppolino et al., 2023; Lai et al., 2024; Abdalla et al., 2025).

1. Human–Machine Interfaces in Renewable Energy Systems

Traditional supervisory systems, such as SCADA platforms, have long been used to monitor and control renewable installations. While these tools provide essential functionalities—data acquisition, alarms, and remote control—they are often limited by rigid architectures and relatively static visualization capabilities. Recent studies emphasize the importance of adaptive and user-centered HMIs, capable of delivering real-time feedback and intuitive interaction. Such interfaces improve operator situational awareness, reduce cognitive workload, and facilitate decision-making during both routine operation and fault conditions. In hybrid systems, where multiple energy sources and storage units must be coordinated, the HMI plays a particularly crucial role in bridging the gap between automated algorithms and human supervision (Javaid et al., 2023; Hossain et al., 2023).

2. Ergonomics and Advanced Supervision

Ergonomics in HMI design goes beyond aesthetics; it encompasses the cognitive and physiological interaction between humans and complex energy systems. Research in this area highlights the need for interfaces that minimize operator fatigue, reduce error probability, and support rapid interventions during emergencies. Advanced visualization techniques, including layered dashboards, interactive graphs, and adaptive alarm hierarchies, have been shown to significantly enhance supervisory efficiency. Integrating ergonomic principles into renewable energy HMIs is therefore recognized as a pathway to improving system resilience, especially under fluctuating resource conditions (Pham et al., 2023).

3. Intelligent and Explainable AI in HMI

Artificial Intelligence (AI) is increasingly embedded within supervisory platforms, enabling predictive maintenance, anomaly detection, and adaptive control recommendations. In particular, explainable AI (XAI) frameworks are gaining attention, as they provide transparency in decision-making processes, which is essential for operator trust and accountability. Studies have proposed AI-driven HMIs capable of adapting visualization layouts based on operator expertise, workload, or contextual priorities. This convergence of intelligent algorithms and human supervision marks a transition from static monitoring interfaces toward cognitive decision-support systems, where humans remain in the loop while benefiting from machine learning insights.

4. Digital Twins and Virtual Supervision

The concept of digital twins—virtual replicas of physical energy assets—has been widely adopted in modern renewable energy management. By synchronizing real-time operational data with simulation models, digital twins enable predictive monitoring, fault diagnosis, and scenario testing without disrupting physical systems. When integrated into HMIs, digital twins provide operators with an enriched supervisory experience: the ability to visualize both present and future states of the hybrid system. Several projects in smart grids and microgrids have demonstrated the potential of digital twins in optimizing resource scheduling, enhancing situational awareness, and supporting resilience under uncertainty.

5. Cybersecurity Challenges in Supervisory HMIs

As supervisory HMIs increasingly rely on interconnected digital platforms, cybersecurity has become a critical concern. Hybrid renewable energy systems are vulnerable to cyberattacks targeting communication links, control algorithms, or operator interfaces. Recent literature emphasizes the integration of intrusion detection, secure communication protocols, and resilience strategies within HMI architectures. A secure interface not only protects system integrity but also safeguards operator decision-making from manipulation or misinformation. Cybersecurity considerations are therefore indispensable for the sustainable deployment of HRES at both local and national scales.

6. Links to Projects and Energy Policies

The development of advanced supervisory HMIs is not only a technical challenge but also a policy-driven priority. International frameworks such as the European Green Deal, the International Energy Agency (IEA) reports, and national smart grid initiatives underline the necessity of robust supervisory infrastructures for large-scale renewable integration. Several recent projects—ranging from community microgrids to utility-scale hybrid plants—have demonstrated the benefits of user-centered supervisory systems in achieving policy goals such as decarbonization, resilience, and affordability. These policy and project-driven initiatives further highlight the timeliness of advancing HMI research in renewable energy contexts.

7. Synthesis

The reviewed literature shows that while significant progress has been made in hybrid energy system modeling and automated control, the supervisory HMI remains a critical and underexplored component. Emerging paradigms in ergonomics, AI-driven adaptability, digital twins, and cybersecurity offer promising directions, yet their integration into renewable HMIs is still at an early stage. This chapter contributes to addressing this gap by presenting a simulation-based supervisory HMI framework for a PV–wind–battery hybrid system, emphasizing user-centered design and adaptive interaction (Romdhane et al., 2023; Allal et al., 2024; Youssef et al., 2023; Lammouchi et al., 2025; Alqunun et al., 2025; Ahmed Baba et al., 2023).

DESCRIPTION AND MODELING OF THE STUDIED HYBRID RENEWABLE SYSTEM

1. System Architecture

The proposed system consists of three primary components: a photovoltaic (PV) source, a wind turbine, and a battery storage unit. The electrical load that is connected to the DC bus via a DC/AC multilevel inverter is represented as a controllable power demand whose value is set by the user (Aloo et al., 2023; Cao et al., 2025; Marzebali et al., 2020).

The wind conversion system is based on a PMSG, an AC/DC converter and a boost DC/DC converter. A fuzzy-based MPPT controller adjusts the generator's output by regulating the duty cycle of the DC–DC converter.

The PV conversion system contains the PV modules and a DC/DC boost converter. An MPPT controller based on fuzzy logic ensures optimal power extraction under variable conditions.

The Battery energy storage system (BESS): The battery is modeled using a dynamic equivalent circuit with state-of-charge (SoC) estimation. It absorbs surplus energy from renewable sources and discharges when generation falls short of demand. Bidirectional power flow is managed through a buck–boost converter with charge/discharge logic.

These components are interconnected via a common DC bus supplying the load, which is variable and partially user-defined through the HMI module.

Figure 1. The hybrid PV-Wind-Storage system.

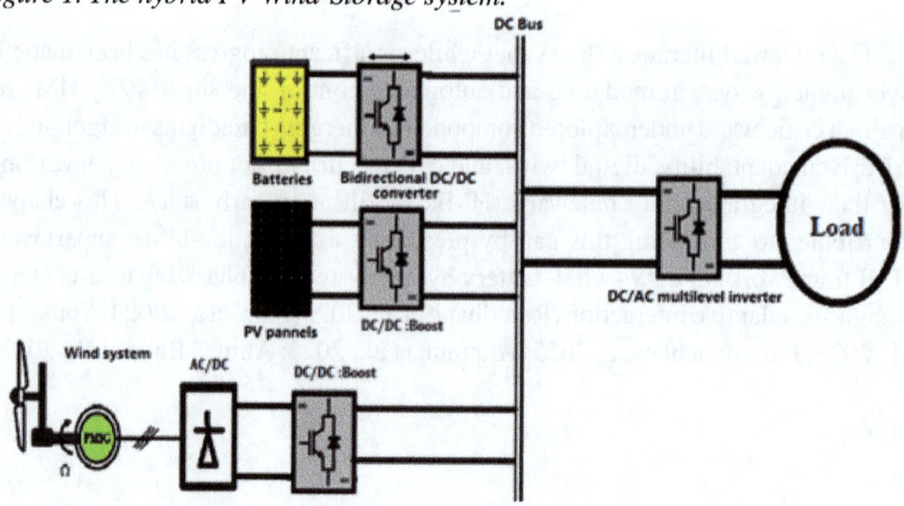

The energy management system (EMS) that governs power flow decisions and mode switching according to a fixed priority logic:

PV Priority: The EMS first utilizes the available solar power.
Wind Contribution: If solar is insufficient, wind power is added.
Battery Support: If both sources cannot satisfy the demand, the battery discharges to make up the difference.

The supervisory HMI interface enables users to monitor real-time behavior, modify operational parameters, and enforce control decisions.

2. Mathematical model of PV System

The single diode model was adopted in this part to describe mathematically the current-voltage characteristic I(V) of a PV cell represented in Fig2.

Figure 2. Equivalent electrical schema of a photovoltaic cell

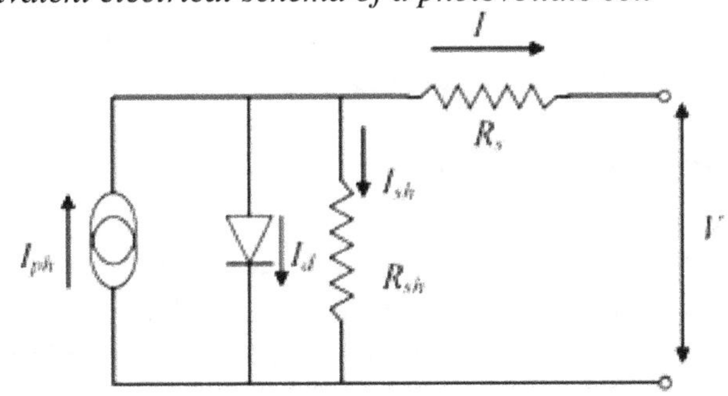

From the equivalent circuit in figure (2.), it is possible to write:

$$I = I_{ph} - I_d - I_{sh} \tag{1}$$

In which Iph is the light generated current, Id is the currents flows across the diode and Ish is

The light generated current Iph depends, particularly on the absorbed sun light, Thus it can be expressed by eq (2).

$$I_{ph} = iccs\frac{G}{G_s}\left(1 - \Delta icc(T_s - T)\right) \tag{2}$$

The current across the diode can be written as:

$$I_d = I_s\left[e^{\left(\frac{V+R \times I}{n \times V_t}\right)} - 1\right] \tag{3}$$

with:

I_s: The diode saturation current can be expressed by the following equation:

$$I_s = \frac{I_{ph-\frac{V_{co}}{R_{sh}}}}{\left[e^{\frac{V_{co}}{n \times Vt}} - 1\right]} \tag{4}$$

Therefore, we can write:

$$I = I_{ph-}I_s\left[e^{\left(\frac{V+R \times I}{n \times V_t}\right)} - 1\right] - \frac{V + R_s \times I}{R_{sh}} \tag{5}$$

3. Wind Energy Conversion System Modeling

The wind conversion system contains a wind generator, diodes based rectifier and a DC-DC boost converter connected to the DC bus.

The Wind Generator Modeling

The wind generator contains a turbine driving an electric synchronous generator (PMSG) via a flexible drive train (Marzebali et al., 2020).

The kinetic energy of the wind is transformed into a mechanical energy that helps to turn the turbine that drives the PMSG to generate, finally, the electric energy.

The power delivered by the wind turbine P_{wt} can be mathematically described by the equations (5).

$$P_{WT} = C_p(\beta, \lambda)P_{wind} = \frac{1}{2}\rho\pi R^2 C_p V_v^3 \tag{6}$$

The power coefficient Cp, given by eq (8) and (9), depends on the pitch angle β and the tip-speed ratio λ described by (7).

$$\lambda = \frac{\omega R_t}{V_v} \tag{7}$$

Where ω is the angular speed (rad/s) and R is the radius of the turbine.

$$C_p = 0.5176\left(\frac{116}{\lambda_i} - 0.4\beta - 5\right)e^{\frac{-21}{\lambda_i}} + 0.0068\lambda_i \quad (8)$$

$$\lambda_i = \frac{1}{\frac{1}{\lambda + 0.08\beta} - \frac{0.035}{\beta^3 + 1}} \quad (9)$$

The PMSG is modeled using the standard d–q transformation equations as shown in (10).

$$\begin{cases} V_d = R_s i_d + L_d \frac{di_d}{dt} - \omega L_q i_q \\ V_q = R_s i_q + L_q \frac{di_q}{dt} + \omega L_d i_d + \omega \varnothing_f \end{cases} \quad (10)$$

Equation (11) describes the electromagnetic torque.

$$C_{em} = \frac{3}{2} p(\varphi_d i_q - \varphi_q i_d) \quad (11)$$

With:
L_d, L_q: The self-inductances of each stator winding.
p: The number of pole pairs.
R_s: The resistance of each stator winding.
The flux vectors, φ_d and φ_q, that are created by the permanent magnet are described in equations (12).

$$\begin{cases} \varphi_d = L_d i_d + \varphi_f \\ \varphi_q = L_q i_q \end{cases} \quad (12)$$

4. Battery Energy Storage System (BESS)

The BESS is modeled as a lithium-ion battery, incorporating **a** state-of-charge (SOC) controller, charge/discharge logic, and overcurrent protection. The battery model follows a Thevenin-based equivalent circuit:

$$V_{bat} = E - R_{int} I_{bat} \quad (13)$$

- E is the open-circuit voltage (OCV) as a function of SOC,
- R_{int} is the internal resistance,
- I_{bat} is the battery current (positive during discharge, negative during charge).

Battery charge/discharge control is based on the bus voltage level and the generation-load mismatch. The BESS operates to maintain the DC bus voltage within acceptable limits and to smooth fluctuations from the PV and wind inputs.

Hybrid Control and Energy Management System

Energy management in hybrid renewable energy systems (HRES) is a critical process aimed at optimizing power flow among various generation sources, energy storage, and loads. In the context of the studied system, which integrates photovoltaics, wind energy, and a battery, the Energy Management System (EMS) ensures continuous power supply, maximizes renewable usage, protects system components, and coordinates with the Human–Machine Interaction (HMI) interface for manual supervision and intervention (Ahmed et al., 2024; Al-Wesabi et al., 2024; Aloo et al., 2023; Bassey, 2023; Cao et al., 2025; Chrifi-Alaoui et al., 2023; Ibrahim, 2024; Marzebali et al., 2020).

This section presents the layered control strategies, including automated rule-based logic, user-in-the-loop decision-making and real-time supervisory adaptation.

The EMS (as shown in Fig.3) is the control core that monitors input generation, battery state, and load demand to make real-time decisions. It includes:

- **Mode selection logic**: Charging, discharging, direct supply, or idle mode;
- **Priority logic**: Based on generation cost, source availability, and user preferences;
- **Battery protection**: Avoids deep discharge or overcharging;
- **Communication layer**: Synchronizes with the supervisory HMI to receive user commands and send status updates.

Figure 3. EMS ALGORITHM.

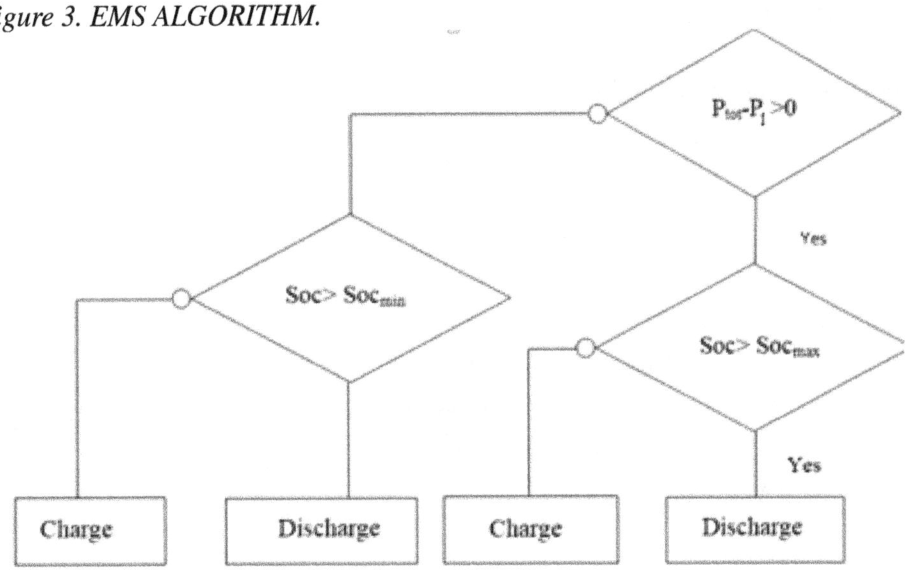

Where

P_{tot} is the total power of the renewable energy resources. $P_{tot}=P_{wind}+P_{PV}$

P_{wind} is the wind generator's power

Ppv is the Photovoltaic power

P_l is the load power

A simple yet effective rule-based controller manages power distribution based on predefined priorities:

1 Prioritize renewable generation to supply the load directly.

2 Store excess energy in the battery if SoC < SOC_{max}.

3 Discharge battery to supply load if renewables are insufficient and SoC > SOC_{min}.

4 If both renewable sources and battery are insufficient, initiate load shedding

The Control of the Proposed System and Simulation Results

Figure 4. PV conversion system control

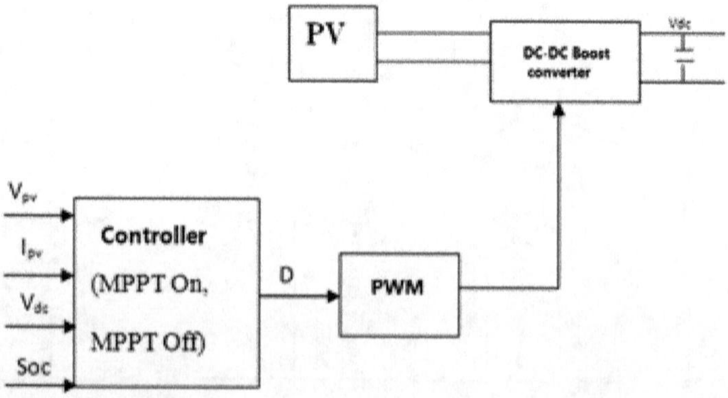

Depending on the state of charge (SOC) of the storage system, The PV conversion system has to operate in two cases in MPPT On case in order to extract the maximal power or in MPPT Off case in order to the power management as shown in fig.4.

The energy management is described by the following algorithm:

- If SOC is inferior then soc_{max} then the MPPT On is active.
- If SOC est superior then soc_{max} the MPPT Off is activated.

The same process is used to control the DC-DC boost converter as the PV conversion system algorithm **(Fig.5).**

- •If SOC is inferior then soc_{max} then the MPPT On is active.
- •If SOC is superior then soc_{max} the MPPT Off is activated.

Figure 5. Wind conversion system control

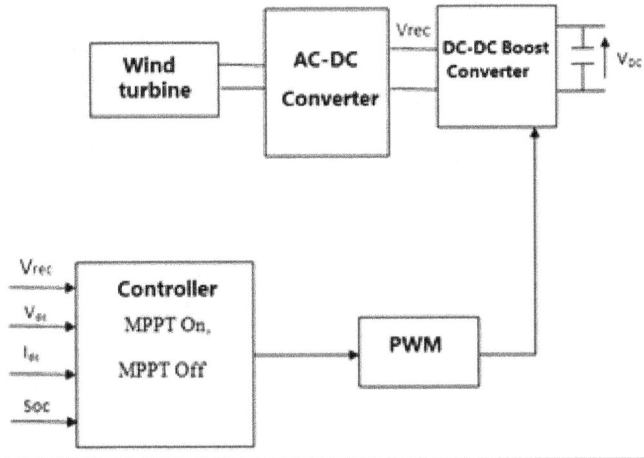

The modeling and the control of the different elements of the hybrid system are done under MATLAB/SIMULINK environment.

Two modes of operation were simulated:

1. Autonomous Mode: No human interaction; decisions based on programmed control logic.
2. Supervised Mode: Operator intervention allowed via HMI interface.

Autonomous Mode

To optimize the PV system as well as the Wind system performances, P&O algorithms are used to maximize the power by controlling the duty cycle of the converters.

The DC bus voltage must be stable and well regulated and adjusted as shown in Fig. 7.

The batteries are used as an auxiliary source in the case of a shortage or absence of production and as a load in the event of overproduction. The load is always connected. Thus, an algorithm is used for energy management.

Indeed, depending on the renewable energy (PV and wind system) production, energy consumption and state of charge of the batteries, the operating scenario of the hybrid system is described by Fig.6.

In case of excess production P_{tot} > PL(Load power)), the batteries are charged. On the other hand, in case of a power deficit (P_{tot} < PL), the batteries are partially discharged to meet the demand as approved by the evolution of the state of charge (see Fig.8).

Figure 6. The operating scenario of the hybrid system.

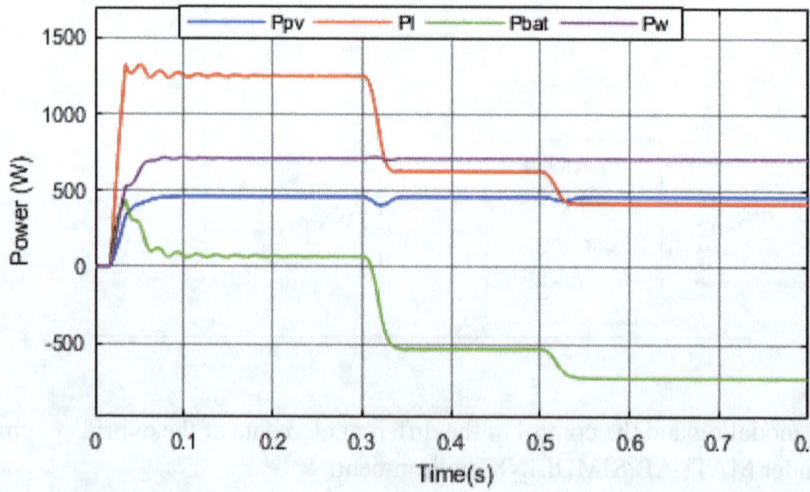

Figure 7. DC bus voltage.

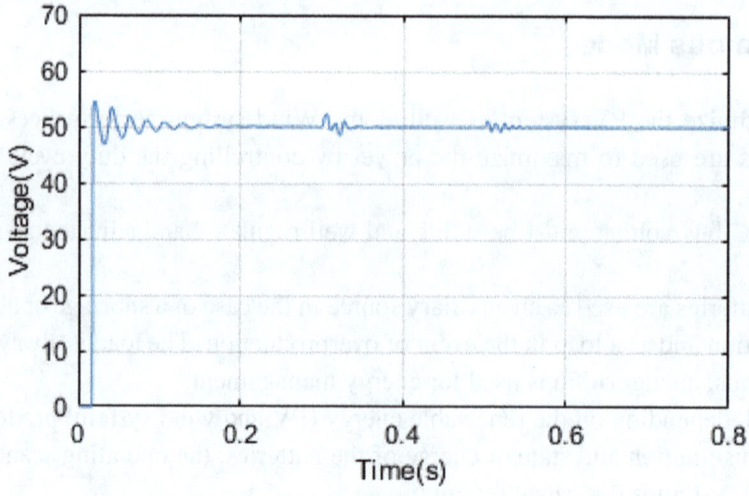

Figure 8. State of charge of batteries.

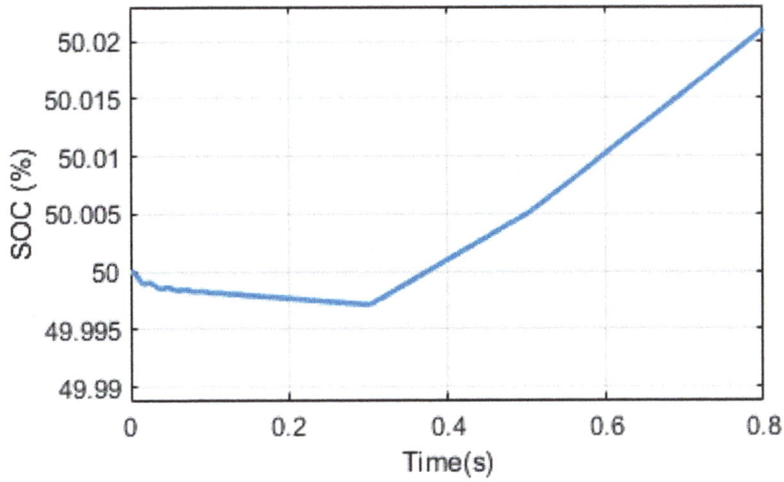

Supervised Mode

The supervisory control is realized through an interface where the user can: Enter the desired power demand (e.g., using a slider or input box) or observe real-time data such as battery state of charge, load current, and voltage.

This supervisory interface, designed in MATLAB App Designer, functions as the core HMI tool. It is connected dynamically to the simulation model built in Simulink and Simscape Electrical. Data flows between the interface and the model are handled using From Workspace and To Workspace blocks, allowing seamless bidirectional communication.

This structure enables a looped interaction: the user inputs a command (P_load), the EMS responds by reallocating energy flows, and the interface reflects the outcome, allowing the user to observe and, if necessary, adjust their input.

User inputs from the HMI (such as requested load demand) are factored into the control logic to override or supplement the automated decisions.

This layer allows users to override EMS decisions, simulate load surges, or study the impact of human intervention in various environmental and operating scenarios.

Integration of HMI in EMS Decision-Making

Unlike conventional EMS systems, this study introduces **user-interactive EMS**, allowing the operator to:

- Override automatic decisions
- Adjust thresholds (e.g., minimum SOC, load priority)
- Inject or remove virtual load
- Choose operational modes (PV priority, wind priority, storage-first, etc.)
- Start/stop MPPT tracking or force fixed voltage tracking

The EMS control blocks include "manual override" flags, monitored by Simulink Dashboard switches. When these are active, the system shifts to semi-automatic mode, where selected user inputs influence decisions while others remain automatic.

A unique feature of this system is the user-tunable load resistor via the HMI.

This feature allows the user to test EMS robustness under changing load scenarios and observe the system's response in real-time.

HUMAN–MACHINE INTERFACE (HMI) DEVELOPMENT USING MATLAB APP DESIGNER

As hybrid renewable energy systems (HRES) evolve toward higher levels of automation and intelligence, the inclusion of human operators in the decision-making loop remains crucial. Human–Machine Interaction (HMI) serves as a bridge between automated control processes and user intuition, enabling the system to operate not only based on sensor input and algorithmic decisions but also on real-time human insight. This section explores the core concepts behind HMI in energy systems and presents design considerations for integrating intuitive, responsive interfaces in supervisory control frameworks (Kiran & Reddy, 2022; Meddeb et al., 2023; Park et al., 2024; Triatmaja et al., 2024).

The supervisory control of the proposed hybrid renewable energy system requires a user-friendly platform that enables real-time monitoring, parameter adjustment, and manual intervention when necessary. To achieve this, a dedicated Human–Machine Interface (HMI) was developed using **MATLAB App Designer**. This HMI provides operators with an intuitive and interactive environment to control the system while visualizing its operational status.

1. Interface Layout and Components

The HMI was designed to integrate both control and visualization functionalities in a single dashboard. The interface consists of the following main components:

1. **Start Button** – Initiates the execution of the Simulink model, establishing communication between the GUI and the simulation environment.

2. **Stop Button** – Terminates the simulation in a controlled manner, ensuring proper data logging and variable reset.
3. **Load Parameter Edit Field** – Allows the operator to manually set or adjust the load value during runtime, enabling flexible testing of various operating scenarios. In this application there is three values of load.
4. **Real-Time Power Graphs** – A multi-axis plotting area displaying instantaneous values of:
   - o Photovoltaic (PV) output power
   - o Wind turbine output power
   - o Battery charging/discharging power
   - o Load power demand
5. **DC Bus Voltage Graph** – Dedicated chart for monitoring the DC link voltage stability, providing immediate insight into system regulation performance.
6. **State of Charge (SOC) Graph** – Shows the battery SOC evolution over time, allowing the operator to monitor storage utilization and depth of discharge.

2. Communication Between the HMI and Simulink Model

The HMI communicates with the Simulink model through MATLAB workspace variables and **set_param / get_param** commands. The process is structured as follows:

- **Initialization**: When the "Start" button is pressed, the app triggers the execution of the Simulink model using the **set_param('ModelName','SimulationCommand','start')** function.
- **Parameter Updates**: Load values modified in the edit field are written directly to the corresponding Simulink block parameter via **set_param** commands, ensuring immediate effect.
- **Real-Time Data Acquisition**: Simulink signals (PV power, wind power, load power, battery power, SOC, and DC bus voltage) are streamed to MATLAB using **To Workspace** or **To App** functions. These variables are then plotted dynamically in the App Designer UIAxes components.
- **Simulation Termination**: Pressing the "Stop" button sends a **set_param('ModelName','SimulationCommand','stop')** command, ensuring a clean shutdown of the simulation process.

3. Advantages of the Developed HMI

The implementation of this HMI provides several operational and research benefits:

- **Enhanced Monitoring** – Real-time display of multiple key performance indicators enables quick assessment of system behavior.
- **Interactive Control** – Operators can adjust load demand during runtime without halting the simulation, supporting rapid scenario testing.
- **Research Flexibility** – The HMI can be easily adapted to include additional controls or visualization panels as the system evolves.

4. Illustration of the Developed HMI

Figures 9 and 10 presents a screenshot of the developed App Designer interface, showing the control buttons, parameter adjustment field, and real-time graphical monitoring.

Figure 9. App Designer interface

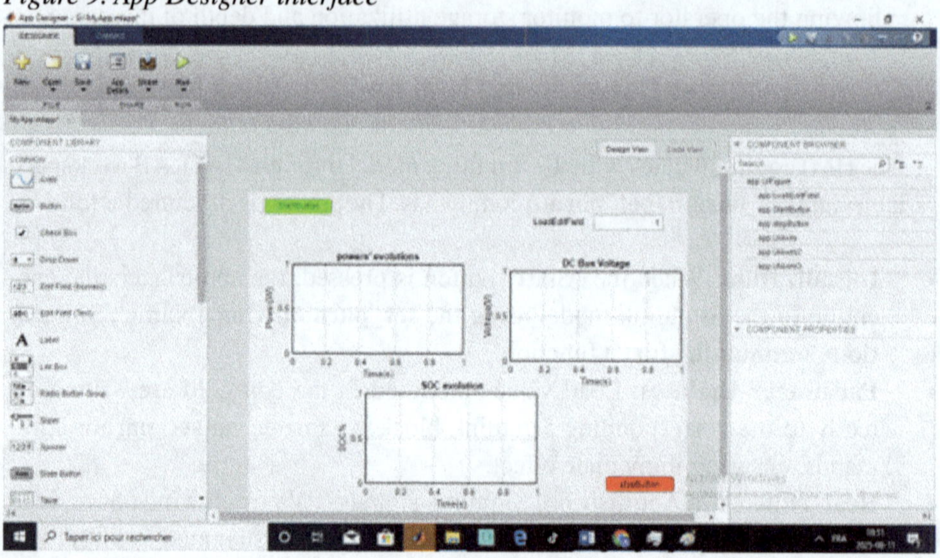

Figure 10. supervisory human–machine interaction (HMI) framework

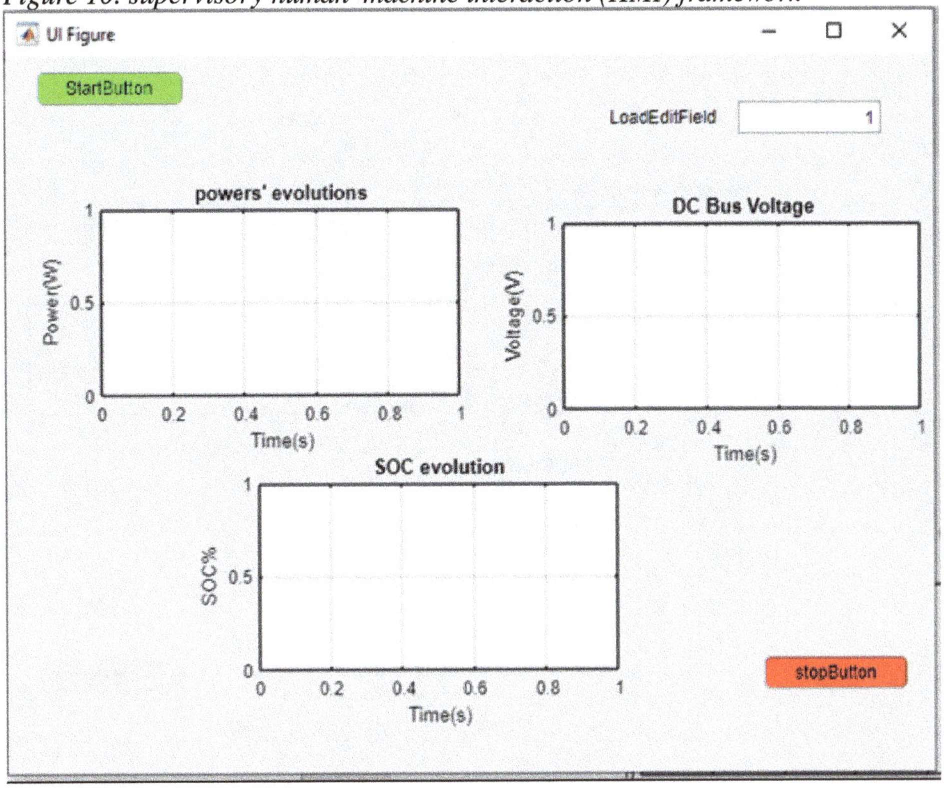

The result of the control via the created interface is shown in Fig.11. As illustrated in the figure, the presented results correspond to the initial stage of the simulation, primarily intended to demonstrate that the designed supervisory interface enables real-time visualization of the system's key parameters—including source power outputs, DC bus voltage, and battery state of charge (SOC)—while allowing user interaction through control buttons and parameter adjustment fields.

Figure 11. The result of the control via the created interface.

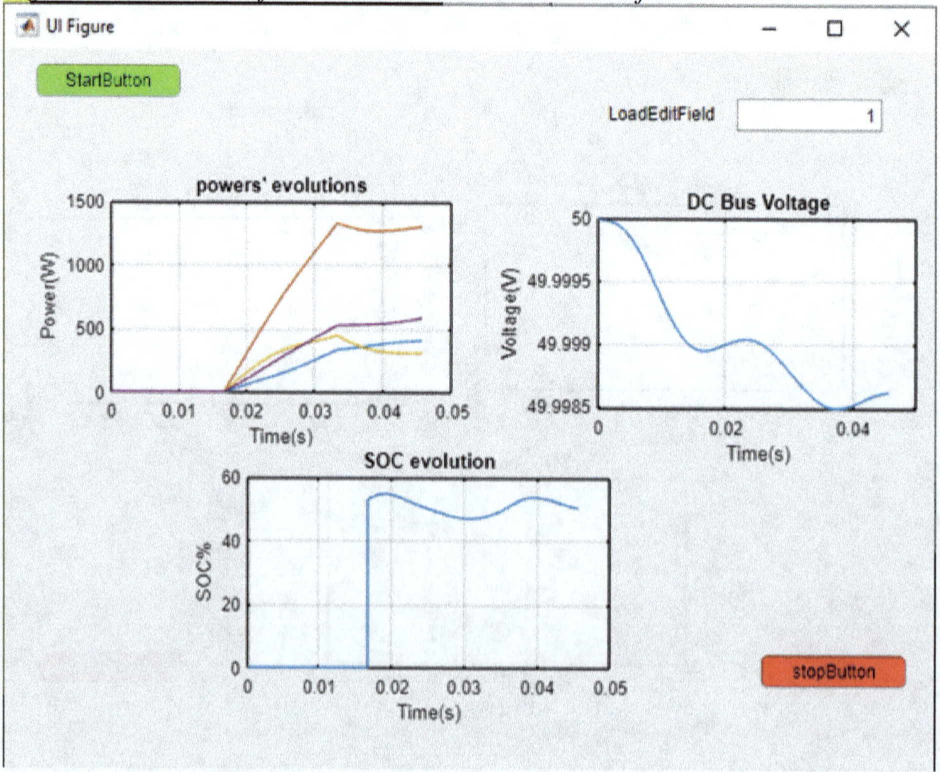

The implementation of a supervisory human–machine interaction (HMI) framework within a hybrid renewable energy system (HRES) reveals significant advantages in both operational efficiency and adaptability. This section synthesizes the insights drawn from the modeling, simulations, and case studies, while also identifying key areas for future research and development.

The success of supervisory HMI frameworks extends beyond technical performance. It touches on themes of energy sovereignty, environmental stewardship, and digital inclusion. By empowering users to participate in energy decisions, these systems help build awareness and trust in sustainable technologies.

Moreover, in energy-insecure regions, such systems can enable community-led energy governance. In such settings, a combination of autonomous control, minimal supervision, and local adaptability ensures resilience against resource scarcity and infrastructure limitations.

CONCLUSION

This chapter has explored the integration of supervisory human–machine interaction within the energy management framework of hybrid renewable energy systems. Beginning with a detailed description and modeling of the hybrid system components, it highlighted the complexity and dynamic behavior inherent to such systems. The development of a hybrid control and energy management strategy was then presented, aiming to optimize energy flow and enhance system reliability.

A key contribution of this work is the design and implementation of a user-friendly Human–Machine Interface built with MATLAB App Designer, tightly coupled with the Simulink model. This interface empowers operators to monitor essential system variables in real time, modify load conditions interactively, and gain immediate feedback on system performance through clear graphical representations. By facilitating seamless communication between the human operator and the automated control system, the HMI plays a crucial role in improving decision-making, operational flexibility, and overall system efficiency.

Looking ahead, future enhancements could include extending the HMI with advanced features such as predictive analytics, adaptive control based on user feedback, and remote access capabilities. Such developments would further strengthen the collaboration between human supervisors and intelligent energy management systems, paving the way for smarter, more resilient hybrid renewable energy solutions.

REFERENCES

Abdalla, S. A., Abdullah, S. S., & Kassem, A. M. (2025). An adaptive frame and intelligent control approach for an autonomous hybrid renewable energy technology consisting of PV, wind, and fuel cell innovation. *Alexandria Engineering Journal, 114*, 279–291. DOI: 10.1016/j.aej.2024.10.057

Ahmad, S., Hasan, S. M. N., Hossain, M. S., Uddin, R., Ahmed, T., Mustayen, A. G. M. B., Hazari, M. R., Hassan, M., Parvez, M. S., & Saha, A. (2024). A review of hybrid renewable and sustainable power supply system: Unit sizing, optimization, control, and management. *Energies, 17*(23), 6027. Advance online publication. DOI: 10.3390/en17236027

Ahmed, M. M. R., Mirsaeidi, S., Koondhar, M. A., Karami, N., Tag-Eldin, E. M., & Ghamry, N. A. (2024). Mitigating uncertainty problems of renewable energy resources through efficient integration of hybrid solar PV/wind systems into power networks. *IEEE Access: Practical Innovations, Open Solutions, 12*, 30311–30328. DOI: 10.1109/ACCESS.2024.3370163

Ahmed Baba, M., Naoui, M., & Cherkaoui, M. (2023). Fault-tolerant control strategy for Hall sensors in BLDC motor drive for electric vehicle applications. *Sustainability, 15*(13), 10430. Advance online publication. DOI: 10.3390/su151310430

Ajiboye, O. K., Ochiegbu, C. V., Ofosu, E. A., & Gyamfi, S. (2023). A review of hybrid renewable energies optimisation: Design, methodologies, and criteria. *International Journal of Sustainable Energy, 42*(1), 648–684. DOI: 10.1080/14786451.2023.2227294

Al-Wesabi, I., Fang, Z., Cai, J., Hussein Farh, H. M., Aboudrar, I., & Dagal, I. (2024). Fast DC-link voltage control based on power flow management using linear ADRC combined with hybrid salp particle swarm algorithm for PV/wind energy conversion system. *International Journal of Hydrogen Energy, 61*, 688–709. DOI: 10.1016/j.ijhydene.2024.02.325

Allal, A., Mansouri, N., & Naoui, M. (2024). Advanced residual optimal mapping approach for precise detection of stator faults in induction motors. *IEEE Access: Practical Innovations, Open Solutions, 12*, 143515–143530. DOI: 10.1109/ACCESS.2024.3442671

Aloo, L. A., Kihato, P. K., Kamau, S. I., & Orenge, R. S. (2023). Modeling and control of a photovoltaic-wind hybrid microgrid system using GA-ANFIS. *Heliyon, 9*(4), e14678. Advance online publication. DOI: 10.1016/j.heliyon.2023.e14678 PMID: 37025884

Alqunun, K., Bechiri, M. B., & Naoui, M. (2025). An efficient bearing fault detection strategy based on a hybrid machine learning technique. *Scientific Reports*, *15*, 18739. Advance online publication. DOI: 10.1038/s41598-025-02439-4 PMID: 40437009

Bassey, K. E. (2023). Hybrid renewable energy systems modeling. *Engineering Science and Technology Journal*, *4*(6), 571–588. DOI: 10.51594/estj.v4i6.1255

Blaabjerg, F., Yang, Y., Yang, D., & Xu, B. (2017). Distributed power-generation systems and protection. *Proceedings of the IEEE*, *105*(7), 1311–1331. DOI: 10.1109/JPROC.2017.2696878

Cao, X., Zheng, Y., Xiao, H., & Xiao, M. (2025). Maximum power point tracking control of offshore wind-photovoltaic hybrid power generation system with crane-assisted. *CMES-Computer Modeling in Engineering & Sciences*, *143*(1), 289–334. DOI: 10.32604/cmes.2025.063954

Chaichan, W., Waewsak, J., Naklua, Y., Ali, F., Mueanmas, C., Nikhom, R., Kongruang, C., & Gagnon, Y. (2025). Sustainability analysis of hybrid renewable-based power generation with battery energy storage system for remote islands: Application to Koh Hang, Thailand. *Cleaner Energy Systems*, *12*, 100203. Advance online publication. DOI: 10.1016/j.cles.2025.100203

Chrifi-Alaoui, L., Drid, S., Ouriagli, M., & Mehdi, D. (2023). Overview of photovoltaic and wind electrical power hybrid systems. *Energies*, *16*(12), 4778. Advance online publication. DOI: 10.3390/en16124778

Coppolino, L., Nardone, R., Petruolo, A., & Romano, L. (2023). Building cyber-resilient smart grids with digital twins and data spaces. *Applied Sciences (Basel, Switzerland)*, *13*(24), 13060. Advance online publication. DOI: 10.3390/app132413060

Faisal, M., Hannan, M. A., Ker, P. J., Hussain, A., Mansor, M. B., & Blaabjerg, F. (2020). Review of energy storage system technologies in microgrid applications: Issues and challenges. *IEEE Access: Practical Innovations, Open Solutions*, *8*, 218685–218693. DOI: 10.1109/ACCESS.2020.3036556

Girija, G. S. P., & Sahoo, A. K. (2025). An adaptive energy monitoring system for a hybrid power plant using renewable energy. *Transactions on Energy Systems and Engineering Applications*, *6*(1), 1–15.

Hossain, M. M., Ortmeyer, T., & Hall, E. (2023). Human-centered design, development, and evaluation of an interface for a microgrid controller. *Advances in Science, Technology and Engineering Systems Journal*, *8*(3), 80–88. DOI: 10.25046/aj080310

Ibrahim, M. M. (2024). Energy management strategies of hybrid renewable energy systems: A review. *Wind Engineering*, *48*(1), 133–161. DOI: 10.1177/0309524X231200010

Javaid, S., Javaid, N., & Alhussein, M. (2023). Towards efficient human–machine interaction for home energy management with seasonal scheduling using deep fuzzy neural optimizer. *Cognition Technology and Work*, *25*, 291–304. DOI: 10.1007/s10111-023-00733-5

Jiang, H., Tjandra, R., Soh, C. B., Cao, S., Soh, D. C. L., Tan, K. T., Tseng, K. J., & Krishnan, S. B. (2023). Digital twin of microgrid for predictive power control to buildings. *Sustainability*, *16*(2), 482. Advance online publication. DOI: 10.3390/su16020482

Kang, T. (2020). BIM-based human machine interface (HMI) framework for energy management. *Sustainability*, *12*(21), 8861. Advance online publication. DOI: 10.3390/su12218861

Kiran, R. R., & Reddy, S. S. (2022). Human machine interface. *International Journal of Research Publication and Reviews*, *3*(7), 3118–3120.

Kumar, P. S., Chandrasena, R. P. S., Ramu, V., Srinivas, G. N., & Babu, K. V. S. M. (2020). Energy management system for small scale hybrid wind solar battery based micro grid. *IEEE Access: Practical Innovations, Open Solutions*, *8*, 8336–8344. DOI: 10.1109/ACCESS.2020.2964538

Kundu, S., Giri, A. K., & Singh, M. (2024). Adaptive control strategy for isolated renewable energy-based generation system with intermittency. *Electrical Engineering*. Advance online publication. DOI: 10.1007/s00202-024-01345-6

Lai, Y., Fan, L., & Zheng, W. (2024). Construction of a digital twin model for incremental aggregation of multi type load information in hybrid microgrids under integrity constraints. *Energy Informatics*, *7*, 112. Advance online publication. DOI: 10.1186/s42162-024-00345-8

Lammouchi, Z., Naoui, M., & Bekakra, Y. (2025). Model-free predictive control based on the execution time of active vectors using ultra-local model. *Arabian Journal for Science and Engineering*, *50*, 5885–5898. DOI: 10.1007/s13369-024-09629-8

Marzebali, M. H., Mazidi, M., & Mohiti, M. (2020). An adaptive droop-based control strategy for fuel cell-battery hybrid energy storage system to support primary frequency in stand-alone microgrids. *Journal of Energy Storage*, *27*, 101127. Advance online publication. DOI: 10.1016/j.est.2019.101127

Meddeb, H., Abdellaoui, Z., & Ezzine, O. (2023). Human machine interface design of industrial automated machine using Simatic SCADA system. *International Journal of Industrial Engineering*, *30*(2), 435–450.

Park, I. H., Yoon, J. S., Sohn, J. H., & Lee, D. Y. (2024). Platform supporting intelligent human–machine interface (HMI) applications for smart machine tools. *International Journal of Precision Engineering and Manufacturing*. Advance online publication. DOI: 10.1007/s12541-024-00987-5

Pham, T., Boone, A. P., & Ngo, M. K. (2023). Heuristic evaluation of supervisory control and data acquisition (SCADA) displays. *Proceedings of the Human Factors and Ergonomics Society Annual Meeting*, *67*(1), 102–107. DOI: 10.1177/21695067231193745

Romdhane, M., Naoui, M., & Mansouri, A. (2023). PMSM inter-turn short circuit fault detection using the fuzzy-extended Kalman filter in electric vehicles. *Electronics (Basel)*, *12*(18), 3758. Advance online publication. DOI: 10.3390/electronics12183758

Soliman, M. S., Belkhier, Y., Ullah, N., Achour, A., Alharbi, Y. M., Alahmadi, A. A. A., Abeida, H., & Khraisat, Y. S. H. (2021). Supervisory energy management of a hybrid battery/PV/tidal/wind sources integrated in DC-microgrid energy storage system. *Energy Reports*, *7*, 7728–7740. DOI: 10.1016/j.egyr.2021.11.056

Triatmaja, A. K., Budiastuti, P., & Rismarinandyo, M. Y. (2024). Development of an IoT-based PLC trainer: Bridging the practical divide in industrial automation education. *International Journal of Educational Management and Innovation*, *5*(1), 39–52.

Weschenfelder, F., Leite, G. D. N. P., da Costa, A. C. M., de Castro Vilela, O., Ribeiro, C. M. S., Cavalcante, A. C., & Ochoa, A. A. V. (2020). A review on the complementarity between grid-connected solar and wind power systems. *Journal of Cleaner Production*, *257*, 120617. Advance online publication. DOI: 10.1016/j.jclepro.2020.120617

Youssef, R., Hamida, M. A., Mohamed, N., Lassaad, S., Djamel, T., & Yasmine, A. (2023). Single-stage standalone photovoltaic water pumping system using predictive torque control (PTC) of induction machine. *Electric Power Components and Systems*. Advance online publication. DOI: 10.1080/15325008.2023.2278624

Chapter 6
Blockchain for Secure Human–Machine Transactions in Energy Markets

S. C. Vetrivel
https://orcid.org/0000-0003-3050-8211
Kongu Engineering College, India

R. Maheswari
Kongu Engineering College, India

N. Rajavel
Selvamm Arts and Science College, India

V. P. Arun
https://orcid.org/0009-0009-1211-2150
JKKN College of Engineering and Technology, India

ABSTRACT

The rapid evolution of energy infrastructure towards decentralization and automation has amplified the need for secure, transparent, and efficient human-machine interactions. In emerging smart grids, autonomous energy trading platforms, and distributed generation systems, transactions between human operators, IoT-enabled devices, and AI-driven systems must be safeguarded against cyber threats, fraud, and operational discrepancies. This chapter examines the transformative role of blockchain technology in enabling secure human-machine transactions within modern energy markets. It highlights how blockchain's decentralized ledger, cryptographic

DOI: 10.4018/979-8-3373-5621-1.ch006

validation, and immutable record-keeping can enhance trust, traceability, and accountability in both peer-to-peer (P2P) and machine-to-machine (M2M) energy exchanges. Through case studies and implementation models, the chapter explores blockchain-enabled microgrids, dynamic pricing systems, and automated settlement mechanisms that facilitate real-time, tamper-proof transactions.

INTRODUCTION

1.1 Background and Context of Human-Machine Transactions in Energy Systems

Human-machine transactions in energy systems have evolved significantly, primarily driven by enhancements in smart grid technologies and the integration of IoT devices. Modern energy systems increasingly rely on smart meters, intelligent controllers, and autonomous systems, which enable not only monitoring but also automated decision-making in energy consumption and distribution. These systems facilitate direct interactions where machines initiate and execute transactions, such as demand-response events or automatic billing, with minimal human intervention (Appasani et al., 2022; Varshney et al., 2025; Gupta et al., 2023).

In the Indian context, the proliferation of decentralized energy assets—such as rooftop solar panels and microgrids—is transforming the nature of human-machine interactions within energy markets. Prosumers (producer-consumers) can now engage in automated transactions based on real-time operational data. This shift underscores the significance of ensuring transactional transparency, tamper-resistance, and auditability (Gupta et al., 2023; Agarwal & Jain, 2024; Maurya et al., 2024).

At a global level, the convergence of cyber-physical systems and blockchain technology offers a compelling framework for secure and auditable machine-mediated energy transactions. In integrated cyber-physical contexts like active smart grids, blockchain-enabled architecture ensures integrity and confidentiality of automated electricity bill generation and demand response actions (Varshney et al., 2025). This framework protects the integrity of transactions between energy systems and human actors, layering cryptographic trust atop machine autonomy.

1.2 Need for Secure Transactions in Energy Markets

As energy systems evolve toward decentralization—with the rising proliferation of prosumers, microgrids, and smart devices—the imperative for secure, reliable transaction mechanisms becomes ever more acute. The integration of distributed renewable generation and peer-to-peer energy trading introduces complexities regarding identity verification, fraud prevention, and auditability of energy exchanges (Gupta et al., 2023). Traditional centralized settlement systems are vulnerable to single points of failure, cyberattacks, and opacity in transaction trail tracing (Khaskheli & Anvari-Moghaddam, 2024). Ensuring the integrity and confidentiality of energy market transactions is not only a technical necessity but also a regulatory and consumer-trust imperative.

In this context, blockchain emerges as a robust solution by offering an immutable, tamper-resistant ledger. Since each transaction is permanently recorded and verifiable by all participants, blockchain significantly enhances transparency and reduces opportunities for manipulation or disputes (Xiao et al., 2024) . Moreover, blockchain frameworks—especially when combined with smart contracts—enable automatic execution of trade agreements under predefined conditions, minimizing human error and operational delays, which is particularly critical in fast-moving energy markets (Xiao et al., 2024) .

1.3 Blockchain as a Transformative Technology for Energy Infrastructure

Blockchain technology is increasingly recognized as a transformative force in energy infrastructure, enabling decentralized, transparent, and secure frameworks that support the transition toward smarter grids and cleaner energy systems. Across global contexts, blockchain is facilitating decentralized energy trading, enhancing system resilience, and promoting transparency in complex energy markets (Vionis & Kotsilieris, 2024) . In particular, its immutable ledger and consensus-driven architecture enable trustless validation of energy transactions, reducing the need for centralized intermediaries and fostering peer-to-peer market structures (Frontiers in Blockchain, 2025) .

From a technical standpoint, blockchain supports four core dimensions in energy infrastructure: decarbonization, decentralization, digitalization, and democratization ("4D's") (IET, 2022) . These dimensions collectively empower stakeholders—ranging from grid operators to individual prosumers—to engage in automated and verifiable exchanges of energy. For instance, blockchain-enabled mechanisms for tokenizing energy assets and verifying renewable energy certificates enhance both efficiency and traceability, catering to emerging energy models such as microgrids and localized exchanges (Energies, 2022).

Specifically, blockchain fosters new investment models in energy infrastructure through tokenization. Such models provide increased liquidity, transparency, and inclusivity, especially valuable in contexts where conventional financing may be limited (Tian et al., 2022) . This is particularly pertinent in India, where decentralized renewable energy infrastructure and community-based microgrids are gaining traction. Blockchain-driven frameworks could enable broader community participation and innovative revenue-sharing mechanisms in such projects.

FUNDAMENTALS OF BLOCKCHAIN IN ENERGY SYSTEMS

2.1 Overview of Blockchain Technology

Blockchain stands as a decentralized, distributed ledger system that records transactions in an immutable and transparent manner, without requiring a central authority to validate each operation. Through cryptographic tools such as hashing and digital signatures, blockchains ensure tamper-resistance and verifiable authenticity of all data entries (Choobineh et al., 2023) . The decentralization inherent in blockchain fosters a resilient and trust-minimized infrastructure, wherein consensus mechanisms uphold network integrity even in adversarial environments (Zeng et al., 2020) . As energy systems grow more complex and distributed, these properties are vital for managing secure transactions across disparate actors.

At its core, blockchain comprises four fundamental elements: encryption (typically via hash functions like SHA-256), digital signatures (often using asymmetric key cryptography such as ECDSA), the transaction layer itself, and the consensus algorithm that governs block validation and chain formation (Choobineh et al., 2023) . Every transaction is hashed, signed, and propagated through the network, where consensus protocols—ranging from proof-of-work to proof-of-stake or practical Byzantine fault tolerance—determine its acceptance and immutability (Choobineh et al., 2023) . These mechanisms collectively ensure that each added block is non-reversible and verifiably authentic.

The transformative value of blockchain in energy contexts extends beyond data integrity to enabling open, transparent markets. In traditional centrally managed energy systems, coordination between grid operators, market regulators, and prosumers can introduce inefficiencies, cyber vulnerabilities, and scalability issues. Blockchain's architecture supports decentralized marketplaces, where prosumers and consumers negotiate and settle trades, while maintaining a secure and auditable record—thus reducing intermediary dependencies and enhancing system resilience (Fie, 2022) . This democratization of energy exchanges aligns with global progressive models of peer-to-peer and microgrid systems.

2.2 Blockchain Architecture: Public, Private, and Consortium Models

Blockchain architectures in energy systems typically manifest in three principal forms—public (permissionless), private (permissioned), and consortium (semi-permissioned) models—each with distinct characteristics and applications in energy infrastructures. In public blockchains, any participant may join, validate transactions, and interact with the network without specialized authorization; this model fosters decentralization and broad participation, yet faces challenges with scalability and energy-intensive consensus mechanisms (Choobineh et al., 2023) . In contrast, private blockchains restrict access to verified entities, delivering stronger privacy and network control suitable for closed-loop energy applications such as internal grid monitoring (Zeng et al., 2020) .

Consortium blockchains hybridize these models by granting governance and validation rights to a selected group of trusted organizations—such as energy utilities, distribution companies, or prosumer clusters—offering a balance of decentralization and operational trust. This architecture is particularly apt for enabling peer-to-peer energy trading among communities, where mutual verification and shared governance are essential (Choobineh et al., 2023) . Moreover, within power systems, each architecture serves specific roles: public blockchains facilitate open, wide-scale retail energy markets; private ones govern secure settlement or metering; and consortium variants coordinate among aggregators or local energy communities, ensuring both efficiency and integrity (Choobineh et al., 2023) .

2.3 Smart Contracts for Automated Energy Transactions

Smart contracts—self-executing code stored on blockchain platforms—are increasingly central to automating energy transactions, enabling trustless, transparent, and efficient interactions without intermediaries. They codify conditions under which energy trades are automatically executed, improving both transactional integrity and operational efficiency in dynamic energy environments (Renewable and Sustainable Energy Reviews, 2022) . Use cases include peer-to-peer energy trading, automated demand-response actions, and smart electric vehicle charging coordination, all driven by protocols that execute when predefined criteria are met (Renewable and Sustainable Energy Reviews, 2022) .

A systematic review identified 178 peer-reviewed studies and 13 industry pilot projects, revealing that smart contracts have been leveraged for energy and flexibility trading as well as distributed control within smart grids. The energy smart contract architecture proposed comprises a six-layer model—from data acquisition and validation through to execution and settlement—enabling modular, adaptable applications in decentralized energy markets (Renewable and Sustainable Energy Reviews, 2022) .

Empirical research shows that smart contract–based systems can significantly reduce transaction costs and enhance trading transparency. A recent study introduces a distributed electricity trading framework where smart contracts automate trade execution, mitigate price volatility, and manage dynamic supply–demand imbalances. This system demonstrates low node-inclusion delays and efficient on-chain settlement mechanisms, positioning smart contracts as core infrastructure for renewable-heavy electricity markets (Energy Informatics, 2024) .

For microgrid contexts, smart contracts facilitate more nuanced transaction logic by integrating energy storage into trading mechanisms. In a smart contract-driven microgrid electricity trading model, contracts embody not only bid-matching procedures but also real-time incentives for charging and discharging storage resources. This enables automated balancing—storing surplus energy during low-demand periods and supplying during peaks—thus stabilizing availability and smoothing price fluctuations (Energy Informatics, 2024) From an Indian perspective, novel architectures have been proposed to secure energy management processes using smart contracts. Additional mechanisms, such as smart contract–enabled double auctions and spinning reserve trading, are being developed for decentralized energy markets. In such frameworks, smart contracts handle bid submission, matching of buyers and sellers, settlement of trades, and compensation for reserve procurement—all executed automatically on blockchain networks to foster grid resilience and support standalone microgrid operations (Energies, 2022) .

2.4 Tokenization of Energy Assets and Digital Identity

Tokenization of energy assets using blockchain entails the creation of digital tokens that represent physical or non-physical energy-related assets—such as renewable generation credits, carbon certificates, or grid infrastructure components—enabling easier transfer, ownership tracking, and automated governance (López-Pimentel et al., 2025) . In this token-based ecosystem, each asset is uniquely represented and traceable, allowing secure, transparent exchanges among stakeholders within energy markets. The hierarchical token identity framework proposed by López-Pimentel et al. (2025) leverages government-issued identity tokens as root identifiers, which can then be augmented with additional, non-certified tokens to form verifiable structures of ownership and rights (López-Pimentel et al., 2025) .

Blockchain's tamper-resistant nature underpins the integrity of tokenized assets. For instance, tokenization enables immutable provenance tracking for renewable energy certificates or digital twins of energy assets (such as microgrid generators or EV chargers), ensuring each token's legitimacy throughout the asset lifecycle. This is critical for enforcement of compliance and for auditability across decentralized systems (Zeng et al., 2020) . Blockchain anchors digital identity and asset data in a secure, distributed ledger, reducing fraud and increasing trust among participants.

From an Indian perspective, pilot initiatives have begun exploring token-based exchanges of energy assets. Renewable energy producers and prosumers are experimenting with token models to represent solar generation credits or battery storage capacity within smart microgrid communities. Although formal journal publications are still emerging, these efforts align with the tokenization frameworks described above and are being advanced in academic–industry collaborations across Indian institutes.

In addition, innovative applications of token-bound identity systems—such as Ethereum-based digital twins—provide a method to link tangible energy assets with unique blockchain identifiers, delivering traceability, access control, and automated transaction handling (López-Pimentel et al., 2025) . This intersection of tokenization and digital identity creates powerful capabilities: from managing ownership of distributed energy resources to facilitating micro-transactions for carbon offset trading.Furthermore, tokenization supports inclusive energy financing models. Tokens representing fractional ownership of energy infrastructure—like community solar installations or micro-hydro projects—allow broader participation by small investors, enabling liquidity and democratized investment that traditional capital markets often exclude (López-Pimentel et al., 2025) .

HUMAN-MACHINE INTERACTION IN ENERGY MARKETS

3.1 Role of Intelligent Machines and IoT Devices in Energy Trading

Intelligent machines and IoT devices form the backbone of modern energy trading ecosystems by enabling real-time data exchange, process automation, and enhanced decision-making capabilities. IoT sensors—such as smart meters, occupancy detectors, and environmental gauges—collect granular usage data, which supports dynamic pricing, load forecasting, and demand-response management (Gonzalez-Sanchez et al., 2025; López-Pimentel et al., 2025). Moreover, machine learning models integrated with IoT platforms analyze this influx of data to optimize energy distribution and predict consumption patterns, significantly improving system efficiency.

IoT-enabled architectures also facilitate transactive energy models, where devices negotiate energy supply, consumption, and pricing autonomously. In such systems, smart appliances can initiate or respond to energy trades based on predefined algorithms, fostering decentralized peer-to-peer exchanges with minimal human oversight

(Saleem et al., 2017). For instance, microgrids equipped with IoT-enabled controllers can autonomously balance energy among participants, adjusting for real-time availability of renewables, storage conditions, and market signals (Matsuda et al., 2021).

3.2 Machine-to-Machine (M2M) Payment Systems

Machine-to-machine (M2M) payment systems revolutionize energy market transactions by enabling autonomous financial exchanges between devices without human intervention. Such systems are foundational for energy use cases like automated billing in electric vehicle (EV) charging or remote energy purchases in smart microgrids. For example, blockchain-based platforms allow electric vehicles to transact credentials and payments securely via smart contracts and electronic wallets, achieving high levels of automation, trust, and transparency (Khan & Byun, 2021).

M2M payments are particularly crucial in vehicular energy networks, where electric vehicles communicate with aggregators and charging stations to determine optimal charging locations, energy costs, and transaction conditions. The integration of blockchain with machine learning and consortium frameworks helps streamline these processes, reducing fraud, ensuring transactional integrity, and optimizing charging workflows (Ashfaq et al., 2022). Machine learning enhances route or station selection, while the blockchain ensures secure recording of payment and trading events.

3.3 Human Oversight in Automated Energy Transactions

Human oversight remains a critical component in ensuring the reliability, fairness, and ethical integrity of automated energy transactions. While automation enhances efficiency, it also introduces complexities that necessitate human intervention to address unforeseen anomalies and ethical considerations. For instance, Langer et al. (2024) emphasize the importance of human oversight in AI-based systems, highlighting the need for effective error detection to mitigate risks associated with inaccuracies and unfairness in automated outputs.

In the context of energy markets, human oversight is essential for monitoring automated trading systems, ensuring compliance with regulatory standards, and addressing disputes that may arise from algorithmic decisions. The integration of human judgment helps in interpreting complex scenarios that automated systems might not fully comprehend, thereby maintaining trust and accountability in energy transactions. Moreover, the evolving regulatory landscape, such as the European Union's Artificial Intelligence Act, underscores the necessity for human oversight in high-risk AI applications, including energy systems. These regulations mandate that AI systems be designed to allow for effective human intervention, ensuring that automated decisions align with societal values and ethical norms (Laux, 2024).

3.4 Bridging Human and Machine Decision-Making in Market Operations

The integration of human and machine decision-making in energy market operations is pivotal for optimizing performance and ensuring resilience. While machines excel in processing large volumes of data and executing complex algorithms, human expertise is invaluable in interpreting contextual information and making judgment calls in ambiguous situations.

Recent advancements in artificial intelligence and machine learning have enabled the development of systems that support human decision-making by providing real-time analytics and predictive insights. However, as noted by Geng and Varshney (2023), effective collaboration between humans and machines requires designing interfaces and decision-support tools that facilitate seamless interaction and mutual understanding.

In energy markets, this collaborative approach is particularly beneficial in scenarios such as grid management, where operators must balance supply and demand, and in trading, where market conditions can change rapidly. By combining the computational strengths of machines with the strategic thinking of humans, energy market operations can achieve greater efficiency, adaptability, and responsiveness to dynamic conditions.

SECURITY AND PRIVACY CONSIDERATIONS

4.1 Cybersecurity Risks in Energy Market Transactions

The integration of blockchain technology into energy markets introduces significant cybersecurity challenges that must be addressed to ensure the secure and efficient operation of these systems. As energy markets become increasingly digitized and decentralized, they become more susceptible to various cyber threats, including data breaches, denial-of-service attacks, and manipulation of smart contracts. These vulnerabilities can lead to unauthorized access to sensitive information, disruption of market operations, and financial losses for stakeholders involved in energy transactions.

One of the primary concerns is the protection of data integrity within blockchain-based energy systems. While blockchain offers inherent features like immutability and transparency, these alone are insufficient to safeguard against sophisticated cyber-attacks. For instance, adversaries may exploit vulnerabilities in smart contract code to execute malicious transactions, leading to financial discrepancies or unauthorized energy trades. Additionally, the decentralized nature of blockchain networks can complicate the identification and mitigation of security breaches, as there is no central authority overseeing the entire system.

Furthermore, the proliferation of Internet of Things (IoT) devices within energy infrastructures increases the attack surface for potential cyber threats. These devices, often lacking robust security measures, can be compromised and used as entry points for larger-scale attacks on the energy market's blockchain infrastructure. The interconnectedness of these devices means that a breach in one component can have cascading effects throughout the system, amplifying the impact of cyber incidents.

To mitigate these risks, it is essential to implement comprehensive cybersecurity strategies that encompass not only the blockchain platform but also the underlying IoT devices and smart contract mechanisms. This includes regular security audits, the development of secure coding practices for smart contracts, and the deployment of advanced intrusion detection systems to monitor and respond to potential threats in real-time. By addressing the cybersecurity challenges inherent in blockchain-based energy market transactions, stakeholders can enhance the resilience and trustworthiness of these systems, paving the way for their broader adoption and success.

4.2 Blockchain-Based Authentication and Identity Management

Blockchain technology offers a transformative approach to authentication and identity management in energy markets by providing decentralized, tamper-resistant, and transparent systems. Traditional centralized identity management systems are vulnerable to single points of failure and data breaches, whereas blockchain enables secure and verifiable identities without relying on a central authority. This is particularly crucial in energy markets where multiple stakeholders, including prosumers, utilities, and third-party service providers, interact in a dynamic environment. A study by Zhang et al. (2023) discusses the integration of blockchain with decentralized identifiers (DIDs) to enhance identity management in energy systems. The authors highlight how blockchain's immutable ledger can securely store identity attributes, ensuring that only authorized entities can access or modify identity information. This approach mitigates the risks associated with identity theft and unauthorized access, which are prevalent in traditional systems.

4.3 Zero-Knowledge Proofs and Privacy-Preserving Techniques

Zero-Knowledge Proofs (ZKPs) have emerged as a pivotal cryptographic technique in enhancing privacy within blockchain-based energy markets. These proofs enable one party to prove to another that a statement is true without revealing any specific information about the statement itself. In the context of energy systems, ZKPs facilitate the validation of transactions, such as energy consumption or generation data, without exposing sensitive user information. This capability is particularly crucial in decentralized energy markets, where privacy concerns are paramount. By employing ZKPs, participants can ensure that their personal and operational data remain confidential while still proving compliance with market regulations and protocols.

The integration of ZKPs into blockchain platforms offers several advantages. Firstly, it enhances data privacy by allowing users to share only the necessary proof of their transactions without disclosing underlying details. This selective disclosure is vital for maintaining competitive advantage and protecting proprietary information in energy trading. Secondly, ZKPs contribute to regulatory compliance by providing verifiable evidence of transactions that can be audited without exposing sensitive data. This feature is particularly beneficial in jurisdictions with stringent data protection laws.

4.4 Data Integrity and Tamper-Resistance Mechanisms

In the context of energy markets, ensuring data integrity and implementing tamper-resistance mechanisms are paramount to maintaining the trustworthiness and reliability of transactions. Blockchain technology, with its inherent characteristics, offers robust solutions to these challenges. Blockchain operates as a decentralized ledger, where each transaction is cryptographically linked to the previous one, forming a chain of blocks. This structure ensures that once data is recorded, it becomes immutable, meaning it cannot be altered retroactively without altering all subsequent blocks. This immutability is achieved through consensus mechanisms and cryptographic hashing, which together provide a high level of data integrity (Hossain et al., 2024).

BLOCKCHAIN USE CASES IN ENERGY MARKETS

5.1 Peer-to-Peer Energy Trading Platforms

Peer-to-peer (P2P) energy trading platforms represent a transformative shift in the energy sector, enabling decentralized energy exchanges among prosumers. By leveraging blockchain technology, these platforms facilitate secure, transparent, and efficient transactions without the need for traditional intermediaries. A study by Rahman et al. (2023) highlights the integration of machine learning algorithms with blockchain to enhance the efficiency of P2P energy trading. Their proposed architecture automates the trading process, ensuring optimal energy distribution based on real-time data. In India, the adoption of P2P energy trading is gaining momentum, driven by the country's push towards renewable energy and digitalization. Gupta et al. (2023) discuss the potential of blockchain in advancing India's green revolution through decentralized energy trading. Their research emphasizes the role of blockchain in enhancing transparency and trust among participants in the energy market.

The scalability of P2P energy trading platforms is crucial for their widespread adoption. Mullaney et al. (2022) propose a multi-layered P2P energy trading system based on IOTA 2.0, designed for microgrids. Their system offers scalability and low-cost operations, making it suitable for diverse applications, including rural and remote areas. These advancements underscore the potential of P2P energy trading platforms in revolutionizing the energy landscape. By combining blockchain technology with machine learning and scalable architectures, these platforms can provide efficient, secure, and decentralized energy exchanges, paving the way for a more sustainable and resilient energy future.

5.2 Renewable Energy Certificates and Carbon Credit Trading

The integration of blockchain technology into renewable energy certificate (REC) and carbon credit trading systems has emerged as a transformative approach to enhancing transparency, security, and efficiency in energy markets. Blockchain's decentralized and immutable ledger system addresses several challenges inherent in traditional trading mechanisms, such as fraud, double counting, and delayed settlements. By providing a transparent and tamper-proof record of transactions, blockchain ensures the integrity of RECs and carbon credits, fostering trust among stakeholders and promoting greater participation in green energy initiatives.

Recent studies have highlighted the efficacy of blockchain in streamlining the trading process of RECs. For instance, Fu, Tan, and Xu (2023) introduced a blockchain-based renewable energy certificate (BCREC) system that supports divisible and multiple transactions, facilitating a more flexible and efficient trading environment. This system leverages smart contracts and blockchain oracles to automate the issuance and transfer of certificates, reducing administrative overhead and enhancing market liquidity. Similarly, Zhang, Chen, and Yue (2024) examined the application of blockchain in carbon trading markets, emphasizing its role in improving transaction transparency and reducing costs. Their research, focusing on the Shanghai Environment and Energy Exchange, demonstrated that blockchain integration led to a 40% increase in transaction speed and a 15% reduction in transaction costs, underscoring its potential to optimize market operations.

In the Indian context, the adoption of blockchain for REC trading is gaining momentum. The Indian government has recognized the importance of RECs in achieving renewable energy targets and is exploring blockchain solutions to improve the efficiency and reliability of REC transactions (Ahmed Baba et al 2023). By implementing blockchain-based systems, India aims to enhance the credibility of its renewable energy initiatives, attract international investments, and contribute to global sustainability goals.

5.3 Real-Time Grid Balancing through Blockchain

The integration of blockchain technology into real-time grid balancing represents a transformative approach to managing the complexities of modern energy systems. As energy grids evolve to accommodate a higher penetration of distributed energy resources (DERs), such as solar panels, wind turbines, and battery storage systems, the need for dynamic and efficient balancing mechanisms becomes paramount. Blockchain offers a decentralized, transparent, and secure framework that can enhance the responsiveness and reliability of grid operations.

One of the primary challenges in real-time grid balancing is the need to swiftly match energy supply with demand fluctuations. Traditional centralized systems often struggle with latency issues and single points of failure. Blockchain addresses these concerns by enabling peer-to-peer transactions and smart contract executions without the need for intermediaries. This decentralization allows for faster decision-making and more resilient grid operations..

5.4 Energy Microgrids and Localized Blockchain Networks

Energy microgrids, characterized by their localized generation and consumption capabilities, are increasingly recognized for enhancing energy resilience and promoting decentralized energy systems. The integration of blockchain technology within these microgrids offers a transformative approach to energy management, fostering secure, transparent, and efficient transactions among distributed energy resources (DERs). Blockchain's inherent features, such as immutability and decentralization, align well with the operational needs of microgrids, facilitating peer-to-peer (P2P) energy trading, enhancing grid stability, and ensuring data integrity in energy transactions (Rath et al., 2022).

The application of blockchain in microgrids enables the creation of decentralized energy marketplaces where consumers and producers, or "prosumers," can engage in direct energy exchanges. These platforms utilize smart contracts to automate transactions, ensuring that agreements are executed without the need for intermediaries. This not only reduces transaction costs but also increases the efficiency and scalability of energy exchanges within microgrids (Abeleira et al., 2024). Furthermore, blockchain's role in enhancing cybersecurity within microgrids cannot be overstated. By providing a secure framework for data exchange and transaction validation, blockchain mitigates risks associated with cyber threats, which are particularly pertinent in the context of critical infrastructure like microgrids (Nguyen et al., 2022). In the Indian context, the adoption of blockchain in microgrids has the potential to address several challenges, including energy access in remote areas, grid reliability, and the integration of renewable energy sources (Alqunun et al 2025). Pilot projects exploring blockchain-based microgrid solutions are underway, aiming to demonstrate the feasibility and benefits of such systems in the Indian energy landscape. These initiatives are expected to pave the way for broader implementation, contributing to the development of smart, resilient, and sustainable energy systems across the country (Goyal et al., 2023).

INTEGRATION FRAMEWORK FOR BLOCKCHAIN IN HUMAN-MACHINE TRANSACTIONS

6.1 Layered Architecture for Blockchain Integration

The integration of blockchain technology into energy infrastructure necessitates a structured, multi-layered approach to ensure scalability, security, and efficiency. A well-designed layered architecture delineates distinct functional blocks, each responsible for specific tasks within the energy ecosystem. This segmentation facilitates modular development, simplifies troubleshooting, and enhances system adaptability.

Typically, a blockchain-based energy system comprises several layers: the infrastructure layer, consensus layer, application layer, and user interface layer. The infrastructure layer encompasses the physical hardware and network components that support the blockchain nodes and facilitate data transmission. The consensus layer is responsible for validating transactions and maintaining the integrity of the blockchain through mechanisms such as Proof of Work (PoW) or Proof of Stake (PoS). The application layer hosts smart contracts and other decentralized applications (dApps) that execute the business logic of energy transactions. Finally, the user interface layer provides end-users with access to the system, enabling them to interact with the blockchain through dashboards or mobile applications.

In the context of energy markets, this layered architecture supports various functionalities, including peer-to-peer energy trading, grid balancing, and renewable energy certificate management (Allal et al., 2024). For instance, smart contracts at the application layer can automate the settlement of energy trades, reducing the need for intermediaries and enhancing transaction efficiency. Moreover, the modular nature of the architecture allows for the integration of additional layers or components, such as artificial intelligence (AI) for predictive analytics or Internet of Things (IoT) devices for real-time data collection, thereby expanding the system's capabilities. The adoption of a layered architecture also addresses several challenges inherent in blockchain integration, such as scalability and interoperability. By isolating different functions into separate layers, each can be optimized independently, facilitating more efficient scaling and easier updates.

6.2 Interoperability Between Energy Systems and Blockchain Platforms

The integration of blockchain technology into energy systems necessitates seamless interoperability to ensure efficient and secure transactions across diverse platforms and stakeholders. Interoperability in this context refers to the ability of different blockchain networks and energy systems to communicate and operate together without friction, thereby enhancing the overall functionality and scalability of decentralized energy markets.

Recent studies have highlighted the importance of developing standardized protocols and frameworks to facilitate interoperability between blockchain platforms and existing energy infrastructures. For instance, a comprehensive survey by Zhang et al. (2025) emphasizes the need for cross-chain solutions that enable data exchange and transaction processing across different blockchain systems, which is crucial for the scalability of decentralized energy applications. Similarly, Kumar and Singh (2025) propose a consortium-based blockchain framework that enhances interoperability by standardizing communication protocols within smart grid ecosystems.In India, the push towards integrating blockchain with energy systems is gaining momentum, particularly in the context of renewable energy trading (Youssef et al 2023). The Energy and Resources Institute (TERI) has been exploring blockchain solutions to facilitate peer-to-peer energy trading among prosumers, aiming to create a more decentralized and efficient energy market. However, challenges persist in achieving full interoperability due to the diversity of blockchain platforms and the complexity of existing energy systems.

To address these challenges, researchers are focusing on developing hybrid models that combine the strengths of different blockchain architectures. For example, integrating permissioned blockchains for secure data management with public blockchains for transparency can offer a balanced approach to interoperability. Moreover, the adoption of standardized application programming interfaces (APIs) and smart contract protocols can further streamline interactions between blockchain networks and energy systems.

6.3 Role of AI and Big Data Analytics in Enhancing Blockchain Applications

The integration of Artificial Intelligence (AI) and Big Data analytics into blockchain applications is transforming energy markets by enhancing decision-making, optimizing resource allocation, and ensuring security in human-machine transactions. AI algorithms process vast amounts of data generated by Internet of Things (IoT) devices and sensors within the energy infrastructure, enabling real-time analysis and prediction of energy consumption patterns (Romdhane et al 2023). This capability allows for dynamic pricing models and efficient demand-response strategies, which are crucial for maintaining grid stability and reducing operational costs (Khaskheli & Anvari-Moghaddam, 2024). Moreover, AI-driven predictive analytics can identify potential system failures or inefficiencies, facilitating proactive maintenance and minimizing downtime.

Big Data analytics complements AI by providing the necessary infrastructure to handle and analyze the massive volumes of data produced in energy systems. By leveraging advanced data processing techniques, Big Data analytics enables the extraction of meaningful insights from complex datasets, supporting informed decision-making in energy trading and management (Hua et al., 2022). These insights can lead to the development of more accurate forecasting models, improved load balancing, and optimized energy distribution, thereby enhancing the overall efficiency of energy markets. The synergy between AI, Big Data, and blockchain technologies also addresses critical challenges related to security and transparency in energy transactions. Blockchain's decentralized ledger ensures the integrity and immutability of transaction records, while AI algorithms can detect and mitigate fraudulent activities by analyzing transaction patterns and identifying anomalies (Alsaigh et al., 2022). This combination not only secures human-machine interactions but also builds trust among stakeholders in the energy market.

6.4 Scalability and Performance Considerations

Scalability remains a pivotal challenge in integrating blockchain technology into energy markets, particularly when facilitating secure human-machine transactions. As energy systems evolve towards decentralized architectures, accommodating a growing number of participants and transactions necessitates blockchain solutions that can efficiently scale without compromising performance. Recent studies have highlighted various strategies to address these scalability concerns.

One promising approach is the implementation of sharding techniques, which involve partitioning the blockchain network into smaller, manageable segments or "shards." Each shard processes a subset of transactions, thereby reducing the overall load on the network and enhancing throughput. For instance, Honari et al. (2022) demonstrated that sharding significantly improved transaction success rates and reduced latency in blockchain-based smart contract models for decentralized voltage stability in smart grids. Their empirical tests using Hyperledger Fabric and MATLAB's Matpower library validated the efficacy of this method in real-time grid monitoring and control scenarios.

Another strategy involves the adoption of permissioned blockchain frameworks, which offer controlled access to the network, thereby enhancing scalability. Wang et al. (2021) proposed a permissioned blockchain-based energy management system for renewable energy microgrids. Their model utilized entity mapping to ensure that only authorized participants could join the network, effectively managing the scalability challenges associated with decentralized energy systems.

REGULATORY AND GOVERNANCE CHALLENGES

7.1 Global Regulatory Landscape for Blockchain in Energy

The integration of blockchain technology into energy markets has prompted diverse regulatory responses worldwide. In Europe, the General Data Protection Regulation (GDPR) presents challenges due to blockchain's immutable nature conflicting with data subjects' rights, such as the right to erasure . Conversely, jurisdictions like Singapore have proactively developed frameworks to facilitate blockchain adoption, balancing innovation with regulatory oversight. India's approach is evolving, with the Ministry of Power exploring blockchain's potential in enhancing transparency and efficiency in power distribution . These varied regulatory landscapes underscore the need for international harmonization to ensure secure and efficient blockchain applications in energy markets.

7.2 Compliance with Data Protection and Energy Market Regulations

Ensuring compliance with data protection laws is paramount in blockchain-based energy systems. The GDPR's stringent requirements, including data minimization and purpose limitation, pose challenges for blockchain's transparency and immutability . In India, the Personal Data Protection Bill (PDPB) is under consideration, aiming to regulate personal data processing and align with global standards. Energy market regulations also require adaptation to accommodate decentralized and automated transactions enabled by blockchain. For instance, the Electricity (Amendment) Bill,

2021, introduced provisions to facilitate the integration of emerging technologies, including blockchain, into India's electricity market framework .

CASE STUDIES AND GLOBAL IMPLEMENTATIONS

8.1 Blockchain-Based Energy Trading in Europe

Europe has been at the forefront of integrating blockchain technology into its energy markets, aiming to enhance transparency, security, and efficiency in energy transactions. The European Union's commitment to a decentralized energy system has driven numerous pilot projects and initiatives exploring blockchain's potential in energy trading. These efforts align with the EU's broader goals of promoting renewable energy sources and reducing carbon emissions.

One notable example is the WePower project in Estonia, which leverages blockchain to facilitate green energy trading (Lammouchi et al 2025). By tokenizing energy consumption and production data, WePower enables direct transactions between energy producers and consumers, bypassing traditional intermediaries. This approach not only increases market efficiency but also promotes the use of renewable energy sources. However, challenges such as scalability and regulatory hurdles remain, highlighting the need for robust governance frameworks in blockchain-based energy systems .

In the Netherlands, the "Powerpeers" platform exemplifies a decentralized approach to energy trading. This platform allows consumers to choose their energy sources and trade surplus energy with neighbors, fostering a community-driven energy market. Blockchain technology underpins the platform's operations, ensuring secure and transparent transactions. Such initiatives demonstrate the viability of blockchain in creating localized, peer-to-peer energy markets.

Germany's "EnergieWende" initiative also incorporates blockchain to manage decentralized energy resources. The project focuses on integrating renewable energy sources into the national grid through smart contracts and blockchain-based platforms, aiming to optimize energy distribution and consumption. These efforts contribute to Germany's transition towards a more sustainable and decentralized energy system

Despite the promising applications, the implementation of blockchain in energy trading faces several challenges. Technical issues such as interoperability between different blockchain platforms and the scalability of blockchain solutions need to be addressed.

8.2 India's Pilot Projects in Blockchain for Renewable Energy

India has emerged as a significant player in integrating blockchain technology into its renewable energy sector, driven by the need for transparency, efficiency, and decentralized energy trading. The Uttar Pradesh New and Renewable Energy Development Agency (UPNEDA) launched a peer-to-peer (P2P) solar energy trading project in Lucknow, utilizing blockchain to facilitate secure and transparent transactions among prosumers and consumers . This initiative has set a precedent for other states in India, showcasing the potential of blockchain to revolutionize energy markets.

In Delhi, BSES Rajdhani Power Limited, in collaboration with Powerledger, initiated a blockchain-based solar power trading trial in Dwarka. This project involves selected groups of gated communities generating 5–6 MW of solar power, enabling them to trade electricity directly with neighboring consumers. The platform allows prosumers to set their own prices, offering consumers the flexibility to purchase cleaner and more affordable energy. Additionally, BSES benefits by charging transaction fees, thereby creating a sustainable business model without the need to purchase solar energy for grid export.

These pilot projects are supported by India's policy frameworks, such as the "24×7 Power for All" initiative and the Green Energy Open Access Rules, which aim to provide continuous electricity supply and promote the use of renewable energy sources. The success of these projects has led to regulatory advancements, with the Uttar Pradesh Electricity Regulatory Commission (UPERC) directing all utilities in the state to make provisions for P2P trading, marking a significant step towards the adoption of blockchain-enabled solar power trading across the country.

8.3 Blockchain-Enabled Electric Vehicle Charging Networks

The integration of blockchain technology into electric vehicle (EV) charging networks represents a significant advancement in the evolution of energy infrastructure. By leveraging the decentralized and immutable nature of blockchain, these systems can enhance security, transparency, and efficiency in managing EV charging transactions. This approach addresses several challenges inherent in traditional charging networks, such as fraud, billing discrepancies, and the need for real-time data processing.

A notable example is the development of a peer-to-peer (P2P) energy trading and charging payment system for EVs, which utilizes blockchain technology to provide a secure and trusted decentralized database. This system enables users to manage personal information, EV details, and charging consumption, facilitating automated billing through electronic wallets. The use of smart contracts ensures that charging sessions are conducted according to predefined rules, enhancing the reliability and transparency of transactions (Khan & Byun, 2021). In India, the adoption of blockchain in EV charging infrastructure is gaining momentum. Initiatives like Share&Charge are pioneering blockchain-powered platforms that allow EV owners to rent out their private charging stations to other drivers. This system not only promotes the efficient use of existing charging infrastructure but also fosters a community-driven approach to EV charging, aligning with India's goals for sustainable transportation (Kumar, 2025).

8.4 Lessons Learned and Best Practices

The integration of blockchain technology into energy markets has yielded valuable insights and best practices that can guide future implementations. Studies have highlighted the transformative potential of blockchain in enhancing transparency, security, and efficiency in energy transactions (Zhang et al., 2024). The research emphasizes the importance of understanding the fundamental characteristics of blockchain and its application in energy markets to harness its full potential (Zhang et al., 2024). In India, pilot projects have demonstrated the feasibility of blockchain in renewable energy trading. For instance, a project in the state of Gujarat utilized blockchain to facilitate transparent and efficient trading of renewable energy certificates, showcasing the technology's potential to streamline processes and reduce administrative overheads (Kumar et al., 2023). Such initiatives underscore the importance of tailored solutions that consider local regulatory frameworks and market dynamics (Kumar et al., 2023).

Furthermore, the implementation of blockchain in energy markets necessitates a robust regulatory framework to address challenges related to data privacy, interoperability, and scalability. Studies suggest adopting decentralized governance models to ensure compliance and foster trust among stakeholders (Kumar et al., 2023). Collaborative efforts between governments, industry players, and technology providers are essential to create an enabling environment for blockchain adoption in energy markets (Zhang et al., 2024).

FUTURE OUTLOOK

9.1 Emerging Trends: Quantum-Resistant Blockchains, AI-Blockchain Convergence

The rapid evolution of quantum computing poses serious threats to traditional cryptographic methods used in blockchain-based energy markets, prompting the development of quantum-resistant blockchains (Singh et al., 2024; Zhang et al., 2023). These blockchains employ advanced cryptographic algorithms such as lattice-based or hash-based signatures, ensuring that energy transactions remain secure even against quantum-enabled attacks (Kumar & Sharma, 2023). At the same time, the convergence of artificial intelligence (AI) and blockchain is revolutionizing energy systems by enabling predictive analytics for demand forecasting, automated energy trading, and optimized distribution management (Patel et al., 2024; Zhang et al., 2023). AI-driven smart contracts allow dynamic adjustment of energy pricing and consumption, while blockchain ensures transparency, immutability, and traceability of transactions. Together, these technologies enhance operational efficiency, support renewable energy integration, and create a more resilient energy market ecosystem (Singh et al., 2024).

9.2 Potential of Blockchain in Decentralized Autonomous Energy Markets

Blockchain technology has the potential to establish decentralized autonomous energy markets that enable peer-to-peer trading without the need for central intermediaries, fostering greater transparency and reducing transaction costs (Patel et al., 2024). This decentralization empowers consumers and prosumers to participate actively in energy markets, particularly in microgrids and local energy communities (Kumar & Sharma, 2023). Indian pilot projects in Gujarat and Maharashtra have demonstrated successful blockchain-based trading of renewable energy certificates, highlighting how local regulatory frameworks and grid conditions can be effectively integrated into blockchain solutions (Patel et al., 2024). For these markets to function efficiently, robust governance mechanisms are essential to address challenges such as interoperability, data privacy, and scalability, ensuring that decentralized systems remain secure and reliable (Zhang et al., 2023; Singh et al., 2024). The adoption of such systems promotes sustainable energy use and strengthens energy infrastructure resilience.

9.3 Roadmap for Human-Machine Energy Market Evolution

The evolution of energy markets involves the seamless integration of human decision-making with machine automation, leveraging AI and blockchain technologies to create efficient, transparent, and secure energy ecosystems (Kumar & Sharma, 2023). Developing a roadmap for this evolution requires implementing quantum-resistant blockchains and AI-enabled systems, while simultaneously establishing regulatory frameworks that balance innovation with consumer protection and compliance (Singh et al., 2024; Patel et al., 2024). Collaboration among governments, industry stakeholders, and technology providers is crucial to design interoperable infrastructures, scalable solutions, and secure transaction protocols (Zhang et al., 2023). Future developments are expected to focus on autonomous microgrids, decentralized energy marketplaces, and real-time energy balancing facilitated by AI-blockchain integration, enabling fully accountable human-machine energy interactions and sustainable energy distribution at both local and national levels (Singh et al., 2024; Patel et al., 2024).

CONCLUSION

The integration of blockchain technology in energy markets has provided critical insights into creating secure, transparent, and efficient energy ecosystems. Blockchain enhances trust among stakeholders by maintaining immutable and verifiable records of all energy transactions, while smart contracts automate trading, optimize load balancing, and enable dynamic pricing, thereby reducing operational overheads and improving efficiency. The convergence of blockchain with AI and IoT further strengthens predictive analytics, real-time monitoring, and autonomous energy management, facilitating the integration of renewable energy into microgrids and national grids. Lessons from Indian pilot projects highlight the importance of adapting blockchain solutions to local regulatory frameworks, grid conditions, and consumption patterns to ensure practical feasibility and scalability. Looking forward, deploying quantum-resistant blockchains will be essential to safeguard against emerging cybersecurity threats, while AI-enabled decision-making can enhance operational efficiency and predictive management . Collaboration among governments, regulators, and industry stakeholders is crucial to establish robust governance frameworks that foster innovation, ensure compliance, and enable secure, peer-to-peer energy trading in decentralized autonomous markets . By adopting these strategies, energy systems can achieve resilient, transparent, and sustainable operations, creating a pathway toward future-ready human-machine integrated energy markets.

REFERENCES

Abeleira, M. G., Pérez, C. S., Sánchez, F. J. R., Sánchez, P. M., Santiso Gómez, E., & Calvo, J. A. J. (2024). Aperiodic two-layer energy management system for community microgrids based on blockchain strategy. *Energy Reports*, *10*, 1234–1245. DOI: 10.1016/j.egyr.2024.01.034

Agarwal, S., & Jain, A. (2024). Blockchain Applications in Smart Grid: Challenges and Opportunities. In *2023 IEEE International Conference on Blockchain and Distributed Systems Security (ICBDS)* (Proceedings). DOI: DOI: 10.1109/ICBDS58040.2023.10346363

Ahmed Baba, M., Naoui, M., & Cherkaoui, M. (2023). Fault-Tolerant Control Strategy for Hall Sensors in BLDC Motor Drive for Electric Vehicle Applications. *Sustainability*, *15*(13), 10430. DOI: 10.3390/su151310430

Allal, . (2024). Advanced Residual Optimal Mapping Approach for Precise Detection of Stator Faults in Induction Motors. *IEEE Access: Practical Innovations, Open Solutions*, *12*, 143515–143530. DOI: 10.1109/ACCESS.2024.3442671

Alqunun, K., Bechiri, M. B., & Naoui, M.. (2025). An efficient bearing fault detection strategy based on a hybrid machine learning technique. *Scientific Reports*, *15*, 18739. DOI: 10.1038/s41598-025-02439-4 PMID: 40437009

Alsaigh, R., Mehmood, R., & Katib, I. (2022). AI explainability and governance in smart energy systems: A review. *arXiv*. https://doi.org//arXiv.2211.00069DOI: 10.48550

Appasani, B., Mishra, S. K., Jha, A. V., Mishra, S. K., Enescu, F. M., Sorlei, I. S., Bîrleanu, F. G., Takorabet, N., Thounthong, P., & Bizon, N. (2022). Blockchain-Enabled Smart Grid Applications: Architecture, Challenges, and Solutions. *Sustainability*, *14*(14), 8801. DOI: 10.3390/su14148801

Ashfaq, T., Khalid, M. I., Ali, G., Affendi, M. E., Iqbal, J., Hussain, S., Ullah, S. S., Yahaya, A. S., Khalid, R., & Mateen, A. (2022). An Efficient and Secure Energy Trading Approach with Machine Learning Technique and Consortium Blockchain. *Sensors (Basel)*, *22*(19), 7263. DOI: 10.3390/s22197263 PMID: 36236363

Choobineh, F., Arif, S., & Hamed, M. (2023). Blockchain technology in energy systems: A state-of-the-art review. *IET Renewable Power Generation*. Advance online publication. DOI: 10.1049/blc2.12020

Energies. (2022). Towards Blockchain-Based Energy Trading: A Smart Contract Implementation of Energy Double Auction and Spinning Reserve Trading. *Energies, 15*(11), 4084. https://doi.org/DOI: 10.3390/en15114084

Energy Informatics. (2024). Optimizing power system trading processes using smart contract algorithms. *Energy Informatics*. Advance online publication. DOI: 10.1186/s42162-024-00457-6

European Commission. (2021). *Regulatory paradigm and challenge for blockchain integration of energy systems*. MDPI., DOI: 10.3390/su15032571

Fie, L. (2022). Blockchain in energy systems: Values, opportunities, and limitations. *Frontiers in Energy Research*. Advance online publication. DOI: 10.1007/s11708-022-0818-8

Frontiers in Blockchain. (2025). *Revolutionizing the energy sector: exploring diversified blockchain applications in production, distribution, trading, and management of energy*. Frontiers in Blockchain., DOI: 10.3389/fbloc.2025.1544770

Fu, S., Tan, Y., & Xu, Z. (2023). Blockchain-Based Renewable Energy Certificate Trade for Low-Carbon Community of Active Energy Agents. *Sustainability, 15*(23), 16300. DOI: 10.3390/su152316300

Geng, B., & Varshney, P. K. (2023). Human-Machine Collaboration for Smart Decision Making: Current Trends and Future Opportunities. *arXiv preprint*. https://doi.org//arXiv.2301.07766DOI: 10.48550

Gonzalez-Sanchez, J., López-Pimentel, J.-C., & Morales-Rosales, L. A. (2025). A Digital Identity Blockchain Ecosystem: Linking Government-Certified and Uncertified Tokenized Objects. *Applied Sciences (Basel, Switzerland), 15*(15), 8577. DOI: 10.3390/app15158577

Goyal, S., Sharma, R., & Kumar, A. (2023). Blockchain-based microgrid solutions for rural electrification in India: Opportunities and challenges. *Renewable & Sustainable Energy Reviews, 167*, 112–123. DOI: 10.1016/j.rser.2023.112345

Gupta, B. B., Tewari, A., & Chaudhary, P. (2023). Blockchain-based identity management systems: A review. *Journal of Network and Computer Applications, 202*, 103404. DOI: 10.1016/j.jnca.2023.103404

Gupta, J., Jain, S., Chakraborty, S., Panchenko, V., Smirnov, A., & Yudaev, I. (2023). Advancing Sustainable Energy Transition: Blockchain and Peer-to-Peer Energy Trading in India's Green Revolution. *Sustainability, 15*(18), 13633. DOI: 10.3390/su151813633

Gupta, J., Jain, S., Chakraborty, S., Panchenko, V., Smirnov, A., & Yudaev, I. (2023). Advancing Sustainable Energy Transition: Blockchain and Peer-to-Peer Energy Trading in India's Green Revolution. *Sustainability*, *15*(18), 13633. DOI: 10.3390/su151813633

Gupta, M., Giri, S., Shanmugam, P. K., Bhaskar, M. S., Holm-Nielsen, J. B., & Padmanaban, S. (2023). Adoption of blockchain platform for security enhancement in energy transaction. *arXiv*. https://doi.org//arXiv.2305.19490DOI: 10.48550

Honari, K., Zhou, X., Rouhani, S., Dick, S., Liang, H., & Miller, J. (2022). A Scalable Blockchain-based Smart Contract Model for Decentralized Voltage Stability Using Sharding Technique. arXiv. https://arxiv.org/abs/2206.13776

Honari, K., Zhou, X., Rouhani, S., Dick, S., Liang, H., Miller Li, J., & Miller, J. (2022). A scalable blockchain-based smart contract model for decentralized voltage stability using sharding technique. *arXiv*. https://doi.org//arXiv.2206.13776DOI: 10.48550

Hossain, M. I., Steigner, T., Hussain, M. I., & Akther, A. (2024). Enhancing Data Integrity and Traceability in Industry Cyber Physical Systems (ICPS) through Blockchain Technology: A Comprehensive Approach. *arXiv*. https://arxiv.org/abs/2405.04837

Hua, W., Chen, Y., Qadrdan, M., Jiang, J., Sun, H., & Wu, J. (2022). Applications of blockchain and artificial intelligence technologies for enabling prosumers in smart grids: A review. *arXiv*. https://doi.org//arXiv.2202.10098DOI: 10.48550

IET. (2022). Blockchain technology in energy systems: A state-of-the-art review. *IET Renewable Power Generation*. Advance online publication. DOI: 10.1049/blc2.12020

Khan, P. W., & Byun, Y.-C. (2021). Blockchain-Based Peer-to-Peer Energy Trading and Charging Payment System for Electric Vehicles. *Sustainability*, *13*(14), 7962. DOI: 10.3390/su13147962

Khan, P. W., & Byun, Y.-C. (2021). Blockchain-Based Peer-to-Peer Energy Trading and Charging Payment System for Electric Vehicles. *Sustainability*, *13*(14), 7962. DOI: 10.3390/su13147962

Khaskheli, S., & Anvari-Moghaddam, A. (2024). Energy Trading in Local Energy Markets: A Comprehensive Review of Models, Solution Strategies, and Machine Learning Approaches. *Applied Sciences (Basel, Switzerland)*, *14*(24), 11510. DOI: 10.3390/app142411510

Kumar, R., & Sharma, S. (2023). Quantum-resilient blockchain frameworks for secure energy trading in India. *Energy Policy*, *170*, 113–127. DOI: 10.1016/j.enpol.2022.114567

Kumar, R., Sharma, S., & Patel, P. (2023). Regulatory challenges and decentralized governance models for blockchain in energy markets. *Energy Policy*, *160*, 112–123. DOI: 10.1016/j.enpol.2022.112123

Kumar, R., & Singh, A. (2025). A blockchain consortium-based framework to enhance interoperability, security, and standardized energy trading mechanisms within smart grid ecosystems. *Energy Reports*, *11*, 1234–1245. DOI: 10.1016/j.egyr.2025.02.014

Kumar, T. (2025). Blockchain in EVs – Securing India's Electric Future with Cutting-Edge Tech. *Green Bharat News*. https://greenbharatnews.com/blockchain-in-evs

Kumar, V., Vyas, S. A. S., & Kumar, R. (2021). Blockchain Technology for Decentralized Energy Transaction in a Smart Grid. IEEE Smart Grid Bulletin. https://smartgrid.ieee.org/bulletins/november-2021/blockchain-technology-for-decentralized-energy-transaction-in-a-smart-grid

Kumar, V., Vyas, S. A. S., & Kumar, R. (2021). Blockchain Technology for Decentralized Energy Transaction in a Smart Grid. IEEE Smart Grid Bulletin. https://smartgrid.ieee.org/bulletins/november-2021/blockchain-technology-for-decentralized-energy-transaction-in-a-smart-grid

Lammouchi, Z., Naoui, M., & Bekakra, Y.. (2025). Model-Free Predictive Control based on the Execution Time of Active Vectors Using Ultra-Local Model. *Arabian Journal for Science and Engineering*, *50*, 5885–5898. DOI: 10.1007/s13369-024-09629-8

Langer, M., Baum, K., & Schlicker, N. (2024). Effective Human Oversight of AI-Based Systems: A Signal Detection Perspective on the Detection of Inaccurate and Unfair Outputs. *Minds and Machines*, *35*(1), 1–19. DOI: 10.1007/s11023-024-09701-0

Laux, J. (2024). Institutionalised distrust and human oversight of artificial intelligence: Towards a democratic design of AI governance under the European Union AI Act. *AI & Society*, *39*, 2853–2866. DOI: 10.1007/s00146-023-01777-z PMID: 39640298

López-Pimentel, J.-C., Gonzalez-Sanchez, J., & Morales-Rosales, L. A. (2025). A Digital Identity Blockchain Ecosystem: Linking Government-Certified and Uncertified Tokenized Objects. *Applied Sciences (Basel, Switzerland)*, *15*(15), 8577. DOI: 10.3390/app15158577

Matsuda, Y., Yamazaki, Y., Oki, H., Takeda, Y., Sagawa, D., & Tanaka, K. (2021). Demonstration of Blockchain Based Peer to Peer Energy Trading System with Real-Life Used PHEV and HEMS Charge Control. *Energies*, *14*(22), 7484. DOI: 10.3390/en14227484

Maurya, V. K., Kothari, R., Sachdev, P., & Rathore, N. S. (2024). Blockchain Technology for Secure Smart Grid Access Control. In Shrivastava, V., Bansal, J. C., & Panigrahi, B. K. (Eds.), *Power Engineering and Intelligent Systems. PEIS 2023. Lecture Notes in Electrical Engineering, 1097*. Springer., DOI: 10.1007/978-981-99-7216-6_28

Mullaney, C., Aijaz, A., Sealey, N., & Holden, B. (2022). Peer-to-Peer Energy Trading meets IOTA: Toward a Scalable, Low-Cost, and Efficient Trading System. arXiv. https://doi.org//arXiv.2210.06427DOI: 10.48550

Nature Research. (2024). Blockchain-based energy consumption approaches in IoT. *Scientific Reports*. Advance online publication. DOI: 10.1038/s41598-024-77792-x PMID: 39543190

Niforos, M., & Rojas, F. (2021). Blockchain governance and regulation as an enabler for market creation in emerging markets. *World Bank Policy Research Working Paper*, (9732). https://doi.org/DOI: 10.1596/1813-9450-9732

Patel, P., Reddy, S., & Verma, A. (2024). AI and blockchain convergence in decentralized energy systems: Insights from Indian pilot projects. *Renewable & Sustainable Energy Reviews*, *182*, 113–132. DOI: 10.1016/j.rser.2024.114321

Pfitzmann, A., & Hansen, M. (2010). A terminology for talking about privacy by data minimization: Anonymity, Unlinkability, Undetectability, Unobservability, Pseudonymity, and Identity Management. *University of Dresden*. Retrieved from http://dud.inf.tu-dresden.de/Anon_Terminology.shtml

Rahman, M., Chowdhury, S., Shorfuzzaman, M., Hossain, M. K., & Hammoudeh, M. (2023). Peer-to-Peer Power Energy Trading in Blockchain Using Efficient Machine Learning Model. *Sustainability*, *15*(18), 13640. DOI: 10.3390/su151813640

Rath, S., Nguyen, L. D., Sahoo, S., & Popovski, P. (2022). Self-healing secure blockchain framework in microgrids. *IEEE Transactions on Industrial Informatics*, *18*(5), 3456–3465. DOI: 10.1109/TII.2022.3156789

Reijsbergen, D., Maw, A., Dinh, T. T. A., Li, W.-T., & Yuen, C. (2022). Securing Smart Grids Through an Incentive Mechanism for Blockchain-Based Data Sharing. *Energy Reports*, *8*, 1320. DOI: 10.1016/j.egyr.2024.08.060

Renewable and Sustainable Energy Reviews. (2022). Smart contracts in energy systems: A systematic review of fundamental approaches and implementations. *Renewable and Sustainable Energy Reviews, 158*, 112013. https://doi.org/DOI: 10.1016/j.rser.2021.112013

Romdhane, M., Naoui, M., & Mansouri, A. (2023). PMSM Inter-Turn Short Circuit Fault Detection Using the Fuzzy-Extended Kalman Filter in Electric Vehicles. *Electronics (Basel), 12*(18), 3758. DOI: 10.3390/electronics12183758

Saleem, Y., Crespi, N., Rehmani, M. H., & Copeland, R. (2017). Internet of Things-aided Smart Grid: Technologies, Architectures, Applications, Prototypes, and Future Research Directions. *arXiv preprint*. https://doi.org//arXiv.1704.08977DOI: 10.48550

Sharma, S., Kumar, S., & Gupta, B. B. (2023). Blockchain-based identity management and access control for energy trading platforms. *Journal of Network and Computer Applications, 202*, 103404. DOI: 10.1016/j.jnca.2023.103404

Singh, A., Mehta, R., & Kumar, V. (2024). Quantum-resistant blockchains for decentralized energy markets: Global perspectives. *Applied Energy, 345*, 125–142. DOI: 10.1016/j.apenergy.2023.123456

Tan, R., Krishna, V. B., Yau, D. K. Y., & Kalbarczyk, Z. (2016). Impact of Integrity Attacks on Real-Time Pricing in Smart Grids. *IEEE Transactions on Smart Grid, 7*(4), 1772–1781. DOI: 10.1109/TSG.2016.2532563

The Energy and Resources Institute (TERI). (2025). Blockchain for Peer-to-Peer Energy Trading: A Case Study. *TERI Energy and Environment Data Diary and Yearbook, 2025*, 45–50. https://www.teriin.org/publications/Blockchain-for-Peer-to-Peer-Energy-Trading

Tian, Y., Wang, C., Asutosh, A., Woo, J., & Adriaens, P. (2022). Blockchain-enabled tokenization for sustainable and inclusive infrastructure investment. *arXiv preprint*. https://doi.org//arXiv.2208.04709DOI: 10.48550

Varshney, N., Sreetharan, V., Vennila, C., Ahila, R., Mane, D. T., Kshirsagar, K., & Pragathi, B. (2025). Blockchain Technology in Energy Management Systems: Enhancing Security and Transparency.

Vieira, G., & Zhang, J. (2021). Peer-to-Peer Energy Trading in a Microgrid Leveraged by Smart Contracts. arXiv. https://arxiv.org/abs/2201.04944

Vionis, P., & Kotsilieris, T. (2024). The potential of blockchain technology and smart contracts in the energy sector: A review. *Applied Sciences (Basel, Switzerland), 14*(1), 253. DOI: 10.3390/app14010253

Vionis, P., & Kotsilieris, T. (2024). The potential of blockchain technology and smart contracts in the energy sector: A review. *Applied Sciences (Basel, Switzerland), 14*(1), 253. DOI: 10.3390/app14010253

Wang, L., Jiao, S., Xie, Y., Mubaarak, S., Zhang, D., Liu, J., Jiang, S., Zhang, Y., & Li, M. (2021). A permissioned blockchain-based energy management system for renewable energy microgrids. *Sustainability, 13*(3), 1317. DOI: 10.3390/su13031317

Wang, N., Chau, S. C.-K., & Zhou, Y. (2021). Privacy-preserving energy storage sharing with blockchain and secure multi-party computation. *Energy, 225*, 120–130. DOI: 10.1016/j.energy.2021.120130

Wang, S., Taha, A. F., Wang, J., Kvaternik, K., & Hahn, A. (2019). Energy crowdsourcing and peer-to-peer energy trading in blockchain-enabled smart grids. *Applied Energy, 235*, 1078–1090. DOI: 10.1016/j.apenergy.2018.11.031

Xiao, Y., Lin, X., Lei, Y., Gu, Y., Tang, J., Zhang, F., & Qian, B. (2024). Blockchain-Assisted Secure Energy Trading in Electricity Markets: A Tiny Deep Reinforcement Learning-Based Stackelberg Game Approach. *Electronics (Basel), 13*(18), 3647. DOI: 10.3390/electronics13183647

Yang, Q., Wang, H., Wang, T., Zhang, S., Wu, X., & Wang, H. (2021). Blockchain-Based Decentralized Energy Management Platform for Residential Distributed Energy Resources in A Virtual Power Plant. arXiv. https://arxiv.org/abs/2105.00174

Youssef, R., Hamida, M. A., Mohamed, N., Lassaad, S., Djamel, T., & Yasmine, A. (2023). Single-Stage Standalone Photovoltaic Water Pumping System Using Predictive Torque Control (PTC) of Induction Machine. *Electric Power Components and Systems*, •••, 1–17. DOI: 10.1080/15325008.2023.2278624

Zeng, Z., Li, Y., Cao, Y., Zhao, Y., Zhong, J., Sidorov, D., & Zeng, X. (2020). Blockchain Technology for Information Security of the Energy Internet: Fundamentals, Features, Strategy and Application. *Energies, 13*(4), 881. DOI: 10.3390/en13040881

Zhang, G., Chen, S. C.-I., & Yue, X. (2024). Blockchain Technology in Carbon Trading Markets: Impacts, Benefits, and Challenges—A Case Study of the Shanghai Environment and Energy Exchange. *Energies, 17*(13), 3296. DOI: 10.3390/en17133296

Zhang, L., Li, X., & Wang, J. (2025). Decentralized EV Charging Network with Tokenized Access and Blockchain Technology. *ScienceDirect*. https://www.sciencedirect.com/science/article/pii/S2773186325001021

Zhang, L., Wang, Y., & Li, J. (2025). Towards Blockchain Interoperability: A Comprehensive Survey on Cross-Chain Solutions. *Computer Networks, 202*, 108–123. DOI: 10.1016/j.comnet.2025.04.015

Zhang, X., Li, Y., & Wang, Z. (2023). AI-integrated blockchain for energy transaction optimization. *Journal of Cleaner Production, 404*, 136–151. DOI: 10.1016/j.jclepro.2023.136789

Zhang, X., Li, Y., & Wang, Z. (2024). Blockchain for energy markets: A comprehensive survey. *Renewable & Sustainable Energy Reviews, 135*, 110–122. DOI: 10.1016/j.rser.2020.110122

Zhang, Y., Liu, Y., & Wang, X. (2023). Blockchain-based decentralized identity management in smart grids. *Journal of Network and Computer Applications, 202*, 103404. DOI: 10.1016/j.jnca.2023.103404

Zhang, Y., & Wang, L. (2023). Exploring the effects of blockchain scalability limitations on performance and user behavior in blockchain-based shared manufacturing systems: An experimental approach. *Applied Sciences (Basel, Switzerland), 13*(7), 4251. DOI: 10.3390/app13074251

Zhao, Z., Liu, Y., Zeng, Z., Chen, Z., & Zhou, H. (2023). Privacy-preserving electricity theft detection based on blockchain. *IEEE Transactions on Industrial Informatics, 19*(5), 3456–3465. DOI: 10.1109/TII.2023.3145678

Zografopoulos, I., Hatziargyriou, N. D., & Konstantinou, C. (2022). Distributed Energy Resources Cybersecurity Outlook: Vulnerabilities, Attacks, Impacts, and Mitigations. *Energy Reports, 8*, 1320. DOI: 10.1016/j.egyr.2024.08.060

KEY TERMS AND DEFINITIONS

Blockchain: A decentralized digital ledger technology that records and verifies transactions across multiple computers, ensuring transparency, immutability, and security in data exchange.

Consensus Mechanism: A method by which blockchain networks achieve agreement among distributed nodes on the validity of transactions, ensuring trust without a central authority.

Decentralized Energy Market: An energy trading system where producers and consumers exchange energy directly through peer-to-peer networks, enabled by blockchain and smart contracts.

Digital Identity: A unique and verifiable representation of an entity—human, machine, or organization—used to authenticate and authorize transactions on digital platforms.

Energy Tokenization: The process of representing energy assets or units of energy as digital tokens on a blockchain, enabling fractional ownership, transparent tracking, and efficient trade.

Human-Machine Interaction: The dynamic communication and coordination between humans and intelligent machines to facilitate autonomous transactions and decision-making processes.

Smart Contract: A self-executing digital agreement embedded with predefined rules that automatically enforces terms once specified conditions are met on a blockchain network.

Sustainable Energy Trading: The exchange of renewable and clean energy resources through transparent, traceable, and efficient digital mechanisms to promote environmental sustainability.

Transaction Security: The protection of digital transactions from fraud, tampering, or unauthorized access using cryptographic and blockchain-based safeguards.

Trustless System: A framework that eliminates the need for intermediaries or centralized trust, relying instead on cryptographic validation and distributed consensus for secure interactions.

Chapter 7
Human-Machine Interaction in Virtual Inertia Control for Smart Grids

Mabrouka Ghiloufi
https://orcid.org/0009-0002-4773-7432
Ecole Nationale d'ingenieurs de Gabes, Université de Gabès, Tunisia

Lassaad Sbita
https://orcid.org/0000-0002-6589-3976
Ecole Nationale d'ingenieurs de Gabes, Université de Gabès, Tunisia

ABSTRACT

The rapid penetration of renewable energy into smart grid has decreased grid inertia due to the substitution of synchronous generators with inverter-based resources. Due to their lack of rotational inertia, renewable energy create a crucial 'inertia gap'. In response to this challenge, virtual inertia control (VIC) has emerged as an important solution, evolving from primitive emulation methods to advanced adaptive and artificial intelligence-based methods. This work investigates the evolution of virtual inertia emulation methods as important countermeasures for smart grids, from their initial fixed-parameter methods through the development into real-time adaptive systems and intelligent control structures. The integration of Human-Machine Interaction (HMI) interfaces is emphasized to enable operators to monitor, interpret, and intervene in inertia management systems. These HMI systems support decision-making by visualizing real-time data, enabling intuitive control of adaptive VIC algorithms, and ensuring human oversight in safety-critical scenarios within low-inertia power grids.

DOI: 10.4018/979-8-3373-5621-1.ch007

INTRODUCTION

Modernization of power systems is essentially based on the incorporation of renewable energy source (RES) such as hydro, solar PV, and wind capacity, to decarbonize electricity production and enable a smooth shift toward smarter, more responsive grid infrastructures. As the global energy system evolves toward sustainability, distributed generation takes an important position in combating climate change through a substantial reduction in carbon dioxide (CO_2) and other greenhouse gas emissions. Beyond to their environmental advantages, large-scale RES integration is also a technological necessity, allowing for advanced grid functions like real-time demand response, decentralization of energy production, and two-way power flow (Khalid, 2024). This transition to the utilization of inverter-based resources (IBRs), from the traditional fossil-fuel-based synchronous generation, is quite naturally transforming the power systems. With this transition, advanced stability concerns arise that need to be addressed carefully to guarantee grid reliability and long-term operational stability (Hossain et al., 2024).

One of the primary disadvantages of RES-dominated grids is that they inherently have low system inertia, which is quite unlike traditional grids where synchronous generators provide high rotational inertia. Inertia is an essential stabilizing factor, mitigating the effects of sudden disturbances by slowing down the rate of frequency deviation and providing ample time for remedial action to be taken (Johnson et al., 2020). In power systems with high IBR penetration, the absence of rotating mass disconnects generation dynamics from the grid frequency response, inducing accelerated transient frequency events and increased susceptibility to cascading failures (Gu & Green, 2022). The progressive loss of effective inertia thus poses a fundamental threat to grid stability, especially in high RES integration conditions, where conventional frequency regulation structures become less effective owing to their dependency on synchronous machine behavior (Bálint Hartmann, 2019).

In order to offset the inherent deficit of inertia in renewable energy-based power systems, virtual inertia emulation (VIE) is becoming a key enabling technology, which attempts to artificially mimic the inertial response characteristics of conventional synchronous generators through the application of advanced power electronic control concepts. By integration of energy storage units (ESU) or converter control system reconfiguration, VIE incorporates efficient inertial response and therefore diminishes the dynamic performance gap between traditional synchronous generation and inverter-based resources (IBRs). Advanced control schemes, such as droop-based inertia emulations and derivative-added frequency support schemes, enable IBRs to enable transient stability (functions formerly reserved for synchronous generators). Nonetheless, VIE deployment needs to be well-planned and integrated with current grid control and regulation systems to avoid destabilization and ensure compatibility and improve overall system robustness (Mathew & Panikkar, 2024).

A number of schemes have been conceived for virtual inertia implementation techniques, classified into three generational categories, each having particular characteristics in terms of dynamic response-ability, energy storage requirements, and compliance with grid code specifications. The initial generation includes conventional control schemes, based on droop control and derivative feed-forward action, offering simplicity and compatibility but low adaptability. The second generation is facilitated by adaptive and optimal tactics such as Particle Swarm Optimization (PSO) and Artificial Bee Colony (ABC) algorithms, facilitating enhanced performance through real-time system identification and parameter change. The future generation comprises machine learning-based control policies and model predictive control (MPC), facilitating enhanced flexibility, accuracy, and predictive response through data-driven modeling and predictability of the system. Together, these new virtual inertia control structures have a foundational role in mitigating the inertia deficit with high-level RESs penetration, hence enabling higher integration of renewable generation without compromising grid stability and operation reliability (Jafari et al., 2024).

This work is organized into three main sections that systematically cover human-machine interaction within the context of virtual inertia control of smart grids.

- The basic concept and theoretical background of inertia shortage are discussed in the initial part, specifically in terms of its effect on power system frequency stability. It discusses why inadequate inertia comes from deploying massive renewable resources to smart grids and why shifting that way may possibly pose operational challenges and stability threats.
- The second section offers a general overview of Challenges in low-inertia grids and its profound effect on power system stability and performance.
- The third part is dedicated to the acquisition of monitoring methods for virtual inertia deployment, ranging from traditional control methods to newer smart methods with artificial intelligence and machine learning. The latest methods have shown better performance and dynamism compared to older methods, with greater observability, precision, and responsiveness for real-time inertia control in dynamic and complex grid systems

FUNDAMENTALS OF POWER SYSTEM INERTIA

In its basic physical sense, inertia is the property of a mass to oppose a velocity change. In power electrical grids, power system inertia is mainly provided by the rotating masses of synchronous machines in the grid. In this context, inertia literally refers to the potential kinetic energy stored in the rotational equipment e.g., rotors and turbines of the synchronous machines. Such stored energy has the inherent buffering function of slowing down the rate of frequency deviation and thus giving sufficient time for the corrective control action to be activated (Fang et al., 2018). Since synchronous generators are the primary source of mechanical inertia in typical power systems, play a crucial role in maintaining frequency stability and ensuring grid resilience (Abolfazl et al., 2025).

The overall kinetic energy E_K can be represented as follows:

$$E_k = \frac{1}{2} J \omega_s^2 \quad (1)$$

Stored kinetic energy is one of the basic dynamic parameters in power system analysis. For one synchronous generator, the energy is characterized by the moment of inertia of its rotating mass and its rotating speed. In typical synchronous-dominant power systems, the inertia constant, represented by H, is calculated through division of storage kinetic energy in the rotating plant with rated apparent generator capacity (Ulbig et al., 2014). The system inertia constant H can be represented mathematically as:

$$H = \frac{E_k}{S_B} = \frac{(½) J \omega_s^2}{S_B} \quad (2)$$

Where:

J Moment of inertia of the rotating mass (kg m2)
ω_s Rated rotor angular speed (rad/s)
E_k Equivalent kinetic energy of SG
H Inertia constant of SG (s)
S_B Rated apparent power of the system (kW)

The swing equation is a basic dynamic model used to describe the electromechanical performance of a synchronous generator during transient faults, e.g., short circuits or sudden load variations. This differential equation models the generator rotor's angular acceleration in relation to a synchronously rotating reference frame, accounting for the equilibrium of the mechanical torque supplied by the prime mover and the electromagnetic torque resulting from the interaction between the rotor magnetic field and the stator currents and damping effects (Brezovec et al., 2015). From Newton's second law of motion in rotational terms, the swing equation is mathematically expressed as:

$$\frac{d\Delta \omega_s}{dt} = \frac{1}{2H}\left(\Delta P_m - \Delta P_e\right) \quad (3)$$

The swing equation is one of the main components of transient stability analysis and is utilized as a major analytical tool to evaluate the capacity of a synchronous generator to stay in synchronism after being subjected to a disturbance. The nonlinearity of the swing equation normally necessitates the use of numerical integration methods, like the Runge–Kutta method, in order to obtain solutions, especially in the presence of intricate and dynamic grid operating scenarios. Correct analysis and interpretation of the swing equation are key to system stability evaluation, control protection scheme design, and ultimately, the robustness and reliability of power systems against potential cascading failure (Hadavi et al., 2025; Ulbig et al., 2014).

Figure 1 is a block diagram model of the swing equation, revealing the dynamic interaction between the electrical and mechanical power in a synchronous generator. The figure has prominent elements, such as summing junctions, integrators, and feedback loops, which collectively simulate the dynamic response of the system. Input signals are equal to electrical power and mechanical power and their algebraic difference as the primary mover for propelling the system dynamics. Inertia constant and damping coefficients are shown as gain factors and integrators utilized in transforming the rotor speed and rotor angle from angular acceleration. Feedback loops in the diagram are responsible for damped torques and synchronization torque, hence exhibit transient response behavior of the system (Ashish Shrestha, 2021). Systematic configuration within this block diagram provides the basic structure for stability studies, simulation of disturbances and development of control measures to maintain generator synchronism under different operating conditions (Baškarad et al., 2018).

Figure 1. Block diagram model of the swing equation

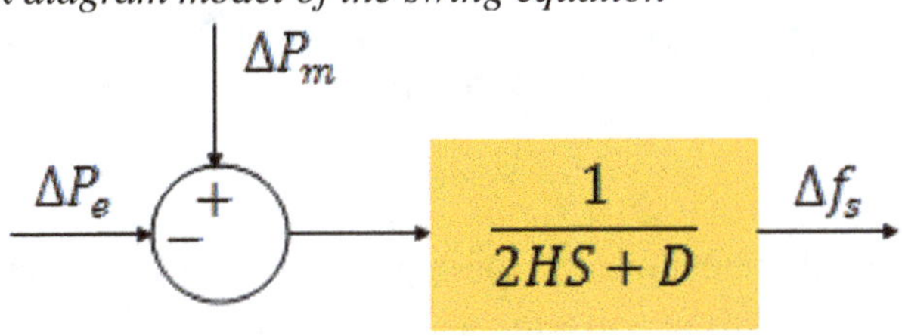

In emergency power imbalances, the power grid uses hierarchical frequency control architecture to enhance system stability and prevent cascading failure propagation. The multilayer control structure operates sequentially across different timescales with corresponding mechanisms and resource allocation techniques used at each layer. The first response is inertial control (0–10 seconds), where synchronous generators (SGs) utilize stored kinetic energy in rotating masses to counter power shortages. The operation slows down rotor speeds, reduces the rate of change of frequency (RoCoF), and mitigates the frequency nadir (the bottom of the frequency deviation curve). Next, primary control (10–30 seconds) activates governor mechanisms that automatically regulate the turbine inputs like steam or fuel flow to increase the mechanical power output and achieve a level of balance between generation and demand. Second, secondary control (15 to 30 minutes) (Rajkumar et al., 2023), controlled by the Automatic Generation Control (AGC), coordinates the dispatch of generating units to finely control power output, restore used reserves into the system, and return the system frequency to its nominal. Lastly, tertiary control (more than 15 minutes) consists of automatic or manual strategic actions such as generation rescheduling, further generating unit commitment, and reserve re-tuning to provide long-term reliability and preparedness for possible future upsets. Combined, coordinated layers of controls ensure the grid's resiliency to react effectively to transient as well as extended disturbances (Deng et al., 2021).

Figure 2. General frequency response and controller involved of a conventional power systems

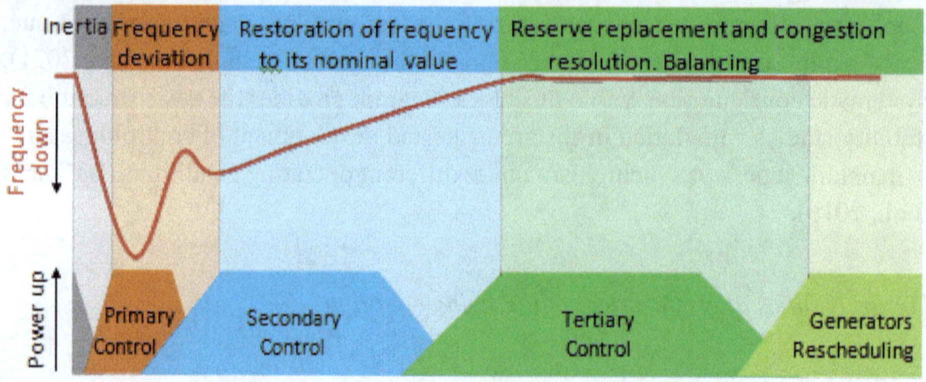

Inertia is a natural maintaining factor in power system frequency stability, providing instant support during a disturbance, restraining the initial effect of power imbalance, and offering precious time for higher-level control to initiate. This makes it an essential requirement for grid operators to ensure continuous monitoring and evaluation of the system inertia that is proactive and adaptive actions which are taken to provide efficient, stable, and reliable power grid operation.

In modern power systems with increased penetration of distributed generation, frequency nadir and rate of change of frequency (ROCOF) are key operating parameters. They are central to control and protection strategies and are vital inputs in estimating grid stability, real-time power balance management, and inertia estimation.

The rate of change of frequency, in mathematical terms being the derivative of the power system frequency $\left(\frac{df}{dt}\right)$ with respect to time, is an important quantity used to estimate grid resilience and dynamic stability (Hongchun SHU, 2025). The first RoCoF measurement records the instantaneous rate of deviation in frequency at the moment a power imbalance occurs (loss of generation or rapid rejection of load) before counteractive control action. This parameter gives valuable information about the inertial response of the system and disturbance resistance without sacrificing stability. RoCoF can generally be estimated from the formula below (Deng et al., 2023):

$$RoCoF = \frac{df}{dt} = \frac{f}{2E_{kinetic}}(P_m - P_e) = \frac{f}{2E_{kinetic}}\Delta P \tag{4}$$

Frequency response dynamics are important in the study of a power system's robustness and stability under a disturbance. The main performance measures are Rate of Change of Frequency, slope at which the system frequency declines immediately after the power mismatch; frequency nadir, the point to which the frequency declines before the recovery process starts; and settling frequency, the final steady-state frequency value when the system has recovered from the disturbance. High values of Large RoCoF indicate high dynamic loading of the grid, and that may lead to the initiation of UFLS schemes in case no corrective action is taken at once. The nadir frequency indicates the severity of insufficiency and inadequacy of primary frequency response, while the settling frequency depends significantly on secondary control actions. These values in aggregate determine the dynamic performance and robustness of the system and directly influence key operation parameters like generator trip levels, inertia adequacy, and renewable energy hosting capacity. Accurate modeling and real-time monitoring of such frequency response parameters are crucial for safe operation of the grid, especially in power systems with rising penetrations of converter-based power plants and low levels of native rotational inertia (Baškarad et al., 2024).

CHALLENGES IN LOW-INERTIA GRIDS

The current energy transition is characterized by the gradual replacement of traditional power generation with renewable generation. This transition presents significant technical and operational challenges. One of the most concerns is the loss of system inertia due to the increasing penetration of inverter-based resources, which decouple generation from the synchronous dynamics of the system. Grid stability has been based on the mechanical inertia of traditional synchronous generators. As they are retired, the demand for alternative means to maintain frequency stability and grid robustness becomes more pressing. Power systems with low inertia are susceptible to larger frequency fluctuations, increased RoCoF, voltage instability, and reduced fault ride-through capability. Therefore, there is a necessary need to develop and install next-generation control paradigms, synthetic inertia solutions, and grid-forming technologies that can maintain system stability, provide reliable performance, and enable the further integration of renewable energy sources into the evolving power grid.

1. Effects of Lower Inertia on Frequency Response

The increased penetration of RES into advanced power systems has resulted in low-inertia grids that introduce severe operational and stability issues. Traditional synchronous generators, which provide mechanical inertia and frequency regulation ancillary services, are being replaced by inverter-interfaced renewable generation with no inherent inertial response. This decrease in system inertia leads to increasingly larger and more rapid frequency deviations after disturbances, thus worsening the grid's ability to sustain stable operation within reasonable frequency limits. Moreover, the absence of damping torque from converter units increases the vulnerability to frequency instability, especially at high RES penetration and abrupt load variations. These dynamics need the development of new control systems and synthetic inertia technologies to regain frequency resilience and enable a reliable grid operation (Liu et al., 2022a).

As shown in Figure 3, decreasing inertia levels lead to a more aggressive and quicker decrease in system frequency after a disturbance, as shown by the steeper initial slope of the frequency deviation curve for lower inertia systems. This increased RoCoF indicates a reduced amount of kinetic energy available to absorb the transient imbalance between generation and load to be absorbed, resulting in larger and more rapid frequency deviations. Conversely, higher-inertia systems exhibit a less rapid and more sluggish frequency response that characterizes the stabilizing influence of greater rotational energy storage (Alsharafi et al., 2018). The difference in frequency response between inertia levels repeats the intrinsic importance of inertia to grid stability, especially in response to sudden disturbances or load alteration. Yet, as inertial capability continues to decline, conventional frequency regulation systems will no longer be able to respond adequately, and sophisticated control methods such as synthetic inertia and primary frequency response with rapid response rates will be required to replace declining indigenous inertial support (Abouyehia et al., 2025).

Figure 3. Frequency response for different levels of inertia H

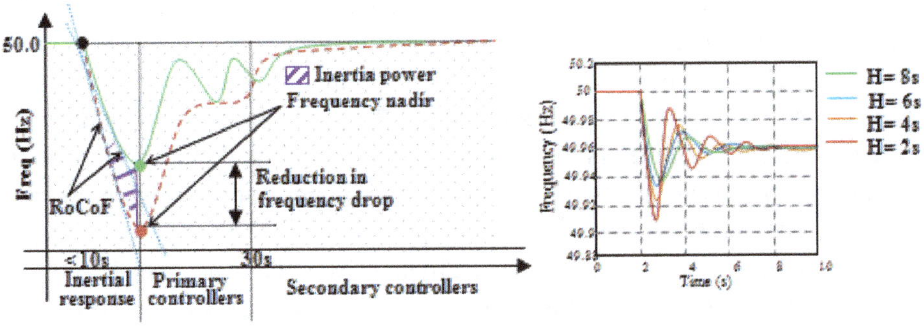

2. Effects of Lower Inertia on Voltage Stability

The reduction of system inertia, which is mostly attributed to the large-scale integration of inverter-interfaced RES, has profound effects not only on frequency stability but also on the voltage stability of modern power systems. Reduced inertia levels are normally associated with reduced short-circuits capacities and poorer grid support features, especially in converter-interfaced dominated systems. This leads to higher sensitivity to sudden load and disturbances with greater voltage swings and longer recovery. Additionally, the lack of native support in the form of reactive power and the voltage regulation ability of synchronous machines further compromises the system's capability to present constant voltage profiles during transients. Therefore, low-inertia systems are more susceptible to voltage instability, particularly with high penetration of renewable and instantaneous load or generation changes. Overcoming these challenges requires advanced voltage control strategies, improved grid-forming inverter capabilities, and flexible resource coordination to guarantee sound voltage stability margins (Elenga Baningobera et al., 2024; Kerdphol et al., 2018).

CONTROL STRATEGIES FOR VIRTUAL INERTIA

The figure 4 depicts a systematic and inclusive VIC Strategy taxonomy, which comprises three general categories: Traditional Strategies, Optimal (Classical) Strategies, and Intelligent Strategies.

Under Traditional Strategies, the focus is on VI-based inverters, which are established techniques such as Virtual Synchronous Machine (VSM), Virtual Synchronous Generator (VSG), and Synchronverter implementations. These methods imitate synchronous machine inertial and damping using power electronic converters. This section also includes Derivative-Based Strategies, an extension of traditional VI control with dynamic extensions for better transient response and stability.

The Optimal Strategies category, under the category of Classical Optimization Techniques, involves sophisticated computational techniques such as Particle Swarm Optimization (PSO) and Artificial Bee Colony (ABC) Optimization to systematically optimize controller parameters to enhance system performance as well as robustness. The techniques are devised for frequency regulation and damping properties in different operating conditions.

Intelligent strategies combine adaptive and data-driven techniques, which represent the cutting edge of VI control. This tier offers Model Predictive Control (MPC) for dynamic real-time optimization based on forecasted system response, and Machine Learning-Based Strategies like Reinforcement Learning (RL), which provide adaptive control capability optimized for complicated, uncertain, and nonlinear grid systems.

This hierarchical framework not only depicts the evolution from classical control schemes to high-level intelligent control systems but also displays the augmented function of optimization and computational intelligence in stabilizing grid. At the core of these strategies are VI-Based Inverters, which provide the working platform for implementing virtual inertia into modern power systems. These inverters bridge the gap between theoretical control systems and practical implementation, facilitating correct incorporation of renewable energy sources without jeopardizing grid stability and frequency (Shobug et al., 2024).

Figure 4. Virtual inertia control strategies

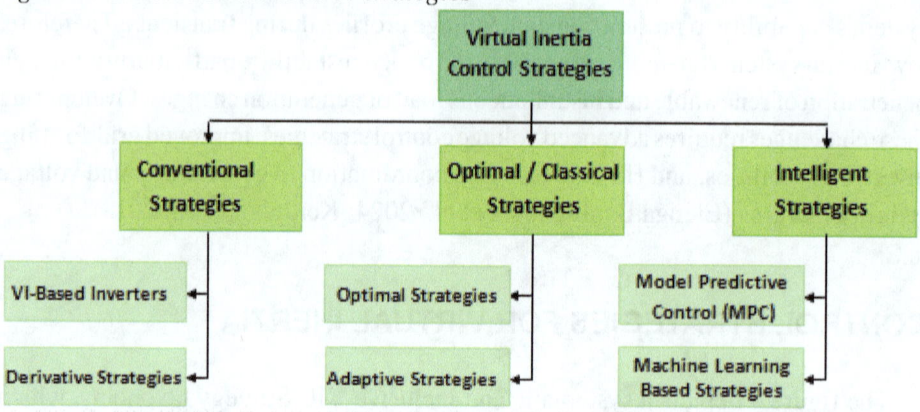

1. Traditional Strategies

a. Implementation of VI-Based Inverters

In principle, VI converters are essentially based on the structural and mathematical paradigms of synchronous machines and synchronous generators, as discussed in classical power systems theory. SG is employed here for both synchronous motors and generators, owing to their duality of operation in the same electromagnetic configuration. Therefore, virtual synchronous machines, virtual synchronous generators, and synchronization converters have the same dynamic concepts because their control models are considered abstracted conceptually and mathematically from physical SG models. A basic SG model with a cylindrical rotor is presented, as shown in Figure 5, which is based on simple electromagnetic equations derived from an approximation of one pole pair per phase ($p = 1$) (Allal et al., 2024). This environment defines a sample bipolar synchronous machine, with a three-phase stator winding configuration and a revolving field excitation system (Yap et al., 2019; Youssef et al., 2023).

Figure 5. A simple schematic of a rotor synchronous generator with distributed three-phase stator windings

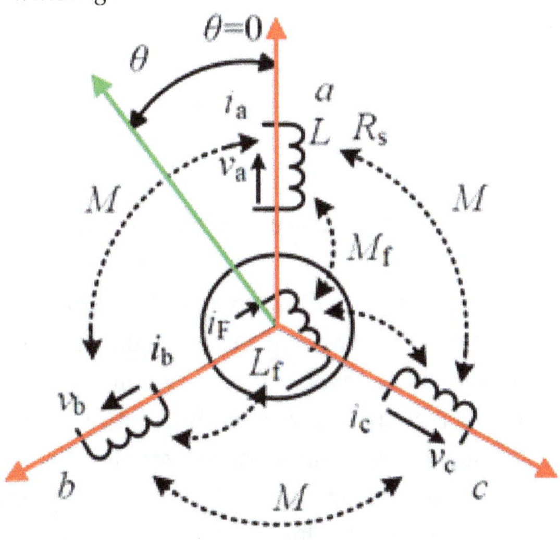

Figure 6 shows the complete single-line model of grid-connected VI-based inverter system. The system begins with a DC source feeding power to a voltage source inverter (VSI), which is controlled by a VI algorithm. This algorithm takes the measured three-phase voltage and current inputs and produces pulse-width modulation (PWM) signals to drive the inverter switches. The output of the inverter is connected to the grid via a filter that consists of a series reactance, an inductive reactance, and a shunt capacitor at the point of common coupling (PCC). The network is extended from PCC to an external power system described as an infinite bus, described by a voltage source with some impedance elements. This infinite bus assumption enables the external system to remain at constant voltage and frequency and provides a stable reference for dynamic performance analysis (Ahmed Baba et al., 2023). The VI algorithm implemented in the controller enables the inverter to mimic the inertial response of a synchronous machine, hence improving the stability of the system and facilitating frequency regulation of high-penetration renewable energy sources (Yap et al., 2019).

Figure 6. Single-line diagram of vi-based inverters interfaced with an infinite-bus power system

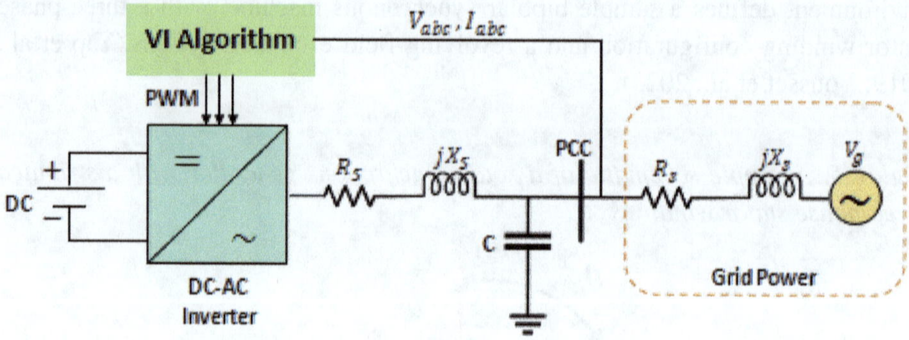

Figure 7 depicts a conceptual structure of a VI-based inverter system, which is supposed to replicate the dynamic response characteristics of SG-type power systems. The design establishes functional equivalence between an aging synchronous generator driven by a prime mover that captures mechanical energy from fossil fuels like natural gas, coal, or oil and an inverter driven by renewable energy sources that mimics artificially the electromechanical phenomenon of an SG. In conventional arrangements, the rotational kinetic energy of the prime mover is passed on to the generator rotor, which in turn develops three-phase alternating current (AC) power to be supplied within the microgrid. This mechanical inertia is intrinsic and plays a crucial role in maintaining frequency stability, thereby enhancing overall system robustness and transient response (Yap et al., 2019).

Conversely, a VI-capable inverter extracts DC power from renewable sources such as wind farms and photovoltaic arrays and mimics the inertial and dynamic response of an SG. It achieves this through advanced control techniques that artificially reproduce the inertial response, allowing the inverter to supply AC power with stabilizing and grid-forming properties. Adding a DC-link capacitor improves the system effective inertia as a storage device, thus providing transient support to sudden load or generation fluctuation. By simulating dynamic performance and rotating mass of the traditional synchronous machines, the inverter offers stable and quick response for microgrid operation without compromising frequency stability and high penetration of the renewable sources and improving overall grid resilience and reliability (Badrudeen et al., 2024).

Figure 7. Comparison of inertia characteristics in renewable energy and conventional power systems for micro grid applications

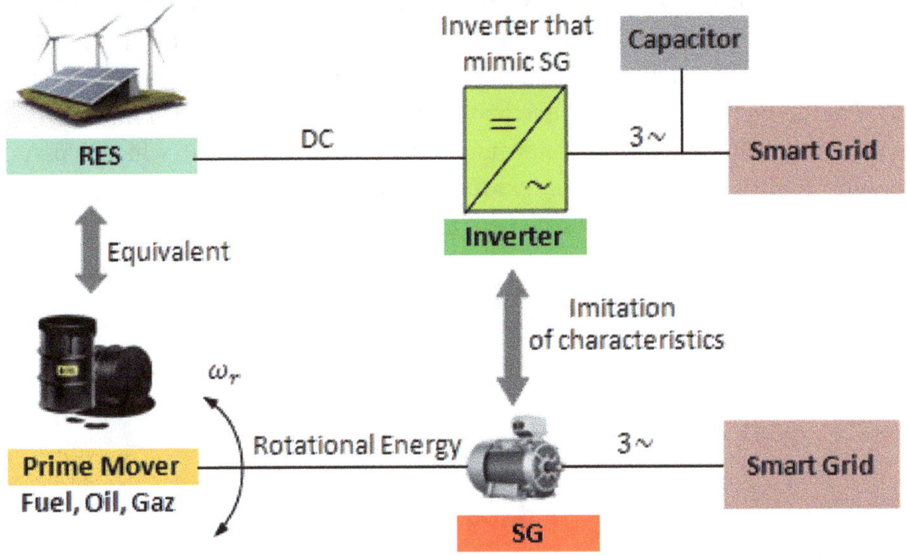

i. Virtual Synchronous Generator (VSG)

VSG heavily depends on traditional inverter hardware supported by a highly intelligent control algorithm to replicate the inertial behavior of a SG under dynamic grid frequency operations. This control mechanism allows grid-side inverters to replicate VI behavior, hence facilitating improved grid stability and dynamic response. VSG concept gained considerable research attention because of its operationally equivalent relationship with SGs, providing a technically viable platform for emission-free integration of renewable energy sources into the grid. By emulating the kinetic energy storage and release dynamics of SGs, the VSG replicates essential electromechanical responses like rotor accelerating and decelerating responses to power unbalance (Zhou et al., 2019).

The inertia constant is of central importance in reducing the RoCoF through a fast dynamic frequency response that depends on the time derivative of system frequency. This role is of special importance to isolated, fragmented, or stand-alone power grids, where there could be a large power imbalance at the moment of an event, which can correspond to large frequency swings. The inertial response is a vital stability mechanism that enhances grid resilience by preventing unnecessary activation of protection relays and associated disconnection events (Yap et al., 2019).

The control topology for VSG technology comprises a dq-frame transformation, and it known as direct-quadrature (d–q) axis conversion, which allows synchronization and control in rotating reference frames. The VSG is based on a current-controlled inverter paradigm, where the dq reference frame is used to implement the current control strategy. The actual references supplied to the current controller are determined based on the reference active and reactive power, as shown in Equation (5). The d-axis current reference is determined based on Equation (6), which is part of the inner control loop for accurate power injection and for mimicking the electromechanical behavior of traditional synchronous generators.

$$P_{VSG} = K_D \Delta\omega + K_I \frac{d\,\Delta\omega}{dt} \tag{5}$$

$$I_d^* = \frac{2}{3}\left(\frac{\left(v_d P_{VSG} - v_q Q\right)}{v_d^2 + v_q^2}\right) \tag{6}$$

Where

$\Delta\omega$the change in angular frequency
$\frac{d\Delta\omega}{dt}$rate of change of $\Delta\omega$
$K_D$damping constant
K_I inertial constant

The figures 8 (a-b-c) illustrate the control structure and operational rules of a VSG, which is designed to mimic the dynamic behavior of a traditional synchronous generator in power electronic networks. Figure 8-a shows a general control strategy of a VSG, indicating its modular core and major functional modules. The system starts with a supply voltage input linked to a filter circuit consisting an inductor and a capacitor, which prevent the high-frequency switching harmonics produced by the Power Electronics Interface (PEI). The filtered output is then conditioned using a Phase-Locked Loop (PLL) to align the internal reference frame of the VSG with the grid voltage, ensuring accurate phase synchronization.

The control block diagram emphasizes the use of a Proportional-Integral (PI) controller for active and reactive power control, as well as a current controller that enables precise tracking of the desired reference current. These references are derived from the power request and the virtual inertia model, which has a vital role in stabilizing the frequency response of the system. d–q frame transformation is utilized to allow decoupled control of active and reactive power components, making real and reactive power flows to be controlled independently.

Figures 8-b and 8-c provide detailed information about the active and reactive power control loops, respectively. Figure 8-b illustrates the active power loop, in which the RoCoF is used to calculate the inertial response, simulating the mechanical inertia of a physical synchronous machine. A proportional gain is employed to regulate the active power injection, counteracting frequency deviations. The reference active power is obtained by measuring the frequency deviation and fed back into the control loop to ensure system stability.

Figure 8-c shows the reactive power loop, whose voltage phase angle and amplitude control ensure synchronization and power factor management. The reference reactive power is obtained from the difference between the reference and measured voltage magnitudes with an Amplitude Measurement block added to increase accuracy. In this case, the PI controller manages the reactive power output to guarantee voltage stability during load variations.

Figure 8. VSG a) The general control strategy, b) The active power loop, and c) The reactive power loop

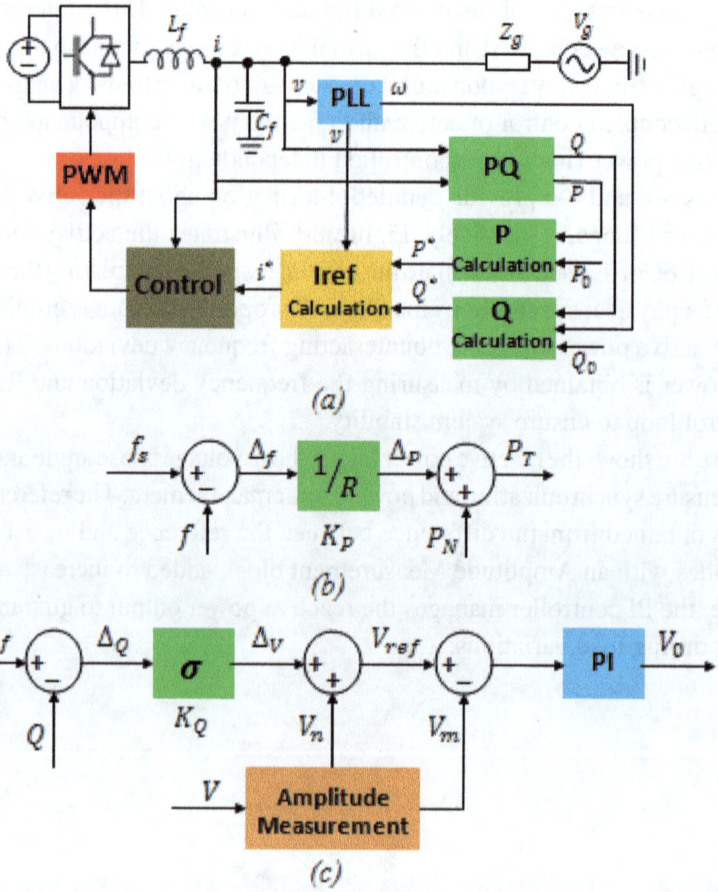

ii. Virtual Synchronous Machine (VSM)

The overall system setup and control structure of a VSM mimic the dynamic performance of an ordinary synchronous generator in power electronic applications. The structure is defined through modular functional components, each of which performs a specific function in achieving precise control of active and reactive power, frequency regulation, and voltage stability.

Central to the system is the Virtual Inertia and Power Control block, which plays a key role in simulating the inertial response of a true synchronous machine. The block receives active power reference and virtual angular velocity inputs through which the VSM can deliver primary frequency response through the simulation of the spinning mass's mechanical inertia. The virtual inertia model enables the system to respond dynamically to frequency deviations, thereby improving grid stability during transient events.

The reactive power control block is employed to regulate the reactive power output, guaranteeing good voltage regulation and the desired power factor. It works in coordination with the Voltage Control block, which generates the voltage reference developed from the grid voltage measurements. The voltage references are then passed through the Current Control block to generate current references and provide accurate tracking of the desired power flow. In this loop, active damping is implemented to suppress oscillations and boost system-damping characteristics, particularly in response to grid disturbances.

The Frequency Droop block is the outer control loop for active power control and frequency assistance. It receives an active power reference, which is combined with a virtual angular velocity reference. The droop controller applies a proportional gain to translate the difference between the VSM's grid frequency and virtual angular velocity into an active power deviation signal (Romdhane et al., 2023). The VSM dynamically responds to frequency deviations by modulating its active power output, thus emulating the inertial response of a real synchronous machine. The droop feature allows the VSM to provide first-response frequency support to help stabilize the grid during transient disturbances (D'Arco et al., 2015).

Figure 9. VSM a) The general control strategy, b) Inertia emulation with power frequency droop

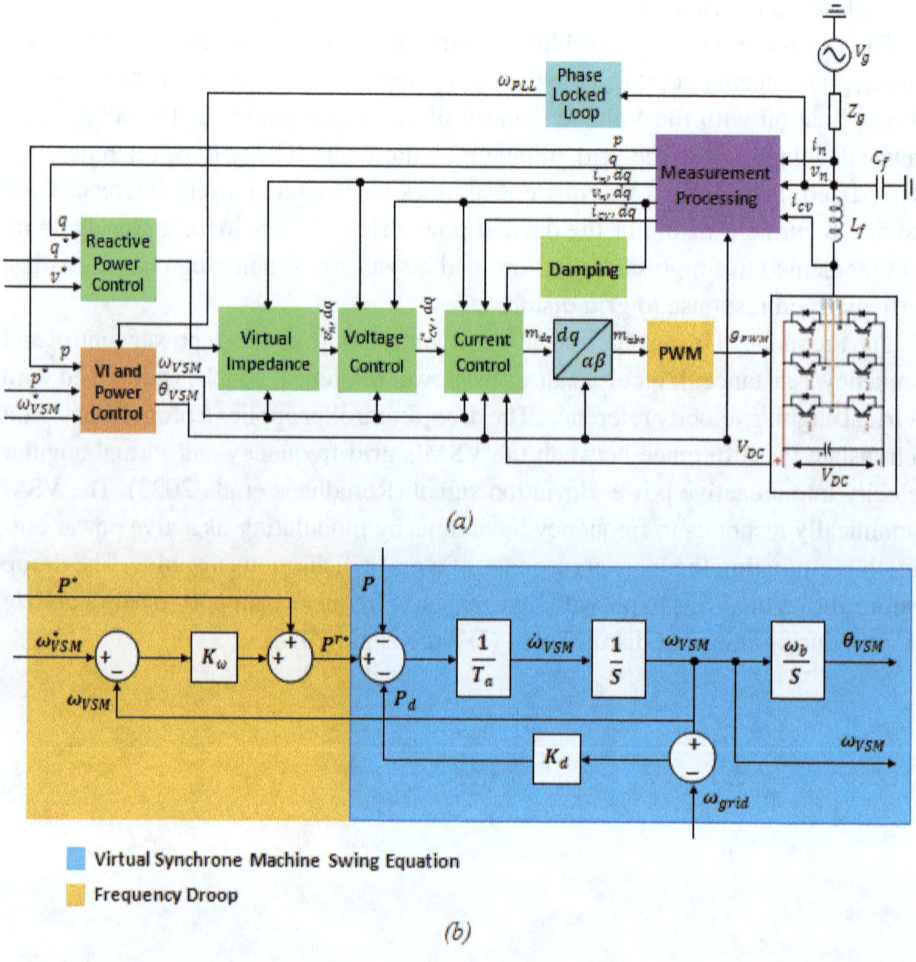

The Virtual Synchronous Machine Swing Equation and frequency Droop are blocks that simulate the mechanical dynamics of a synchronous generator. This block has a first-order model of inertia represented by the time constant, which controls the rate of change of the virtual angular velocity of the VSM in response to active power imbalances. The swing equation used in the program is linearized concerning the rotor angular speed, allowing the calculation of inertial acceleration based on the instantaneous power mismatch in the system (Barać et al., 2021). The linearized representation allows one to have a direct correlation between the mechanical power input deviation and the rate of change in rotor speed that accumulates, represented by the following equation:

$$\frac{d\omega_{VSM}}{dt} = \frac{P^{r*}}{T_a} - \frac{P}{T_a} - \frac{P_d}{T_a} \tag{7}$$

Where

P^{r*} Virtual mechanical input power
P measured electrical power flowing from the VSM into the grid
P_d Damping power
T_a Mechanical time constant

The general operations of a generic VSM are the following:

Step 1: The voltages at point of common coupling (PCC) between the output of VSM and grid are measured to compute the phase currents of the VISMA in real-time.
Step 2: The computed currents are utilized as reference currents for a current-controlled inverter.
Step 3: The pulse width modulation (PWM) modulator drives the inverter.

iii. Synchronverter

A synchronverter is an advanced control structure aimed to bestow grid-inverters with dynamic synchronous machine behavior, including their inertial response and damping torque. Synchronverters emulate the swing equation of a synchronous generator, enabling VSIs to provide synthetic inertia to the power system and thus improve frequency stability in power systems dominated by renewable sources. This control strategy enables self-consequent operation by dynamically controlling the output active and reactive power based on system frequency and voltage operating conditions. Additionally, synchronverters allow seamless mode switching between grid-connected and islanded modes with better transient stability and voltage regulation. Consequently, they are a critical technology for preserving grid resilience and reliability in today low-inertia power grids with decreased rotational mass and increasing variability in generation and load dynamics (Vasudevan et al., 2020).

The feature of power in the synchronverter is the physical interface between the AC grid and the DC source. At its core is a bidirectional synchronverter, which is the central unit for energy conversion. The synchronverter is directly coupled to a direct current (DC) bus, on which a capacitor is used to regulate the DC voltage V_{dc}. This DC bus is powered by a source in the form of a renewable source like a photovoltaic panel or wind turbine, indicated by the current source on the diagram. The VSI is PWM modulated to offer high precision in regulating the output AC voltage V_{abc} and current i_{abc} supplied to the grid. The AC bus of the inverter is directly connected to the grid via an LCL filter, consisting of an inductor L_f and a capacitor that suppresses high-frequency switching harmonics and provides smooth power transfer. The use of a transformer implies the possibility of voltage level transformation between the inverter and the grid for the sake of harmonic-free integration with conventional power systems (Zhong & Weiss, 2010).

The control part of the synchronverter, shown in the lower half of the figure, is designed to emulate the dynamic response of a synchronous generator, hence providing synthetic inertia and frequency support to the grid. Central to this control framework is the application of the swing equation, which governs the interaction between mechanical torque T_m, electrical torque T_e, rotor angle θ_g, and rotor speed ω_g. The mechanical reference torque T_m^* is supplied by a proportional-integral (PI) controller for DC bus voltage V_{dc} control to stabilize the system. The electrical torque is computed using the relationship between the grid voltages and output current of the inverter from equations 8, 9, and 10 in the figure. To simulate the inertial response of a synchronous machine, the system includes a first-order integrator that models the rotor inertia and damping coefficient, enabling the rotor speed to dynamically respond to changes in torque. In addition, the control strategy also comprises droop control mechanisms of active power control and reactive power control such that the active power reference P^* is tied with frequency deviation and the reactive power reference Q^* is tied with voltage deviation. The double-loop control structure allows the synchronverter to control both frequency and voltage independently without any external perturbation, thereby improving the overall grid stability (Yap et al., 2019).

Figure 10. Power and control parts of a synchronverter

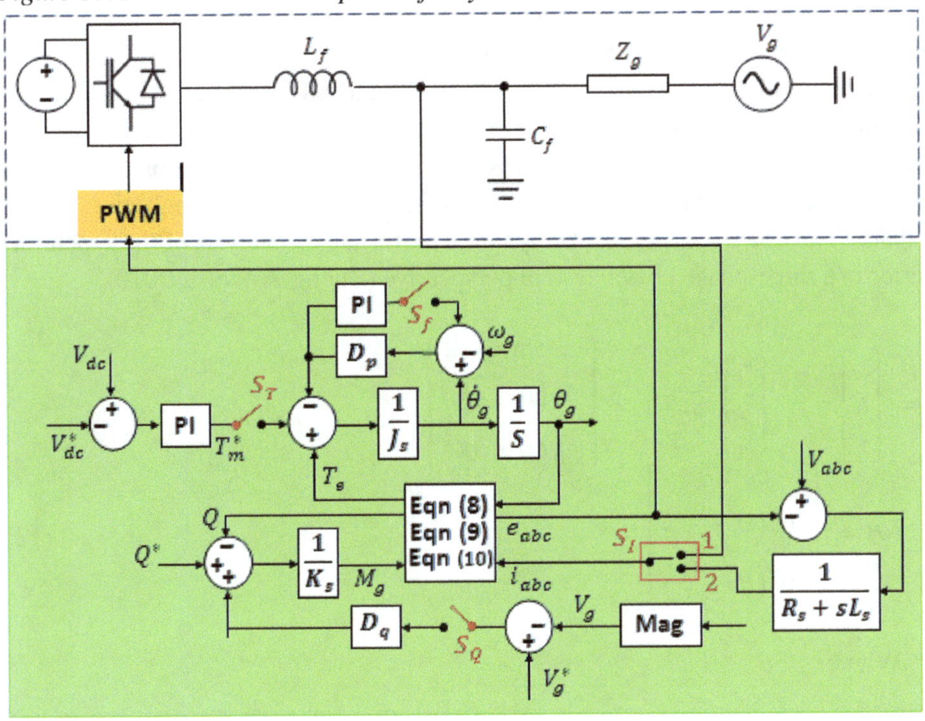

The synchronverter is represented by the swing equations of a synchronous generator (SG) and the dynamic model. Equations (8), (9), and (10) constitute the core of this mathematical model and offer a rigorous representation of SG dynamics. Specifically, Equation (9) determines the electrical torque T_e from the mutual inductance M_f, fundamental current i_f, and rotor angle θ. Equation (8) gives the electromotive force e produced by the SG and is a function of the same variables. Equation (10) gives the reactive power as a function of the rate of change of the rotor angle, mutual inductance, fundamental current, and cosine of the rotor angle $\cos\theta$. Collectively, these equations are the basis upon which the synchronverter inertial response and frequency regulation capability are simulated and tested (Zhong & Weiss, 2010).

$$e = \dot{\theta} M_f i_f \widetilde{\sin}\theta \tag{8}$$

$$T_e = M_f i_f \langle i, \widetilde{\sin}\theta \rangle \tag{9}$$

$$Q = -\dot{\theta} M_f i_f \langle i, \widetilde{\cos}\theta \rangle \tag{10}$$

$$P = \dot{\theta} M_f i_f \langle i, \widetilde{\sin}\theta \rangle \tag{11}$$

Equation (12) presents the three-phase stator flux, voltage, and current vectors, as well as the trigonometric components cosθ and sinθ. The equation presents these values within a mathematical framework considering the spatial distribution of the three-phase system. In particular, the voltage vector and current vector are column matrices with their respective phase components ($v_a \ v_b \ v_c$) and ($i_a \ i_b \ i_c$). Additionally, the cosine and sine terms are matrices that represent phase shifts in the language of the three-phase set-up with the angles shifted in multiples of 2π/3 radians. This is a necessity for the dynamic performance analysis of the synchronverter in a three-phase power system perspective (Zhong & Weiss, 2010).

$$i = \begin{bmatrix} i_a \\ i_b \\ i_c \end{bmatrix} v = \begin{bmatrix} v_a \\ v_b \\ v_c \end{bmatrix} \widetilde{sin}(\theta) = \begin{bmatrix} \sin(\theta) \\ \sin\left(\theta - \frac{2\pi}{3}\right) \\ \sin\left(\theta - \frac{4\pi}{3}\right) \end{bmatrix}$$

$$\widetilde{\cos}(\theta) = \begin{bmatrix} \cos(\theta) \\ \cos\left(\theta - \frac{2\pi}{3}\right) \\ \cos\left(\theta - \frac{4\pi}{3}\right) \end{bmatrix} \tag{12}$$

Where

M_f Mutual-inductance between stator and field winding
T_e Electromagnetic torque
P, Q Active power, reactive power
e Electromotive force (EMF) or synchronverter inner generated voltage

Equations (13) and (14) determine the frequency and voltage droop phenomena, respectively. In this context, Mechanical torque T_m is considered to be the control input, while the electromagnetic torque T_e is determined by the coupling between the current and the rotor angle. Accordingly, the state variables of the synchronverter are the inductor current, capacitor voltage, virtual rotor position, and virtual angular velocity. These variables preserve the dynamic response of the synchronverter, enabling it to mimic the inertial response and frequency regulation capabilities of a traditional synchronous generator (Zhong & Weiss, 2010).

$$\dot{\theta} = \frac{1}{J_s}\left[T_m - T_e + D_p(\dot{\theta}_r - \dot{\theta}_m)\right] \tag{13}$$

$$M_f i_f = \frac{1}{K_s}\left[Q_{SET} - Q + D_q(V_r - V_m)\right] \tag{14}$$

Where

D_p Damping factor for active power loop
D_q Damping factor for reactive power loop
K_s Regulator coefficient of reactive power
Q_{SET} Set point of reactive power
$\dot{\theta}_r$ Reference angular velocity
$\dot{\theta}_m$ Measured feedback angular velocity
V_r Reference r.m.s voltage
V_m Measured feedback grid voltage

b. Power Synchronization Control

Power synchronization control is a fundamental mode in modern power systems, allowing grid-fed inverters to control their output as a function of the local bus voltage, frequency, and phase angle, thus providing stable and coherent power transfer. In virtual inertia systems, the aforementioned control mode performs a critical function in its operation by simulating the native synchronizing torque of synchronous generators during frequency variations. Through dynamic active power injection control relative to grid frequency and phase angle variation, power synchronization control enables inverter-based resources to effectively imitate the inertial response of conventional units. The service supports system stability by damping swift frequency excursions and enabling transient recovery, coupled with virtual inertia control techniques. Power synchronization control thus contributes to improved damping capabilities and frequency regulation, making it a fundamental component for the maintenance of grid stability under low-inertia renewable energy-driven systems (Liu et al., 2022b).

Equation (15) is the mathematical representation of a Power Synchronization Control (PSC) controller, whose function is to control active and reactive power in grid-side inverters. Its control strategy functions by varying the virtual rotor angle and reference voltage magnitude V_m as a function of reference and actual power differences. Particularly, the first of Equation (15) provides the correlation between the virtual rotor position and the active power error $(P^* - P)$. The second equation controls the correction of the reference voltage magnitude with respect to the reactive power error $(Q^* - Q)$, where K_q is the reactive power control gain and V_{S_max} is the maximum value of the voltage reference. Interestingly, this term deliberately omits the damping torque term and instead emphasizes the direct proportionality of power deviations and respective adjustments in the voltage reference and rotor angle. This has the advantage of minimizing the structure of control without compromising synchronization and power control in low-inertia power systems.

$$\begin{cases} \theta = \frac{K_p}{S}(P^* - P) + \omega_I t \\ V_m = \frac{K_q}{S}(Q^* - Q) + V_{S\_max} \end{cases} \quad (15)$$

Where

K_p, K_q Gain coefficients in active and reactive power loops
P^*, Q^* Reference value of P and Q
V_{S_max} Nominal amplitude of the MMC phase voltage
ω_I Fundamental voltage angle generated by the PSC controller
Vm Magnitude of PSC output phase voltage

The diagram shows the power-synchronization control structure of grid-interfaced voltage source converters (VSCs) meant to provide coherent and stable operation in synchronism with the AC grid. The control system is formed by multiple interacting loops with functions to regulate active and reactive power exchange, and grid synchronization (Zhang et al., 2010a; 2010b).

In the center of the control structure is the current-reference control block, which produces reference currents i_d^*, i_q^* from the frequency u_f and phase angle $\Delta\theta_v$ of the grid. The reference currents are obtained from the output of the current control loop, whose function is accurate tracking of the required current components in the $\alpha\beta$ frame. This conversion from the $\alpha\beta$ to the three-phase (*abc*) frame is done via a coordinate transformation block, allowing three-phase voltage references (v_a^* v_b^* v_c^*) to be generated for the control of the PWM modulator of the VSC.

There are two main control feedback loops within the control loop: AC-voltage control and active-power control. For the AC-voltage control loop, the reference AC voltage U^* is compared with the measured grid voltage U_f, and the error signal ΔV is passed through a transfer function $K_{12}(s)$ to produce an input signal to the current-reference control block. The loop prevents the VSC from deviating from correct voltage regulation and grid synchronization.

Active-power control loop is used to control the active power sent to the grid. In this case, reference active power P^* and active power are compared and the resulting error is passed through a transfer function $K_{11}(s)$. The output of the loop is added with the phase angle deviation $\Delta\theta_V$ to control the virtual rotor angle, and consequently active power output of the VSC. This process allows the VSC to respond in a dynamic manner to grid condition changes or load requests in a stable way.

Figure 11. Power synchronization control

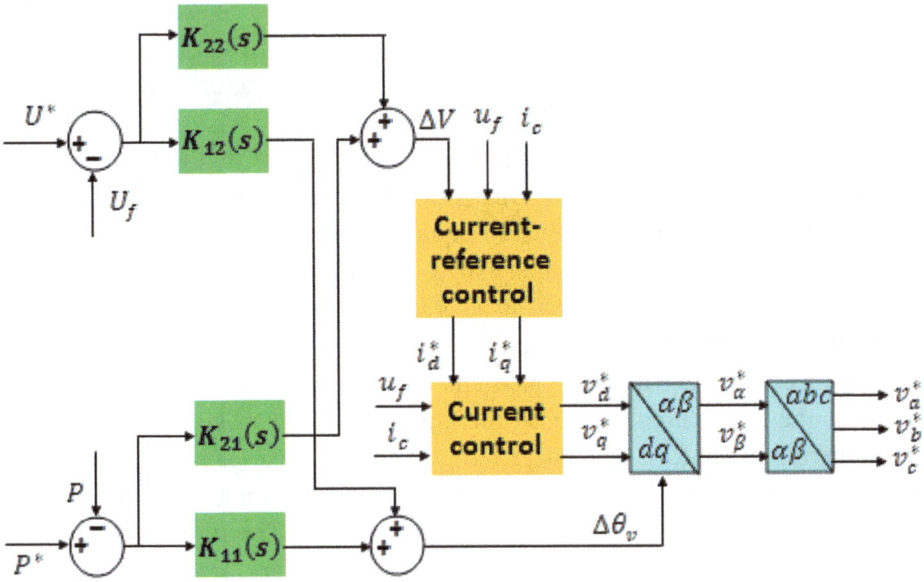

c. Derivative Strategies

Derivative control strategies constitute a high-technology control paradigm for virtual inertia supply in power systems, utilizing differential control action dynamic response properties to improve frequency stability. These strategies are based on the concept of information retrieval from the rate of change of the variables of a system, specifically frequency deviation and its derivatives, to synthesize an appropriate inertial response. By including a derivative term in the control loop, such methods trigger faster correctives during rapid power imbalances, thus enhancing transient performance as well as reducing extreme frequency swings. Virtual inertia emulation-based control through derivatives is able to dampen oscillations more effectively and boost robustness under varying operating conditions. Thus, these methods are important for maintaining the dynamic stability of low-inertia power grids, specifically those with high-penetration renewable energy power grids, where traditional inertia is greatly reduced.

The figure 12 demonstrates the dynamic structure of a derivative technique-based virtual inertia control scheme intended to simulate both virtual damping and inertia for power systems. Central to this scheme is the Virtual Rotor (Inertia Control) block, consisting of the most important elements responsible for simulating inertial response. The system input is frequency deviation *Δf*, which passes through a derivative technique block. This block computes the rate of change of frequency $^{d}/_{dt}$ and provides critical information regarding the transient response of the system. The output of the derivative method is then fed into the Virtual Damping block, which carries out a damping effect in a bid to suppress vibration and stabilize the system response (Kerdphol et al., 2019).

After the virtual damping, the signal is then multiplied by the virtual inertia constant J_{VI}, a fake rotational inertia that simulates the response of a synchronous generator. The result is added to another input, most likely external control signals or reference values, prior to being filtered by a first-order low-pass filter $\left(1 + sT_{ESS}\right)$. This filter simulates the dynamics of an inverter-based energy storage system (ESS) in order to best impose the virtual inertia response with practicality in mind in real-world applications like bandwidth restrictions.

Lastly, the filter output is passed through a limiter to make sure the created virtual inertia power ΔP_{VI} lies within the specified boundaries (P_{ESS_min} P_{ESS_max}). The limiter prevents the injection and absorption of high amounts of power from the ESS, thus ensuring operational reliability and safety. The system design illustrates how derivative methods can be applied to improve grid stability by offering timely and adaptive virtual inertia assistance for low-inertia power grids with high penetration of renewable energy.

Figure 12. The dynamic model of derivative strategy-based virtual inertia control

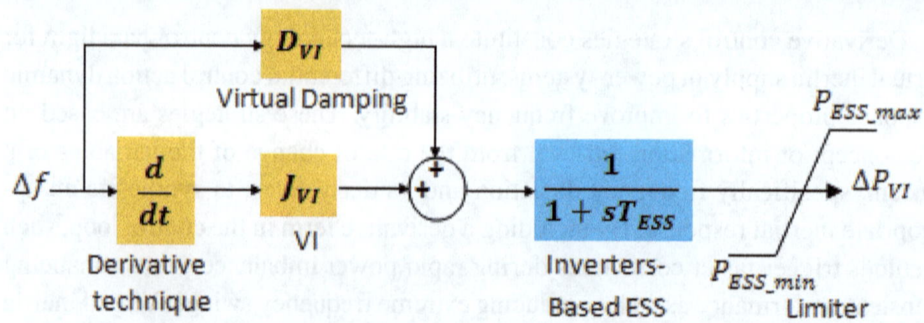

Equation (16) mathematically presents the virtual inertia control principle of power systems in derivatives. It represents the dynamic relationship between frequency deviation Δf and synthesized virtual inertia power ΔP_{VI}, the key component that helps establish grid stability during low-inertia situations. The equation is derived below:

$$\Delta P_{VI} = \frac{J_{VI} + D_{VI}}{1 + sT_{ESS}}(\Delta f) \tag{16}$$

Where

J_{VI}Virtual inertia constant
D_{VI}Virtual damping constant
T_{ESS}Time constant of inverter-based ESS

This equation reveals that the virtual inertia power ΔP_{VI} is directly proportional to the frequency deviation Δf, with the magnitude and phase of the response being a function of the virtual inertia constant J_{VI}, virtual damping constant D_{VI}, and the ESS time constant T_{ESS}. With the inclusion of these parameters, the equation yields a uniform model for generating an adaptive and responsive inertial action, in which power systems can be maintained stable under transient conditions such as sudden load changes or generation imbalance. It is most importantly applied in modern grids with high penetrations of renewable sources of energy, where traditional mechanical inertia is significantly less.

2. Classical Strategies

a. Particle Swarm Optimization

Classical optimization methods have been in vogue for decades to improve the efficiency of virtual inertia control systems in new power electronics-based power systems. One of the most powerful and effective metaheuristic methods is Particle Swarm Optimization (PSO), which has been extensively employed to optimize virtual inertia controller parameters for grid-connected voltage source converters (VSCs). PSO works by iteratively improving a population of potential solutions known as particles that move in the search space through personal and global best experiences. The method has a rapid convergence, it is capable to provide optimal or near-optimal solutions. Additionally, it is well-suited to real-time applications where there is a need for dynamic response and frequency stability. In VIC, PSO is used to adjust control coefficients like damping gains and virtual inertia constants to improve system robustness against abrupt load and disturbance changes. The ability of PSO to solve nonlinear, multi-dimensional optimization problems without relying on gradient information further underlines its suitability in conditions of power systems. In summary, PSO is an effective and well-established computational intelligence method that bridges conventional control theory and advanced optimization paradigms in contemporary energy conversion systems.

The Particle Swarm Optimization is an iterative algorithm that modifies the velocity vector of every particle within the swarm and subsequently modifies its position by adding on the calculated velocity. It is the fine-tuning mechanism that is based on two vital cognitive parameters: the global best solution, (the lowest cost obtained over the entire swarm), and the local best solution (the lowest cost obtained by any individual particle in its search history). The behavior of convergence of the algorithm is decided based on the interaction between the two solutions. If the local best solution is of lower cost compared to the currently available global best solution, it takes the role of being the new global best and thus compels the swarm towards more promising areas in the search space (Maulidhia et al., 2022) (Han et al., 2024).

Swarm Size: The choice of population or herd size for swarm intelligence algorithms is typically problem-dependent. Experimental results indicate that efficient swarm sizes tend to be in the range of 20-50 particles. Historically, it has been found through research that PSO can frequently operate with smaller population sizes than other forms of evolutionary computation in converging to an optimum or near-optimum solution.

Acceleration Coefficients: c1 and c2, acceleration coefficients that control the individual and social cognitive components, respectively, in standard PSO implementations, are typically initialized with classical values of 2.0. However, can be independently varied based on the particular requirements of the optimization problem at hand. Different configurations have experimented with different studies, and the typical values range between 0 and 4. Symmetrical values of c1and c2 are used in an attempt to provide a well-balanced exploitation capability.

Inertia Weight: The PSO algorithm's search process is regulated by three important components: cognitive (personal experience), social (influence from society), and inertial (momentum) terms. These three terms regulate collectively the trade-off between global exploration and local exploitation in the search space. The inertia weight, nonetheless, dominantly regulates the velocity update mechanism and, consequently, influences the convergence behavior and diversification ability of the algorithm.

The optimization of the VIC parameter with PSO flowchart starts with the formation of an objective function that minimizes error. This is the performance measure index for the optimization system. The PSO algorithm is then initialized by specifying its operational parameters, including population size, inertia weight, and acceleration coefficients. A set of particles for which the solution for each particle is virtual inertia (Jv) and virtual damping (Dv) is then generated at random within the possible search space. The algorithm evaluates each particle sequentially based on the constructed objective function and propels the swarm toward the optimal values of the parameters. The optimized Jv and Dv parameters are then saved to Simulink and used in the control system model to facilitte real-time simulation and performance checks.

Figure 13. Flowchart of VIC parameters optimization with PSO

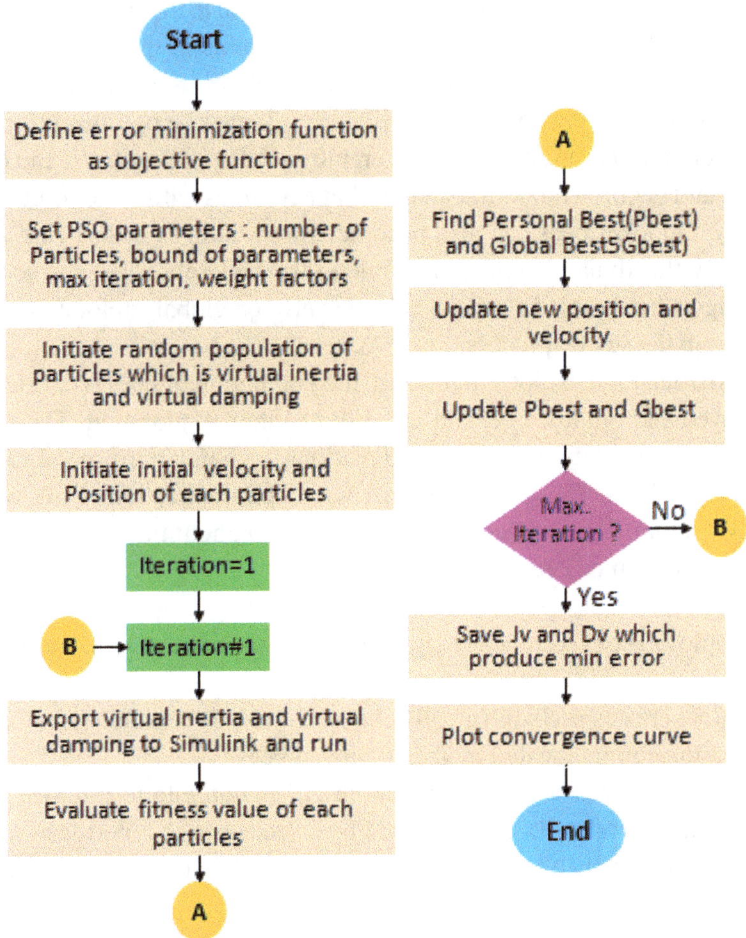

b. Artificial Bee Colony Optimization Algorithm

The Artificial Bee Colony (ABC) algorithm is a bio-inspired computational intelligence algorithm that was formulated by Karaboga in 2005. ABC mimics the foraging behavior of honeybees in a mathematically abstracted environment to solve hard optimization problems. The population of bees in the algorithm involves three types: employed bees, onlooker bees, and scout bees. There exists a task for exploited bees to explore the search space for the first time by producing random candidate solutions represented as nectar positions (NPs). Once a viable NP is found, an employed bee evaluates its quality in terms of fitness value and communicates this information to the rest of the colony via a symbolic "dance" within the hive. The duration or the intensity of the dance is proportional to the fitness (profitability) of the discovered solution. Observer bees that are stable in the hive are those that decode these dances and choose more promising NPs to search probabilistically, favoring more fit solutions. If a single NP can no longer be optimized further in a finite number of iterations, its respective employed bee relinquishes it and becomes a scout bee and begins a random search to find new areas within the solution space. Particularly, onlooker bees can also become employed bees if they find a new or better NP. By this dynamic interaction among working, observer, and scout bees, the ABC algorithm is able to strike a balance between global exploration and local exploitation in the search process.

In the Artificial Bee Colony algorithm, the population of bees is divided evenly into employed and onlooker bees in order to maintain exploration. The source of food, or candidate solutions (or Nectar Positions, or NPs), correspond in number to the employed bees. The algorithm starts with the initialization phase, where the scout bees randomly create the initial solutions throughout the domain. Three key steps are executed in each cycle:

1. The employed bees search the neighborhood of the current NPs and assess their quality.
2. The onlooker bees probabilistically choose good NPs depending on the quality information communicated by the employed bees.
3. When an employed bee's corresponding NP is not able to improve within a limited number of trials, it is replaced by a scout bee that performs a random search to find a new NP, thereby improving the global search capability of the algorithm and avoiding premature convergence.

ABC optimization is presently a promising technique in virtual inertia control of power systems to maximize frequency stability in power grids with a high penetration of renewable energy in the present era. Through adaptive tuning of control parameters of virtual inertia emulators, e.g., droop coefficients and time constants, the ABC algorithm is able to effectively model the exploration and exploitation stage of bees' foraging behavior to locate the optimal solution within a high-dimensional search space. This algorithm provides benefits in terms of convergence rate, stability, and capability for global searches that allow the algorithm to be highly suitable for real-time implementation where changing system conditions require prompt and precise parameter adjustment. Experiments have demonstrated that virtual inertia control algorithms based on ABC can significantly enhance frequency response performance, damp oscillations suppression, and enhance overall system robustness under different loads and generation conditions (Aluko et al., 2020).

Once the objective function is defined, there has to be a systematic sequence of steps executed in order to implement the Artificial Bee Colony (ABC) algorithm into the model. This is a procedural step-by-step framework that ensures the implementation of the algorithm and is graphically explained by way of the flowchart presented in Figure 14.

Figure 14. Flowchart of ABC optimization algorithm

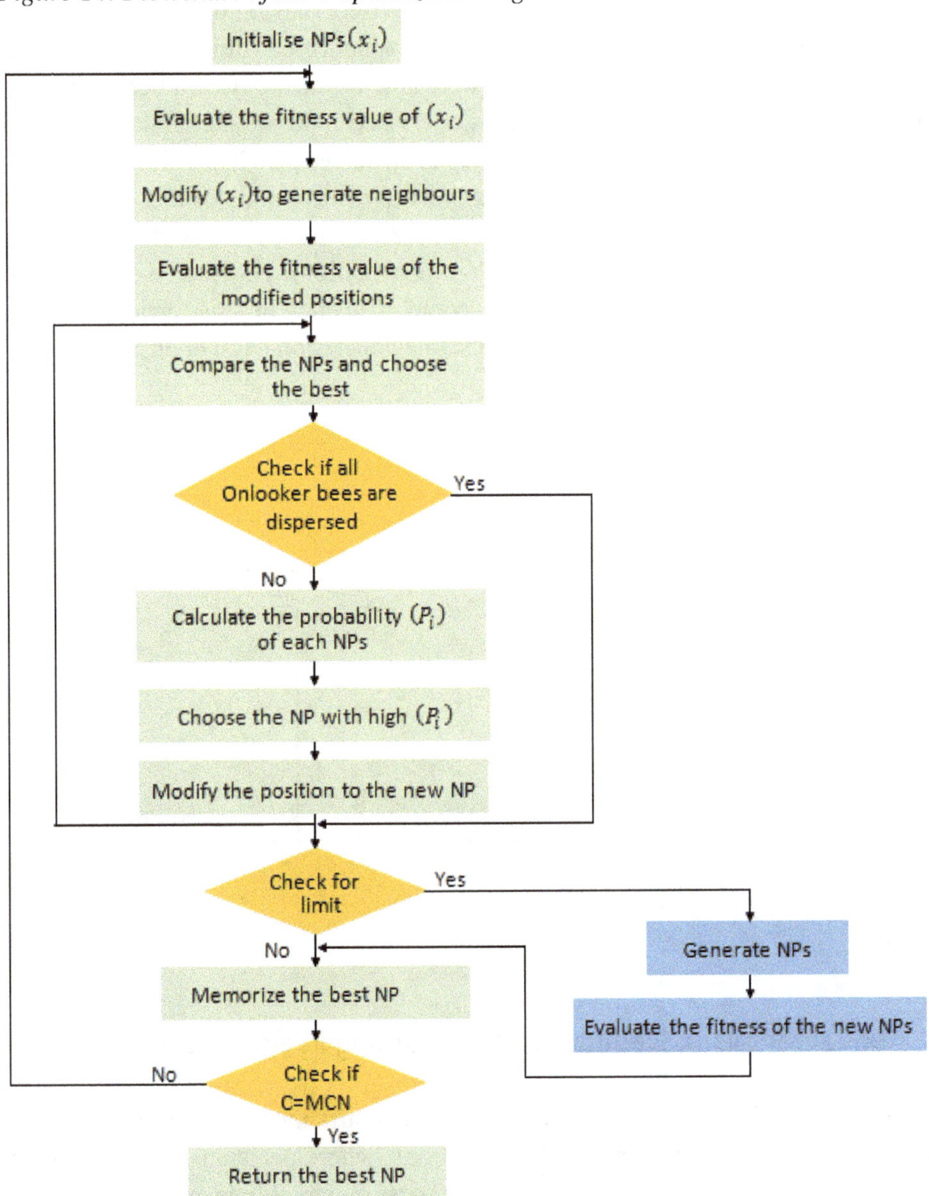

3. Intelligent Strategies

Machine learning (ML) has become a major intelligent approach in the VIC of smart grids, and it is a solid foundation for system stability and dynamic performance (Alqunun et al., 2025). VI, an essential element in new power systems, is utilized to imitate the inertial response of traditional synchronous generators and thus suppressing frequency deviations caused by sudden loads or generator alternations. Machine learning methods, such as supervised, unsupervised, and reinforcement learning provide data-driven and adaptive solutions that support real-time decision-making and optimization. Supervised algorithms are able to model system behavior from historical information, while the use of unsupervised approaches enables patterns to be identified and anomalies to be uncovered in advanced grid dynamics. Reinforcement learning, conversely, allows agents to learn control policies that are optimal using iterative interaction in the environment in a way that virtual inertia controllers can dynamically respond to changing operating conditions. Together, these machine learning paradigms enhance the resilience and efficiency of smart grids and also, facilitate the seamless integration of renewable generation (You et al., 2020).

Figure 15. Machine learning methods

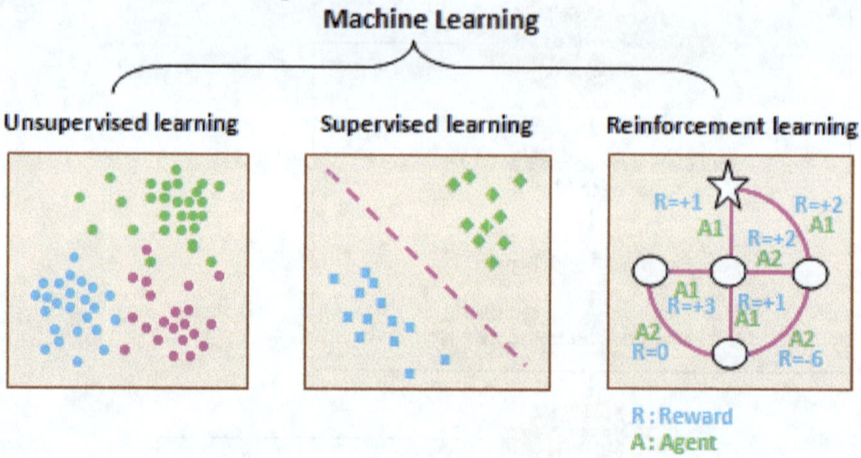

a. Fundamentals of Reinforcement Learning in Smart Grids

Markov Decision Processes (MDPs) are a robust theoretical framework for modeling sequential decision-making situations, where decisions are made in an evolving environment over time. One of the most remarkable features of MDPs is that they can model the temporal dependencies in decision situations, under which action taken in any particular phase may affect not only the immediate or instant state but even future states and rewards. This fundamental property introduces a trade-off between short-time and long-time rewards, as effective policies frequently have the task of balancing immediate gains with future benefits and in certain circumstances it may be beneficial to select an action that appears sub-optimal at the short time because such policies can lead to better overall performance over extended time horizons. This complex interplay between current and future problems highlights the strategic richness and practical relevance of MDPs as an approach to solving intricate decision-making problems (Brown et al., 2024).

Reinforcement learning (RL) is a branch of machine learning that allows an agent to learn optimal decision-making policies through interaction with a changing environment by virtue of controlling reward signals that capture the success of its actions. In smart grid networks, RL has emerged as a powerful technology for ensuring grid stability and reactivity, especially through the utilization of virtual inertia control. In contrast to conventional synchronous generators that normally provide inherent physical inertia, renewable energy resources incorporated into contemporary power systems normally do not possess this natural stabilizing feature and contribute additional frequency instability with loading or generation imbalance oscillations. Virtual inertia control imitates the inertial response using energy storage systems or controllable loads to compensate for the problem, and RL algorithms serve as effective tools to optimize this process. By continuously monitoring and updating of control action based on real-time system behavior, RL can effectively balance energy supply and demand, as well as frequency stability, under varying operating conditions. Such an adaptive aspect of RL makes it extremely attractive for decentralized and dynamic smart grid systems where traditional model-based controllers can be dominated by uncertainty and complexity.

Figure 16. RL a) The basic structure b) The control model

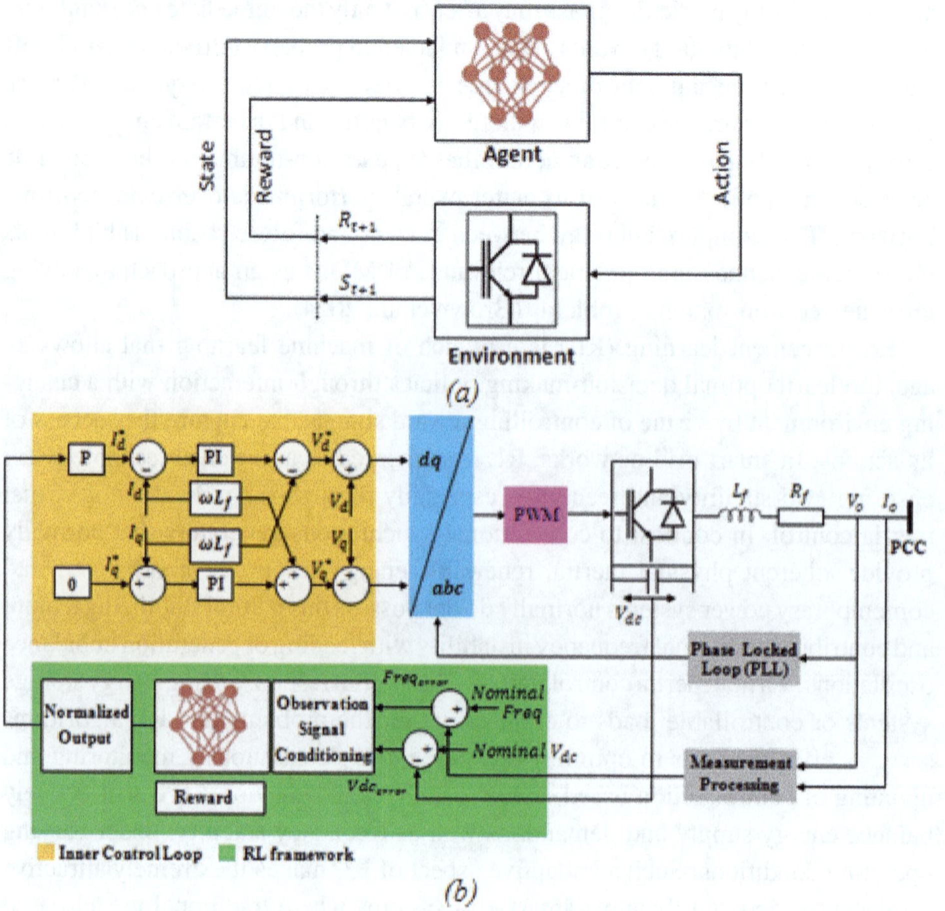

b. Model Predictive Control

Model Predictive Control (MPC) is a advanced control approach that has gained a significant interest in managing and optimizing smart grid systems. Essentially, MPC is focused on a dynamic model of the system for predicting future actions over a defined time horizon to support anticipatory decision-making through optimization of control actions within operating constraints (Lammouchi et al., 2025). In smart grids, in which uncertainty and complexity result from the integration of renewable energy sources, energy storage devices, and smart loads, MPC presents a strong framework for real-time control and resource allocation. By periodically updating forecasts using arrival measurements and at each time step optimizing a constrained finite-horizon optimization problem, MPC guarantees stability and performance under changing grid conditions. Its capacity to handle multivariable systems with direct constraints renders it particularly suitable for applications such as demand-side management, frequency regulation, and voltage management in modern power systems (Kerdphol et al., 2017).

Figure 17. MPC a) The general concept b) The control model

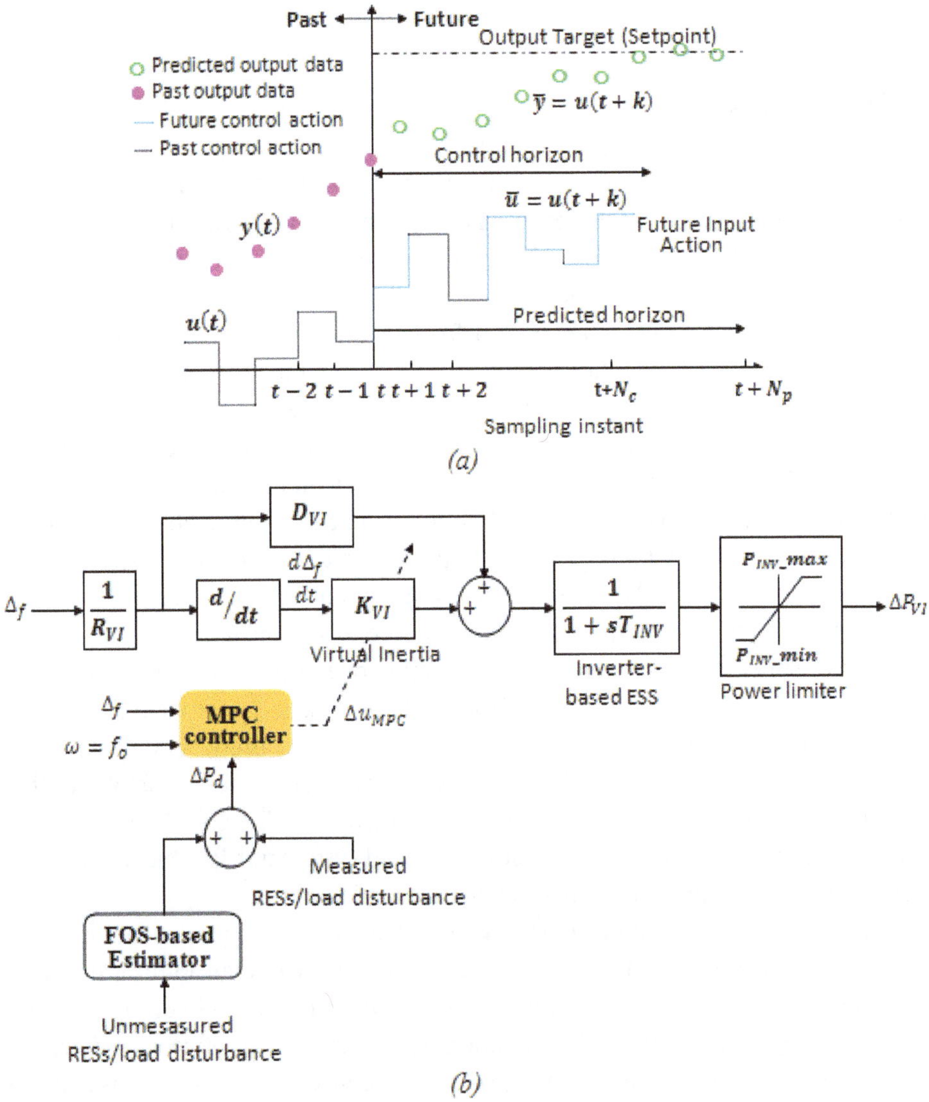

With the aim of improving system stability under high penetration of renewable energy and dynamic loading changes, an MPC method is incorporated into the virtual inertia control loop. This method calculates an optimal virtual inertia block input sequence in real time. Furthermore, the VI model dispatches an appropriate level of synthetic inertia power into the microgrid and allows the frequency deviation to closely trace a zero-reference path under transients. This allows for efficient control of frequency variations within a specified operating range on acceptable frequency deviations and the inertia rate of change of power.

CONCLUSION

The incorporation of renewable energy sources into the grid has materially changed its operating dynamics, lowering the system inertia appreciably and disrupting conventional frequency regulation mechanisms. In this context, virtual inertia control (VIC) has become central in facilitating stability in low-inertia power grids, primarily through the employment of grid-forming inverters that emulate the inertial response conventionally supplied by synchronous generators. In this work, a detailed review of virtual inertia control techniques has been presented, from the traditional methods like Virtual Synchronous Generators (VSG), Synchronverters, and power synchronization control, to sophisticated optimization and machine learning-based methodologies such as Reinforcement Learning and Model Predictive Control.

A critical but frequently neglected component of the effective implementation of such control actions is Human-Machine Interaction (HMI). As control systems become more sophisticated and autonomous, the contribution of human operators to monitoring, supervising, and intervening in control operations becomes correspondingly more significant. Well-designed HMI systems enhance situational awareness, enable timely decision-making, and facilitate the adaptive tuning of control parameters to account for changing grid conditions. Moreover, the integration of intuitive visualization functions, real-time feedback control, and explainable AI components into HMI design promotes operator confidence and system transparency, especially in machine learning algorithm-driven smart control systems.

This chapter emphasizes that the future of virtual inertia control depends not only on developing sophisticated technology but also on how a human abilities interface with automated systems without fatigue. Data-driven and adaptive control schemes need to be supported by robust HMI solutions to bridge the gap between the intricate power system dynamics with the human cognitive capacity. As smart grids become increasingly advanced, a consistent combination of cutting-edge control algorithms and smart user-focused human-machine interface design will be vital to sustain grid resilience, operational performance, and long-term sustainability. Future study should focus on co-design control systems and HMI solutions that are not only technically sound but also cognitively aligned with the needs of grid operators, thereby leading to safer, and more dependable and human-sensitive power systems.

REFERENCES

Abolfazl Hadavi, M. T. (2025). Critical inertia thresholds for frequency stability in renewable Energy-Integrated power systems. *International Journal of Electrical Power & Energy Systems*, *169*, 110733.

Abouyehia, M., Egea-Àlvarez, A., & Ahmed, K. H. (2025). Evaluating inertia estimation methods in low-inertia power systems: A comprehensive review with analytic hierarchy process-based ranking. *Renewable & Sustainable Energy Reviews*, *217*, 115794.

Ahmed Baba, M., Naoui, M., & Cherkaoui, M. (2023). Fault-tolerant control strategy for Hall sensors in BLDC motor drive for electric vehicle applications. *Sustainability*, *15*(13), 10430.

Allal, A., Lammouchi, Z., Khechekhouche, A., Mohamed, N., Alqunun, K., & Al-ghadhban, A. (2024). Advanced Residual Optimal Mapping Approach for Precise Detection of Stator Faults in Induction Motors. *IEEE Access: Practical Innovations, Open Solutions*, *12*, 143515–143530.

Alqunun, K., Bechiri, M. B., Naoui, M., Khechekhouche, A., Marouani, I., Guesmi, T., & Allal, A. (2025). An efficient bearing fault detection strategy based on a hybrid machine learning technique. *Scientific Reports*, *15*(1), 18739. PMID: 40437009

Alsharafi, A. S., Besheer, A. H., & Emara, H. M. (2018). Primary frequency response enhancement for future low inertia power systems using hybrid control technique. *Energies*, *11*(4), 699.

Aluko, A. O., Dorrell, D. G., Pillay Carpanen, R., & Ojo, E. E. (2020). Heuristic optimization of virtual inertia control in grid-connected wind energy conversion systems for frequency support in a restructured environment. *Energies*, *13*(3), 564.

Ashish Shrestha, F. G.-L. (2021). Parametric Sensitivity Analysis of Rotor Angle Stability Indicators. *Energies*, *14*(16), 5023.

Badrudeen, T. U., Nwulu, N. I., & Gbadamosi, S. L. (2024). Low-inertia control of a large-scale renewable energy penetration in power grids: A systematic review with taxonomy and bibliometric analysis. *Energy Strategy Reviews*, *52*, 101337.

Bálint Hartmann, I. V. (2019). Effects of decreasing synchronous inertia on power system dynamics—Overview of recent experiences and marketisation of services. *International Transactions on Electrical Energy Systems*, *29*(04).

Barać, B., Krpan, M., Capuder, T., & Kuzle, I. (2021). Modeling and initialization of a virtual synchronous machine for power system fundamental frequency simulations. *IEEE Access: Practical Innovations, Open Solutions*, *9*, 160116–160134.

Baškarad, T., Holjevac, N., & Kuzle, I. (2024). A new perspective on frequency control in conventional and future interconnected power systems. *International Journal of Electrical Power & Energy Systems*, *156*, 109731.

Baškarad, T., Kuzle, I., Đaković, J., & Ilak, P. (2018, November). Analysis of the wind generation impact on inertial and primary frequency response of the croatian electric power system. In *Mediterranean Conference on Power Generation, Transmission, Distribution and Energy Conversion (MEDPOWER 2018)* (p. 93). Stevenage UK: IET.

Brezovec, M., Kuzle, I., & Krpan, M. (2015). Detailed mathematical and simulation model of a synchronous generator. *Journal of Energy: Energija*, *64*(1-4), 102–129.

Brown, S., Sinha, S., & Schaefer, A. J. (2024). Markov decision process design: A framework for integrating strategic and operational decisions. *Operations Research Letters*, *54*, 107090. PMID: 38560724

D'Arco, S., Suul, J. A., & Fosso, O. B. (2015). A Virtual Synchronous Machine implementation for distributed control of power converters in SmartGrids. *Electric Power Systems Research*, *122*, 180–197.

Deng, H., Wu, J., Luo, J., Yan, R., Zhang, C., Li, P., & Zheng, R. (2021). Research on a Power Grid Cascading Failure Prevention and Control Method considering WSN. *Wireless Communications and Mobile Computing*, *2021*(1), 9439977.

Deng, X., Mo, R., Wang, P., Chen, J., Nan, D., & Liu, M. (2023). Review of RoCoF estimation techniques for low-inertia power systems. *Energies*, *16*(9), 3708.

Elenga Baningobera, B., Oleinikova, I., Uhlen, K., & Pokhrel, B. R. (2024). Challenges and solutions in low-inertia power systems with high wind penetration. *IET Generation, Transmission & Distribution*, *18*(24), 4221–4244.

Gu, Y., & Green, T. C. (2022). Power system stability with a high penetration of inverter-based resources. *Proceedings of the IEEE*, *111*(7), 832–853.

Han, Y., Lin, M., Li, N., Qi, Q., Li, J., & Liu, Q. (2024). DCWPSO: Particle swarm optimization with dynamic inertia weight updating and enhanced learning strategies. *PeerJ. Computer Science*, *10*, e2253. PMID: 39314689

Hongchun, SHU, H. D. (2025). Estimation of inertia demand in power systems considering wind power virtual inertia. *International Journal of Electrical Power & Energy Systems*, *167*, 110597.

Jafari, M., Gharehpetian, G. B., & Anvari-Moghaddam, A. (2024). On the role of virtual inertia units in modern power systems: A review of control strategies, applications and recent developments. *International Journal of Electrical Power & Energy Systems*, *159*, 110067.

Johnson, S. C., Rhodes, J. D., & Webber, M. E. (2020). Understanding the impact of non-synchronous wind and solar generation on grid stability and identifying mitigation pathways. *Applied Energy*, *262*, 114492.

Kerdphol, T., Rahman, F. S., Mitani, Y., Hongesombut, K., & Küfeoğlu, S. (2017). Virtual inertia control-based model predictive control for microgrid frequency stabilization considering high renewable energy integration. *Sustainability*, *9*(5), 773.

Kerdphol, T., Rahman, F. S., Phunpeng, V., Watanabe, M., & Mitani, Y. (2018). Demonstration of Virtual Inertia Emulation Using Energy Storage Systems to Support Community-Based High Renewable Energy Penetration. *IEEE Global Humanitarian Technology Conference (GHTC)*, (pp. 1-7). San Jose, CA, USA.

Kerdphol, T., Rahman, F. S., Watanabe, M., Mitani, Y., Turschner, D., & Beck, H. P. (2019). Enhanced virtual inertia control based on derivative technique to emulate simultaneous inertia and damping properties for microgrid frequency regulation. *IEEE Access: Practical Innovations, Open Solutions*, *7*, 14422–14433.

Khalid, M. (2024). Smart grids and renewable energy systems: Perspectives and grid integration challenges. *Energy Strategy Reviews*, *51*, 101299.

Lammouchi, Z., Naoui, M., Bekakra, Y., Khechekhouche, A., & Sbita, L. (2025). Model-free predictive control based on the execution time of active vectors using ultra-local model. *Arabian Journal for Science and Engineering*, *50*(8), 5885–5898.

Liu, D., Zhang, X., & Chi, K. T. (2022a). Effects of high level of penetration of renewable energy sources on cascading failure of modern power systems. *IEEE Journal on Emerging and Selected Topics in Circuits and Systems*, *12*(1), 98–106.

Liu, W., Teodorescu, R., Kerekes, T., & Dragicevic, T. (2022b, October). Advanced power synchronization control of modular multilevel converter in stiff grid. In *IECON 2022–48th Annual Conference of the IEEE Industrial Electronics Society* (pp. 1-6). IEEE.

Mathew, R., & Panikkar, P. P. K. (2024). Inertial issues in renewable energy integrated systems and virtual inertia techniques. [IJPEDS]. *International Journal of Power Electronics and Drive Systems, 15*(1), 466.

Maulidhia, A. N. A., Asfani, D. A., Priyadi, A., & Setiadi, H. (2022, November). Frequency stability analysis on optimization of virtual inertia control (VIC) capacitor energy storage (CES) controller settings using particle swarm optimization. In *2022 international conference on Computer engineering, network, and intelligent Multimedia (CENIM)* (pp. 359-364). IEEE.

Md Tonmoy Hossain, M. Z. (2024). Next generation power inverter for grid resilience: Technology review. *heliyon, 10*(21), e39596.

Rajkumar, V. S., Ştefanov, A., Presekal, A., Palensky, P., & Torres, J. L. R. (2023). Cyber attacks on power grids: Causes and propagation of cascading failures. *IEEE Access: Practical Innovations, Open Solutions, 11*, 103154–103176.

Romdhane, M., Naoui, M., & Mansouri, A. (2023). PMSM Inter-Turn Short Circuit Fault Detection Using the Fuzzy-Extended Kalman Filter in Electric Vehicles. *Electronics (Basel), 12*(18), 3758.

Shobug, M. A., Chowdhury, N. A., Hossain, M. A., Sanjari, M. J., Lu, J., & Yang, F. (2024). Virtual inertia control for power electronics-integrated power systems: Challenges and prospects. *Energies, 17*(11), 2737.

Ulbig, A., Borsche, T. S., & Andersson, G. (2014). Impact of low rotational inertia on power system stability and operation. *IFAC Proceedings Volumes, 47*(3), 7290-7297.

Vasudevan, K. R., Ramachandaramurthy, V. K., Babu, T. S., & Pouryekta, A. (2020). Synchronverter: A comprehensive review of modifications, stability assessment, applications and future perspectives. *IEEE Access: Practical Innovations, Open Solutions, 8*, 131565–131589.

Yap, K. Y., Sarimuthu, C. R., & Lim, J. M. Y. (2019). Virtual inertia-based inverters for mitigating frequency instability in grid-connected renewable energy system: A review. *Applied Sciences (Basel, Switzerland), 9*(24), 5300.

You, S., Zhao, Y., Mandich, M., Cui, Y., Li, H., Xiao, H., & Zhang, Y. (2020, November). A review on artificial intelligence for grid stability assessment. In *2020 IEEE International Conference on Communications, Control, and Computing Technologies for Smart Grids (SmartGridComm)* (pp. 1-6). IEEE.

Youssef, R., Hamida, M. A., Mohamed, N., Lassaad, S., Djamel, T., & Yasmine, A. (2023). Single-stage standalone photovoltaic water pumping system using predictive torque control (PTC) of induction machine. *Electric Power Components and Systems*, •••, 1–17.

Zhang, L., Harnefors, L., & Nee, H. P. (2010a). Modeling and control of VSC-HVDC links connected to island systems. *IEEE Transactions on Power Systems*, *26*(2), 783–793.

Zhang, L., Harnefors, L., & Nee, H. P. (2010b). Interconnection of two very weak AC systems by VSC-HVDC links using power-synchronization control. *IEEE Transactions on Power Systems*, *26*(1), 344–355.

Zhong, Q. C., & Weiss, G. (2010). Synchronverters: Inverters that mimic synchronous generators. *IEEE Transactions on Industrial Electronics*, *58*(4), 1259–1267.

Zhou, B., Meng, L., Yang, D., Ma, Z., & Xu, G. (2019). A Novel VSG-Based Accurate Voltage Control and Reactive Power Sharing Method for Islanded Microgrids. *Sustainability*, *11*(23), 6666.

Chapter 8
Modeling and Simulation of Triple Junction Solar Cell-Based Photovoltaic Water Pumping:
Modeling and Simulation of Photovoltaic Water Pumping

Beni Hamed Salah
ENSIT, Tunisia

Ben hamed Mouna
https://orcid.org/0000-0002-8331-3728
ISSIG, Tunisia

ABSTRACT

In this work, a high-efficiency photovoltaic water pumping system is conceived, modelled, implemented and validated in Matlab/Simulink environment. The system is made up of a triple junction photovoltaic generator, a centrifugal pumping system, and a three-phase permanent magnet synchronous motor (PMSM) supplied with a three-phase inverter chosen as a pump driver. As a motor drive decoupling strategy, the field oriented control approach is adopted. Three control loops make up the traditional vector control strategy: a speed loop control, two loops for the stator current components of the PMSM. In most cases, the reference quadrature stator flux is obtained from the speed controller. However, the reference direct stator flux is typically set to zero values in many applications. However, issues arose if the application required the high-speed mode. Consequently, a vector control technique and a field weakening region algorithm are developed in this study. The

DOI: 10.4018/979-8-3373-5621-1.ch008

designed system and all control algorithms are implemented in Matlab/Simulink environment. At various speeds, including the two working modes at standard test conditions, significant results are presented and analyzed.

INTRODUCTION

All around the world, there is a critical need for clean drinking water. The world's energy future is about to alter as a result of the energy storm that has been caused by rising energy resource prices, declining oil reserves, and growing environmental concerns. In fact, the overuse and careless exploitation of electricity is the fundamental foundation for industrial development and the enhancement of life. On the other side, there are currently 7 billion people living on the planet, and although population growth is slowing down, 10 billion people are predicted to live there by 2050 (Foster, Ghassemi, & Cota, 2009). Non-renewable energy resources including coal, natural gas, oil, and uranium are used to provide a sizable portion of the world's electricity (Muhsen, Khatib & Nagi, 2017). The main issue is the scarcity and non-renewability of these energy resources, which can take millions of years to replenish (Li, Jin, Akram & Chen X, 2017). However, excessive use of these resources has negative effects on the ecosystem, including CO2 emissions, global warming, air pollution, deforestation, and general environmental degradation. Indeed, between 1973 and 2015, the amount of primary energy consumed globally increased from 6101Mtoe to 13.647Mtoe, causing a steady rise in CO2 emissions from 15.458Mt to 32.294Mt and a 0.6C increase in average global temperature compared to the period between 1900 and 2010. Our globe currently faces formidable obstacles that put the world's ability to meet energy requirements and address environmental issues in danger. As a result, communities will also need to substitute additional renewable, accessible, and environmentally beneficial resources for these ones (Vafaeipour, Zolfani, Varzandeh, Derakhti & Eshkalag, 2014). Only energy from the sun currently meets these criteria of both abundance on the surface of the earth and infinite regeneration. Mono-junction solar cells are the first ones that had been used (Miguel Angel, 2021). As less sunlight is converted into electric energy, the conversion efficiency stays low. The multi-junction solar cell is subsequently created as a remedy for this problem (Aleksandra, Monika & Ireneus, 2021). The converted solar cell is composed of several semiconductors with different wavelengths, which allows for the conversion of a considerable part of sunlight to energy (Aleksandra, Monika & Ireneus, 2023). It therefore outperforms mono-junction solar cells (Shafidah, Mohd, Lin & Kyungkon, 2025). The fundamental power conversion component of a photovoltaic system is the PV module (Anika, Syed, Mohammad, Kaiser & Abu, 2025). To meet the need for electric energy, photovoltaic modules should be installed in parallel or serial arrangements (Marwa, Doaa, Aref, Belal, Mahmoud & Abou, 2025). A PV generator is then obtained. To supply drinking water in urban and rural regions as well as for agriculture, the majority of developing nations rely heavily on rainfall. However, its periodic rains frequently fall short of meeting the country's expanding water needs, which has a negative impact on agriculture yields. Communities used water pumps to suit their demands and lessen reliance on rain. Nowadays, people rely primarily on rain and suffer if there isn't enough water. To achieve their water needs, the majority of nations mostly rely on seasonal rains. One of the most potential uses of solar energy, given the lack of electricity in rural regions and the high costs of delivery, is photovoltaic pumping. A degree of autonomy is also provided by these systems to numerous remote settlements without access to the energy grid. Additionally, diesel pumps typically cost 2 to 4 times more to operate, maintain, and replace than solar pumps (Yang, Linen, Xiaoshun, Hongchun, Tao, Haofei, Lin & Liming, 2019). For this, some nations have implemented solar pumping systems, mostly for irrigation and the delivery of drinking water.

The structure of this chapter is as follows. Modeling of the triple junction photovoltaic generator is shown in the second section. In the third section, the modeling of the Moto-pump system is presented. The fourth section presents and discusses the simulation's findings. The conclusion and some possible futuristic viewpoints are presented in section four.

MODELLING OF THE TRIPLE JUNCTION PHOTOVOLTAIC GENERATOR

The triple-junction photovoltaic solar cell is composed of three sub-cells connected in series, as shown in Fig. 1 (Hamed et al., 2023). Each sub-cell is consisted of a light source, a diode, a parallel-connected resistor, and a series-connected resistance (Anika, Syed, Mohammad, Kaiser & Abu, 2025). The electricity flow is denoted by (1).

$$I_{pv} = I_{phj} - I_{Dj} - I_{Rshj} \tag{1}$$

In which I_{phj} represents the light generated current, I_{Dj} denotes the currents intensity across the diode and I_{Rshj} shows the current intensity moving into the shunt resistor.

For the top sub-cell j=1, j=2 for the medium sub-cell and j=3 for the bottom sub-cell.

When incident light is absorbed, a stream of liberated electrons starts to flow, resulting in an electrical flow. This is referred to as light-generated current. According to equation (2), it depends on how much sunlight is absorbed.

$$I_{cj} = \frac{G_s}{G_{s\,STC}}\left[I_{ccj} + \gamma\left(T_{pcj} - T_{pvSTC}\right)\right] \quad (2)$$

Where T_{pvSTC} is the temperature of a solar cell under typical test conditions in degrees Celsius., The actual short circuit current's temperature coefficient, measured in A/°C, is γ, The solar radiation and the solar radiation under standard test conditions, expressed in W/m², are, respectively,.$G_s$and$G_{s\,STC}$.

Figure 1. Electrical equivalent circuit of multi-junction InGap/InGaA/Ge photovoltaic solar cell (Hamed et al., 2023).

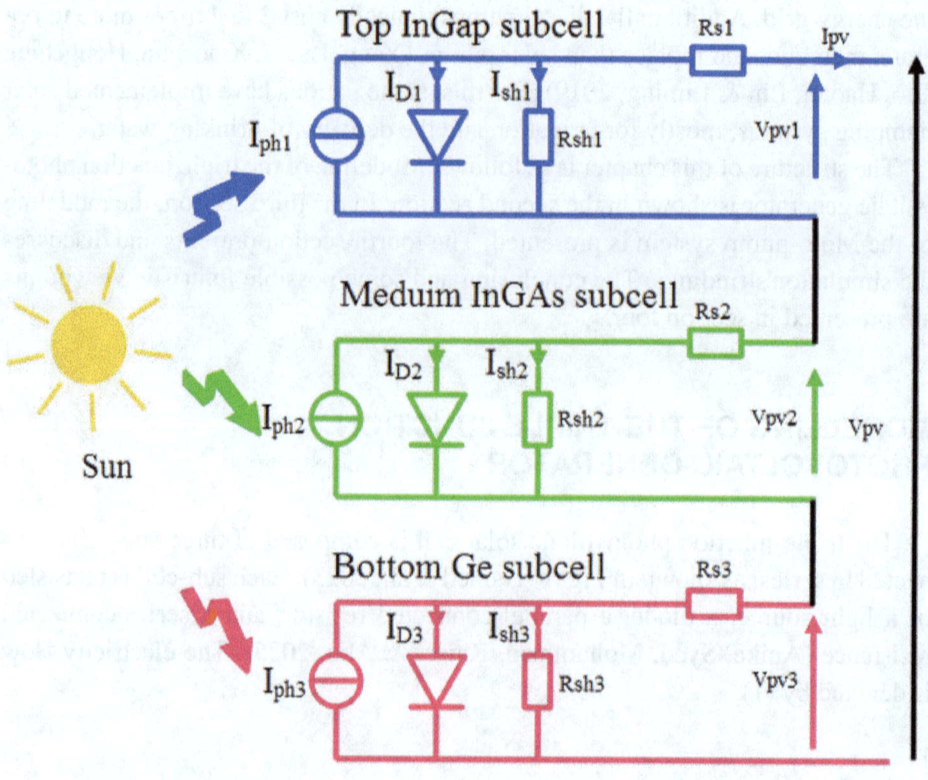

The current intensity expression of the diode is given as (3).

$$I_{dj} = I_{satj}\left[\exp\left(\frac{q\,V_{DJ}}{n_j C_B T_{pcj}}\right) - 1\right] \quad (3)$$

The mathematical behavior of the electrical flow caused by the thermal creation of electron-hole pairs inside the diode's depletion zone and the diffusion current caused by minority carriers flowing across the depletion region is described by Equation (4) (Desideri, Francesc, Alicia, Jérôme, Montse & Joan, 2025).

$$I_{satj} = K_j T_{pcj}(\frac{3+\delta_j}{2})\exp(\frac{-E_{gj}}{n_j C_B T_{pcj}}) \tag{4}$$

Where q is the electron charge, n_j is the diode ideality factor, C_B is the Boltzmann's constant, E_{gj} is the band-gap energy, K_j and δ_j are constant and T is the absolute temperature.

Equation (5) illustrates the entire electric potential of the triple-junction solar cell.

$$V_{pv} = \sum_{i=1}^{3}\frac{n_i BT}{q} Ln(\frac{I_{phi} - I_{pvg} - I_{Rshi}}{I_{sati}} + 1) - R_s I_{pv} \tag{5}$$

MODELLING OF THE MOTO-PUMP SYSTEM

The electrical behavior of the PMSM in the the (d - q) reference frame, may be represented by a nonlinear differential equation as it is shown with (6) (Salah, Franck & Alain, 2007).

$$\begin{cases} U_{fd} = R_{sf} \times i_{fd} + L_{fd} \times \frac{d i_{fd}}{dt} - \omega_f \times L_{fq} \times i_{fq} \\ U_{fq} = R_{sf} \times i_{fq} + L_{fq} \times \frac{d i_{fq}}{dt} + \omega_f \times L_{fd} \times i_{fd} + \omega_f \times \lambda_{ap} \end{cases} \tag{6}$$

In which, the direct and quadratic stator voltage and current components are represented respectively withU_{fd}, U_{fq}andi_{fd}, i_{fq} represent respectively. The stator resistance is designed withR_{sf}. The d and q axis self-stator inductances are represented withL_{fd} andL_{fq}, synchronous speed is given withω_f and λ_{ap}designs the permanent magnet flux linkage.

The motor's mechanical response to applied forces is indicated with equation (7).

$$J_{st}\frac{d\omega_f}{dt} = n_{pp}\left(T_{mot} - T_{res}\right) - f\omega_f \tag{7}$$

In whichT_{mot},J_{st}, $f$and n_{pp} are the electromagnetic torque, the total moment of inertia, the viscous friction coefficient and the number of pair poles respectively.

The load torque is applied with the centrifugal pump. It varies in accordance with the rotational PMSM shaft speed. It is defined with (8).

$$T_{res} = K \times \omega_f^2 \tag{8}$$

Which K is the um constant.

The electromagnetic torque is expressed with (9).

$$T_{mot} = \frac{3}{2} \times n_{pp} \times \left(\lambda_{ap} \times i_{fq} + (L_{fd} - L_{fq}) \times i_{fq} \times i_{fd}\right) \tag{9}$$

Based on the vector control strategy, equation (9) begins

$$T_{mot} = \frac{3}{2} \times n_p \times \phi_a \times i_{qs} \tag{10}$$

For fixed flux value, the PMSM drive behavior is equivalent to that of the separately excited DC motor (Hamed et al., 2023). The field-oriented control approach integrates three control loops: one outer loop for speed control and two inner loops for current components control. The target direct component of the current is a value derived from the field weakening algorithm as it is detailed in (Flah & Sbita, 2013; Hamed et al., 2023) and the target quadratic component value is the output of the speed controller. In the three loops, the proportional and integral controller is used.

THREE PHASE VOLTAGE INVERTER MATHEMATICAL DYNAMIC MODEL

The Moto-pump is connected the triple junction photovoltaic generator by using a three phase voltage inverter. It makes it possible to convert alternating three-phase symmetric output voltage of a desired magnitude and frequency from direct current voltage measured at the output of the triple junction photovoltaic generator (Mohamed et al. 2024). The inverter's output voltages are generated according to the three pulse width modulation control signals. The three phase inverter output voltages feed the PMSM coupled to the centrifugal pump. These output voltages can be expressed as indicated in equation (11) (Shafqat & Taosif, 2025).

$$\begin{bmatrix} U_{as} \\ U_{as} \\ U_{as} \end{bmatrix} = \frac{U_{pvg}}{3} \begin{bmatrix} 2 & -1 & -1 \\ -1 & 2 & -1 \\ -1 & -1 & 2 \end{bmatrix} \begin{bmatrix} S_a \\ S_b \\ S_c \end{bmatrix} \tag{11}$$

WhereS_a, S_b and $S_c$are the control signals of the three-phase inverter's semi-conductors.

SIMULATION RESULTS

A computer simulation built on the Matlab/Simulink software suite was used to simulate the designed high efficiency photovoltaic pumping system. The simulation is carried out in such a way to study the system behavior in a wide speed range including the field weakening region. Indeed, a suitable speed trajectory is selected for pumping system as it is illustrated with Fig. 2. The chosen target shaft speed Started from zero to a speed less than a rated one and then exceeds the nominal speed value. The global control scheme based on the vector control strategy is composed of four essential blocs. The first is the field weakening scheme used to generate the adaptive reference stator flux. The three other blocs are used for shaft speed and stator current components control. A proportional and integral controllers are used to shaft speed and stator current components control loops. The PI controller's parameters are obtained by trial and test error. The obtained simulation results at standard test conditions are displayed in Fig. 2 to Fig. 11. Simulation results are testified under load appliance. The evolution of the load torque applied with the centrifugal pump is presented in Fig. 3. In Fig. 4 and 5, the simulation results of the reference stator current components are presented. In this test, both the field weakening operation and load torque disturbance are considered. That is the direct stator current component is required to reach a zero value when the speed is less than its rated value and the flux is weakened with a target negative value of the direct stator current component. The triple junction photovoltaic generator electric curves at direct connected operating mode are given in Fig. 8 to Fig. 11. Fluctuations are observed in voltage, power and current although constant metrological conditions. As it is indicated in Fig. 11, the operating point is far away from its optimal one.

Figure 2. Evolution of PMSM speeds responses.

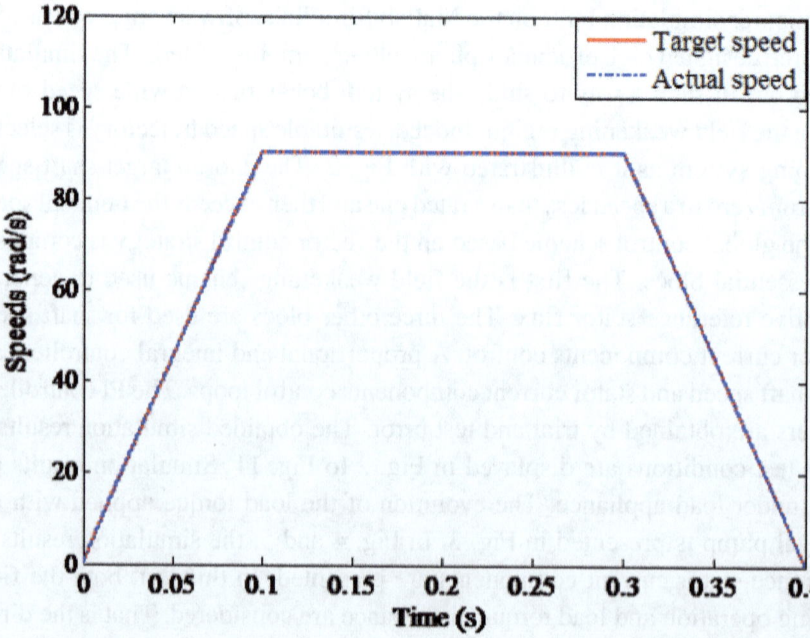

Figure 3. Evolution of the load torque response.

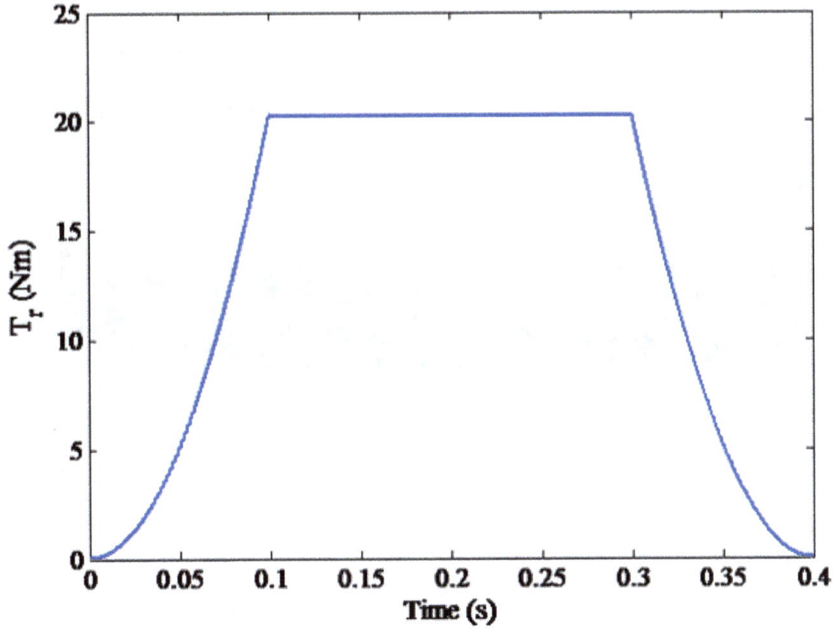

Figure 4. Evolution of the actual and target direct components of the PMSM stator current responses.

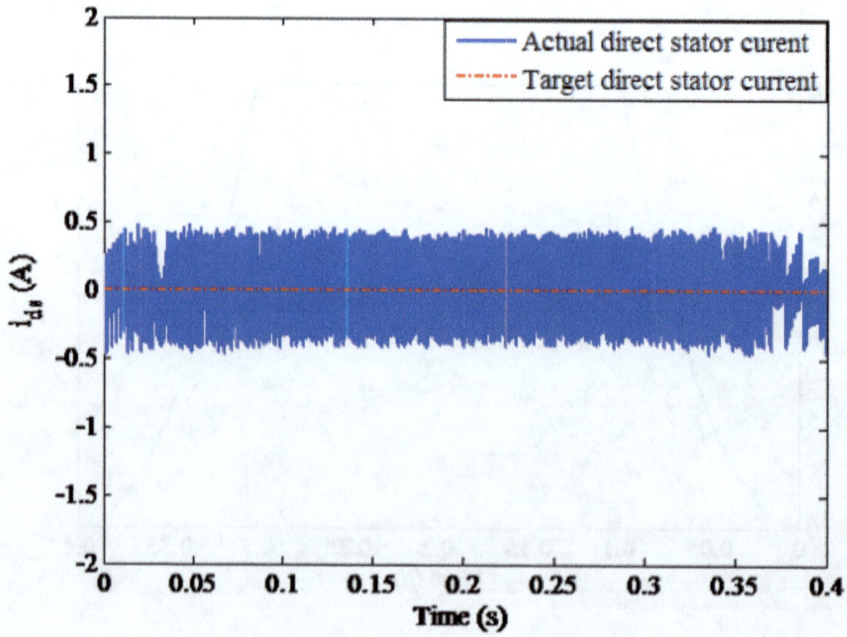

Figure 5. Evolution of the actual and target reverse components of the PMSM stator current response.

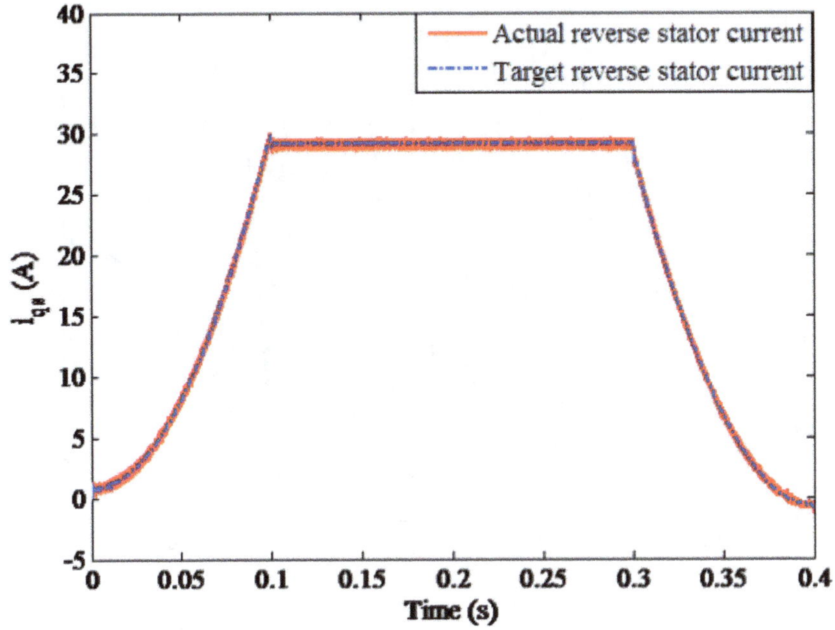

Figure 6. Evolution of the PMSM stator winding's currents responses.

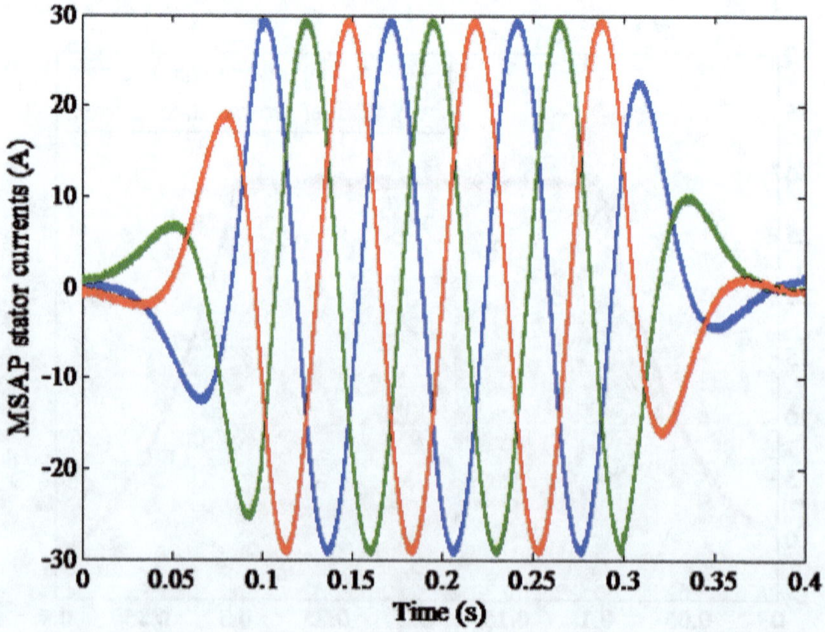

Figure 7. Evolution of the electromagnetic torque response.

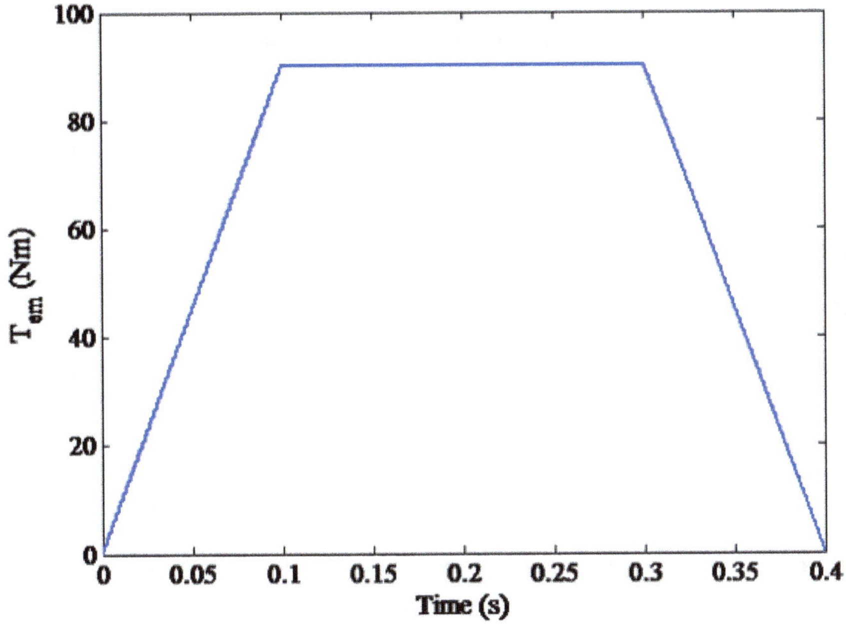

Figure 8. Evolution of the triple junction photovoltaic generator current response.

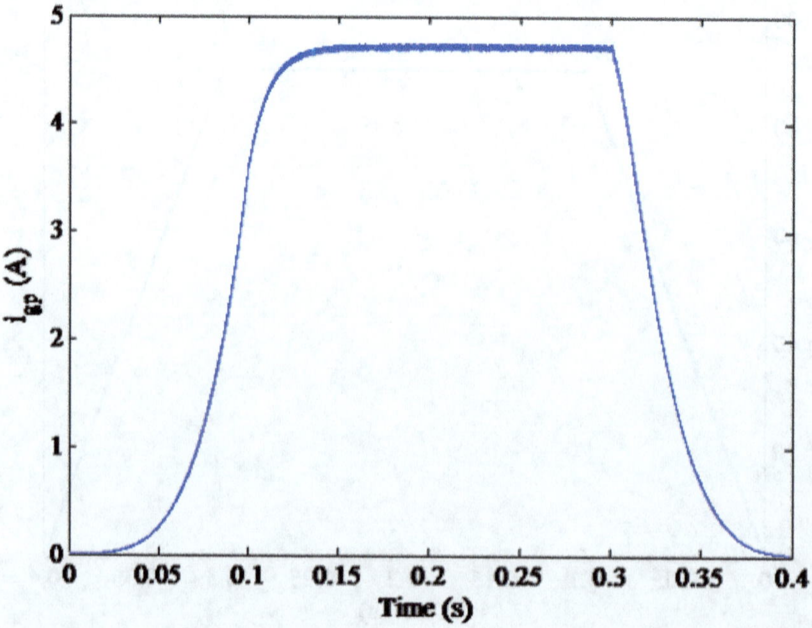

Figure 9. Evolution of the triple junction photovoltaic generator voltage response.

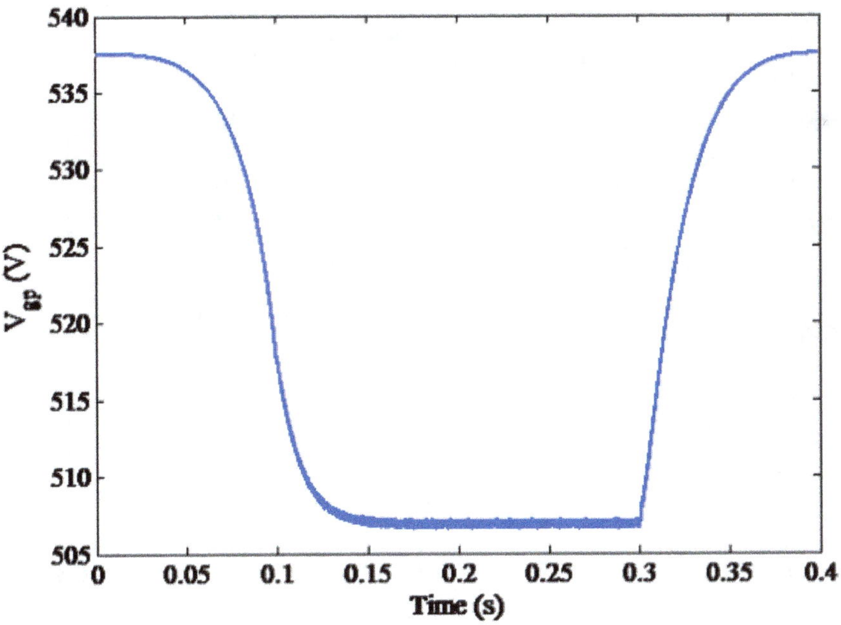

Figure 10. Evolution of the triple junction photovoltaic generator power response.

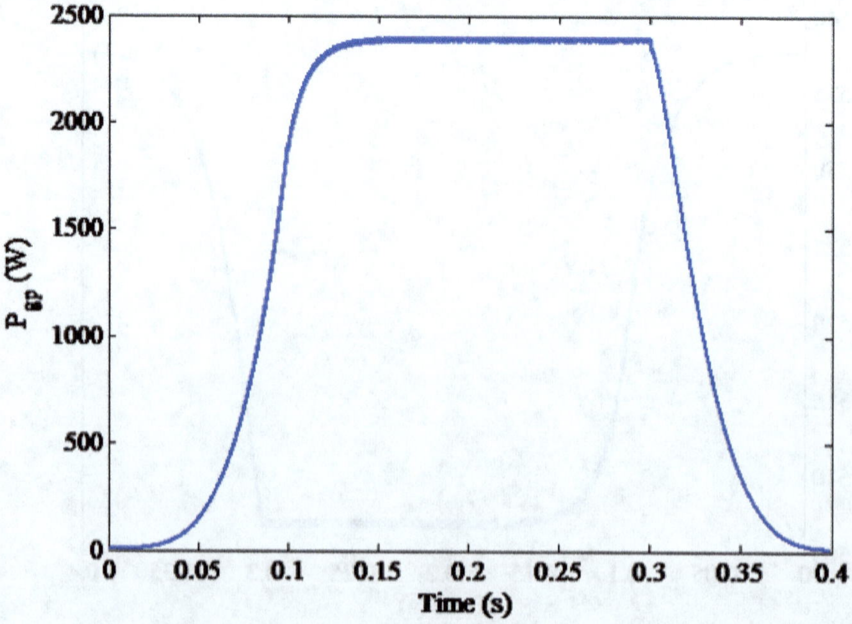

Figure 11. Pumping system operating point at direct connection behavior.

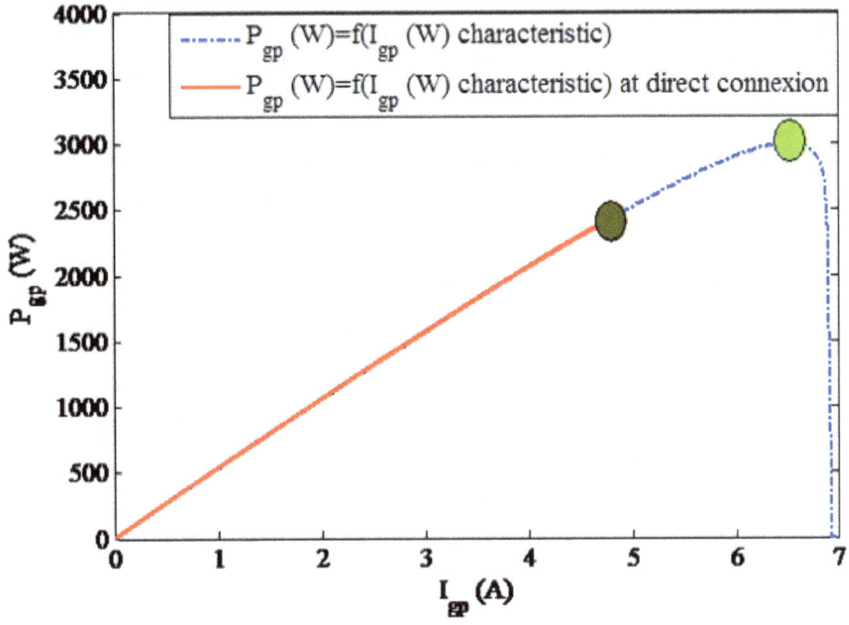

CONCLUSION

In this paper, a high efficient photovoltaic system conceived around a triple junction solar cell is developed. It is used for feeding the water pumping system. The photovoltaic generator is coupled to the centrifugal pump with a three phase permanent synchronous motor. The indirect field oriented control strategy is used in the flux and torque decoupling associated with a field weakening algorithm. Different operating conditions are considered including a wide shaft speed range.

The performance of the investigated system is shown by the simulation results produced using the Matlab and Simulink software packages.

Despite the excellent performance achieved, the system as designed for direct connection faces significant issues. The most common operational point is a sub-optimal one. The maximum power point tracking algorithm can be used to solve this issue. Future research will also focus on testing the entire photovoltaic water pumping system in real-world scenarios. The aforementioned trend toward futuristic applications will be the subject of the author's future work.

REFERENCES

Aleksandra, Z., Monika, P., & Ireneus, Z. (2021). Application of LCA to Determine Environmental Impact of Concentrated Photovoltaic Solar Panels—State-of-the-Art. *Energies*, *14*(31), 1–20.

Aleksandra, Z., Monika, P., & Ireneus, Z. (2023). Cooling of Concentrated Photovoltaic Cells—A Review and the Perspective of Pulsating Flow Cooling. *Energies*, *16*(2), 1–23.

Anika, T., Syed, N. S., Mohammad, J. H., Kaiser, A. R., & Abu, K. (2025). Advances in multijunction solar cells: An overview. *Solar Energy Advances*, *5*, 1–14.

Desideri, R., Francesc, M. P., Alicia, C., Jérôme, B., Montse, V., & Joan, R. U. (2025). Enhancing efficiency of dense array CPV receivers with controlled DC-DC converters and adaptive microfluidic cooling under non-uniform solar irradiance. *Solar Energy Materials and Solar Cells*, *279*, 1–12.

Foster, R., Ghassemi, M., & Cota, A. (2009). *Solar energy: renewable energy and the environment*. CRC Press.

Hamed, S. B., Abid, A., Hamed, M. B., Sbita, L., Bajaj, M., Ghoneim, S. S., & Kamel, S. (2023). A robust MPPT approach based on first-order sliding mode for triple-junction photovoltaic power system supplying electric vehicle. *Energy Reports*, *9*, 4275–4297.

Li, G., Jin, Y., Akram, M., & Chen, X. (2017). Research and current status of the solar photovoltaic water pumping system–a review. *Renewable & Sustainable Energy Reviews*, *79*, 440–458.

Marwa, M. E., Doaa, M. A., Aref, Y. E., Belal, A. A. Z., Mahmoud, S., & Abou, O. (2025). Multijunction solar cell parameter estimation based on metaheuristic algorithms. *Results in Engineering*, *25*, 1–12.

Miguel Angel, R. (2021). Quo Vadis Solar Energy Research? *Applied Sciences (Basel, Switzerland)*, *11*(30), 1–11.

Mohamed, S., Mohsen, Z., Miral, S., & Islam, M. (2024). Optimized and Sustainable PV Water Pumping System with Three-Port Converter, a Case Study: The Al-Kharijah Oasis. *Electricity*, *5*(2), 227–253.

Muhsen, D., Khatib, T., & Nagi, F. (2017). A review of photovoltaic water pumping system designing methods, control strategies and field performance. *Renewable & Sustainable Energy Reviews*, *68*, 70–86.

Salah, L., Franck, P., & Alain, G. (2007). Higher order sliding mode control based on integral sliding mode. *Automatica*, *43*(3), 531–537.

Shafidah, S., Mohd, N. H., Lin, X., & Kyungkon, K. (2025). Predicting high-performance perovskite solar cells using AI-based machine learning models. *Materials Today Sustainability*, *18*, 1–14.

Shafqat, A., & Taosif, I. (2025). Sensorless induction motor drive using coupled inductor based three phase inverters for solar pumping applications. *Results in Engineering*, *25*, 1–15.

Vafaeipour, M., Zolfani, S., Varzandeh, M., Derakhti, A., & Eshkalag, M. (2014). Assessment of regions priority for implementation of solar projects in Iran: New application of a hybrid multi-criteria decision making approach. *Energy Conversion and Management*, *86*, 653–663.

Yang, B., Linen, Z., Xiaoshun, Z., Hongchun, S., Tao, Y., Haofei, L., Lin, J., & Liming, S. (2019). Novel bio-inspired memetic salp swarm algorithm and application to MPPT for PV systems considering partial shading condition. *Journal of Cleaner Production*, *215*, 1203–1222.

Chapter 9
Smart Charging of Electric Vehicles:
A Comprehensive Review from a Human-Machine Interaction Perspective

Mabrouka Romdhane
https://orcid.org/0009-0005-5779-6518
University of Gafsa Faculty of Sciences of Gafsa, Tunisia

Mohamed Naoui
https://orcid.org/0000-0001-9117-3510
National Engineering School of Gabes, Tunisia

Ali Mansouri
Higher School of Applied Sciences and Technology of Gafsa, Tunisia

ABSTRACT

The global transition from conventional vehicles to electric mobility is accelerating, driven by environmental concerns, regulatory frameworks, and advances in energy infrastructure. As electric vehicles (EVs) become increasingly embedded within smart grid, intelligent energy management systems are required to address variable demand, battery aging, charging variability, and grid stability. Artificial Intelligence (AI) has become a key enabler by providing efficiency and reliability. This chapter reviews AI-driven strategies for battery charging and bidirectional energy exchange, focusing on Computational Intelligence (CI) and Machine Learning (ML). Their applications in state estimation, charging control, optimization, anomaly detection, and smart grid coordination are analyzed. Particular attention is given to V2G and G2V architectures and the critical role of Human-Machine Interaction (HMI),

DOI: 10.4018/979-8-3373-5621-1.ch009

which translates AI-generated insights into understandable, user centered charging decisions. By comparing CI and ML approaches and identifying open challenges, this chapter offers a structured perspective on intelligent and HMI integrated EV charging systems.

INTRODUCTION

Given the accelerating urgency of current energy and environmental challenges, the integration of electric vehicles (EVs) into modern smart grids has emerged as a pivotal strategy for optimizing energy distribution, enhancing system flexibility, and reducing greenhouse gas emissions (Lammouchi et al. 2025)(Li et al. 2016). Electrified transportation represents a critical component of global decarbonization efforts, yet its successful deployment depends not only on the proliferation of EVs but also on the development of advanced charging infrastructures capable of responding to highly dynamic and uncertain operating conditions (Bechiri et al. 2024). These conditions include fluctuating renewable energy availability, varying user charging behavior, network congestion, battery degradation, and the increasing complexity of bidirectional energy exchanges (Sallán et al. 2009). In this context, artificial intelligence (AI) has become a transformative enabler, offering innovative tools to intelligently manage the charging and discharging cycles of EV batteries and to support a seamless transition toward sustainable, resilient, and human-centric energy systems(H. Zhang et al. 2018)(Faria et al. 2019).

The adoption of AI in EV energy management is driven by its unique ability to analyze large and heterogeneous data streams, learn nonlinear relationships, and autonomously adapt to rapidly changing environments. Numerous studies have demonstrated the growing importance and effectiveness of AI in this domain. For example, Sinha, Neena, et al. provide an extensive literature review based on Scopus and the PRISMA methodology, offering a well-structured roadmap for the integration of AI techniques into smart mobility and smart grid frameworks. Their work underscores the fundamental role of AI in predictive modeling, decision support, and real-time optimization across multiple layers of the EV ecosystem. Similarly, other authors have conducted comprehensive analyzes of AI-based approaches for lithium-ion battery State of Charge (SoC) estimation, comparing the performance of classical machine learning models with more advanced architectures such as convolutional neural networks (CNNs), recurrent neural networks (RNNs), and generative models. These studies reveal that AI significantly outperforms traditional estimation techniques, particularly in scenarios involving complex battery dynamics or noisy measurement conditions(Youssef et al. 2024). In parallel, recent research emphasizes the necessity of optimizing EV battery performance given the rapid and widespread adoption of electric mobility worldwide. Predictive machine learning (ML) techniques have been shown to provide accurate forecasts of critical battery health indicators including State of Charge (SoC), State of Health (SoH), State of Function (SoF), and Remaining Useful Life (RUL) thereby enhancing battery safety, extending lifecycle, and improving user experience (Romdhane and Naoui 2023). These studies encompass a wide range of learning paradigms, including supervised, unsupervised, and deep learning algorithms, each evaluated in terms of accuracy, robustness, and computational efficiency. Within this broader research landscape, the present chapter aims to provide a structured and in-depth overview of AI applications in EV battery charging and energy management. We propose a clear classification of AI approaches into two main categories: Computational Intelligence (CI) and Machine Learning (ML)(Tarusan et al. 2020). Computational Intelligence encompasses heuristic and bio-inspired methods capable of dealing with uncertainty, nonlinearity, and real-time decision-making, while Machine Learning focuses on data-driven modeling, statistical inference, and predictive analytics. By examining how these techniques operate individually and in combination, we shed light on their practical applicability to key EV energy-management functions, including Grid-to-Vehicle (G2V) and Vehicle-to-Grid (V2G) operations, smart-grid integration, state estimation, and charging optimization (Fabbri et al. 2014)(Hajer and Aymen n.d.).

Furthermore, this chapter presents a comparative analysis of CI and ML methods, outlining their respective strengths, limitations, and suitability for different operational contexts. Such a comparison is crucial, as it guides researchers, engineers, and policymakers in selecting appropriate AI strategies tailored to specific challenges within EV energy systems. Ultimately, by situating AI at the intersection of electric mobility, intelligent computational energy infrastructures, and human–machine interaction, this chapter contributes to an understanding of how advanced methods can drive the evolution of sustainable, efficient, and resilient energy ecosystems.

HYBRID VEHICLES AND PURE ELECTRIC VEHICLES

Electric vehicles (EVs) are generally classified into two main categories: pure battery-electric vehicles (BEVs) and hybrid electric vehicles (HEVs). The essential difference lies in the presence of an internal combustion engine (ICE), which is included only in hybrid configurations. In both cases, the propulsion system integrates an energy source, a battery pack, power-electronics converters, and an electric drive. These components require a high-performance control platform capable of supervising energy flow, regulating motor torque, and coordinating auxiliary systems(Naoui et al. 2017).

Each vehicle category presents distinct benefits and limitations. BEVs are inherently clean and produce no direct emissions, aligning with global efforts to reduce pollution and greenhouse gases. This has motivated substantial research aimed at improving battery technology, optimizing motor sizing, and enhancing charging solutions. BEVs rely solely on electrical energy for propulsion and on-board systems, whereas HEVs combine two or more energy sources or storage technologies. According to the International Electrotechnical Commission (IEC), a vehicle qualifies as a hybrid as long as at least one of its energy sources supplies electrical power for propulsion(Grunditz and Thiringer 2016).

Figure 1 represent a general classification of electrical vehicles, where hybrid configurations are classified based on the combination of energy sources:

- Battery Electric Vehicle (BEV): powered exclusively by a battery.
- Fuel-Cell Electric Vehicle (FCEV): combines a fuel cell with a high-capacity battery.
- Hybrid Electric Vehicle (HEV): integrates both an ICE and an electric battery.
- Plug-in Electric Vehicle (PEV): battery powered with the ability to charge from the electrical grid.

Figure 1. General EVs classification

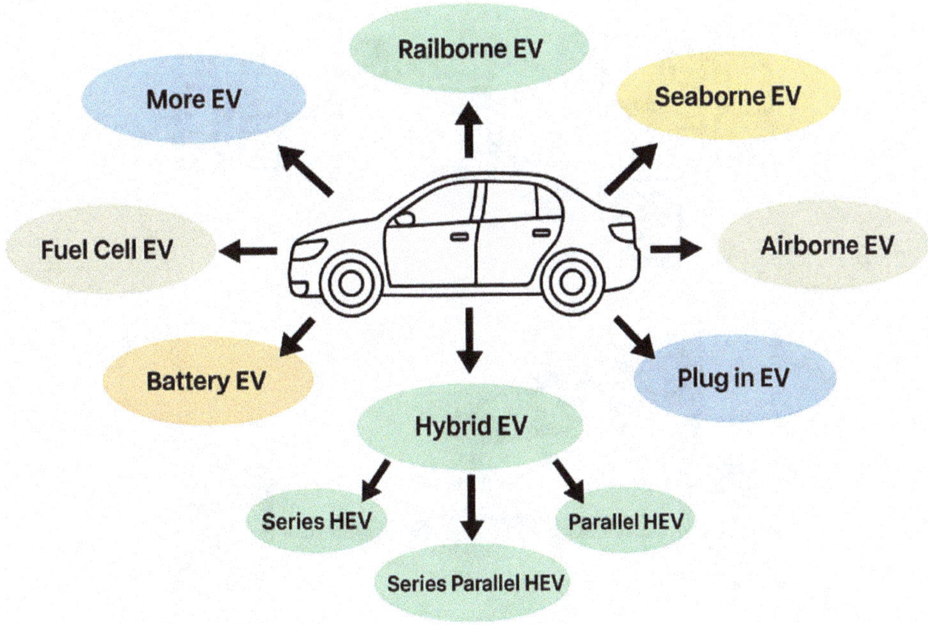

1. Hybrid Vehicle Architectures

The architecture of an HEV defines how power flows from the combustion engine and the electrical subsystem to the drivetrain. Three main topologies are widely recognized in the literature: series, parallel, and series-parallel (power-split). Each configuration has distinct structural features and operational characteristics(Naoui et al. 2023).

a. Series Hybrid Architecture

In a series hybrid, shown in Figure 2, the ICE is mechanically isolated from the wheels. Instead, it drives a generator that produces electrical power either to recharge the battery or to supply the electric propulsion motor. The traction motor alone provides the vehicle's driving torque (Wang et al. 2020).

Figure 2. The series architecture of the hybrid vehicle

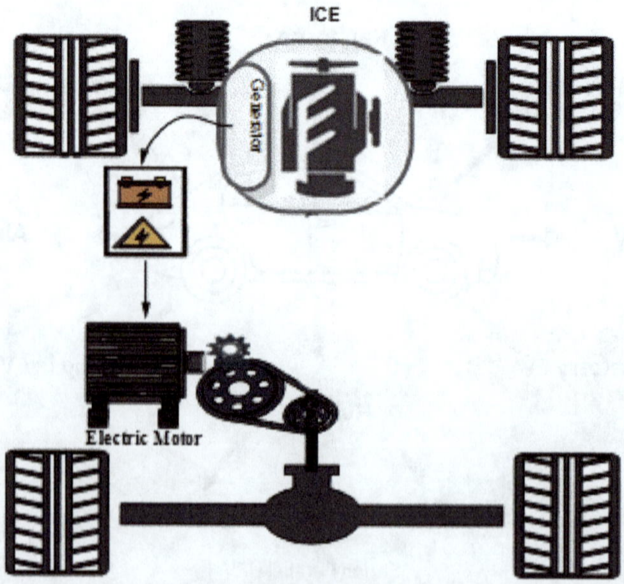

This arrangement offers several advantages:

- Excellent low-speed efficiency, especially in urban environments,
- Optimal and stable operation of the ICE near its maximum efficiency point,
- Flexible placement of the generator set.

However, series hybrids suffer from:

- Reduced efficiency at higher speeds due to multiple energy-conversion stages,
- The need for a high-power traction motor,
- Limited availability of purely mechanical driving modes.

Despite these constraints, series systems offer low emissions and fuel savings of up to 15–30% under urban driving conditions.

b. Parallel Hybrid Architecture

In a parallel hybrid, shown in Figure 3, both the ICE and the electric motor can deliver torque directly to the wheels. The mechanical coupling can be achieved using belts, pulleys, gears, or a single-shaft configuration. Depending on the design, the system may operate in torque-addition or speed-addition mode(C. Zhao et al. 2020).

Figure 3. Double-shaft and single-shaft parallel hybrid configuration

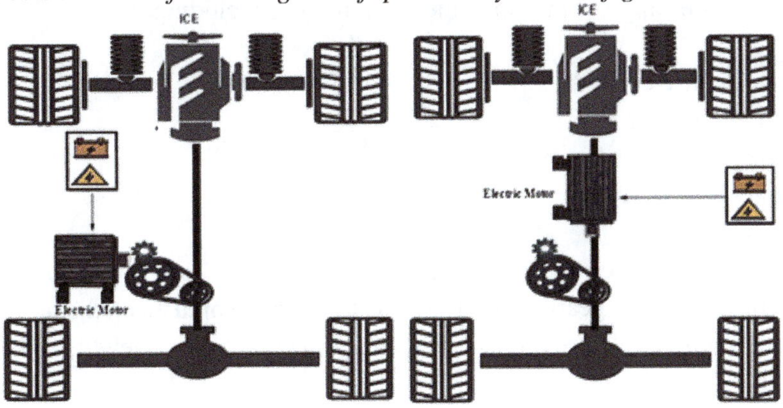

This type of architecture has several significant assets, such as:

- High energy efficiency at medium and high speeds,
- The ability to operate in full electric or full thermal mode (depending on design),
- Fewer modifications to the conventional drivetrain.

However, the parallel arrangement also presents some limits, we can mention:

- Complex coordination between ICE and electric torque contributions,
- Increased control complexity,
- Potential reduction in transient performance due to ICE dynamics.

Even so, parallel hybrids remain popular due to their simplicity and cost efficiency.

c. Series-Parallel (Power-Split) Hybrid Architecture

The series-parallel hybrid, also known as a power-split architecture illustrated in Figure 4, combines the strengths of both previous topologies. It employs a planetary gear set or similar mechanism to divide power between mechanical and electrical paths. This allows the ICE, traction motor, and generator to share responsibilities under different driving conditions (Cheng et al. 2015).

A well-known example of this configuration is the Toyota Prius design, which uses two electric machines one assisting traction and one adjusting engine speed. The architecture supports multiple operating modes, including:

- Pure electric drive,
- Pure thermal drive,
- Series hybrid mode,
- Parallel hybrid mode.

This architecture delivers high efficiency and smooth torque delivery without clutches or multi-ratio gearboxes. However, it requires sophisticated control algorithms and multiple electrical machines, which increase cost, weight, and system complexity.

Figure 4. Typical power-split hybrid arrangement

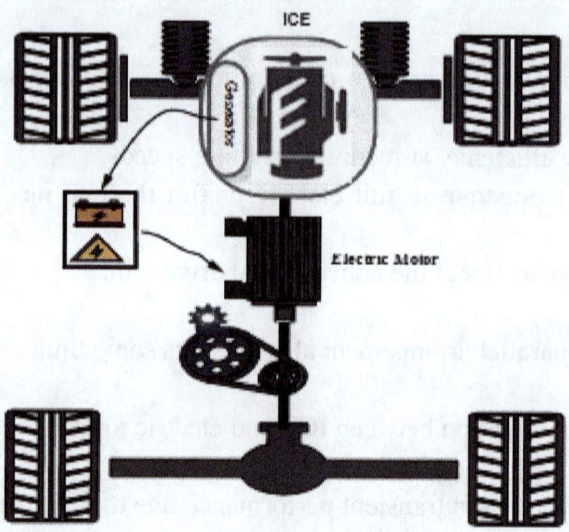

d. Comparison of Propulsion Architectures

Table 1 presents a comparative overview of the three main HEV architectures, summarizing their key strengths and limitations (Naoui et al. 2023)

Table 1. Comparative overview of the three main HEV architectures

Feature / Aspect	Series Hybrid	Parallel Hybrid	Series-Parallel Hybrid
Power to wheels	Electric motor only	ICE + electric motor	Electric motor + ICE + power split
Efficiency – city	High	Moderate	High
Efficiency – highway	Low	High	High
Mechanical complexity	Low	High	Very high
Control complexity	Moderate	Low	Very high
Component sizing	Large motor and battery	Smaller motor	Balanced sizing
Cost	Moderate	Low	High

2. Pure Electric Vehicle (PEV)

A Pure Electric Vehicle (PEV), also known as a Battery Electric Vehicle (BEV) shown in Figure 5, relies entirely on electrical energy stored in rechargeable batteries to power an electric traction motor. Unlike hybrid architecture or conventional internal combustion engines (ICEs), PEVs do not incorporate any form of combustion engine or fuel-based propulsion system. Consequently, they produce zero tailpipe emissions and offer a highly efficient and environmentally friendly mode of transportation (Bansal 2015). A pure electric vehicle (PEV) architecture is composed of several key subsystems that work together to ensure efficient propulsion and energy management. The electric traction motor converts electrical energy into mechanical torque to drive the wheels, with induction motors, permanent magnet synchronous motors (PMSMs), and switched reluctance motors (SRMs) being the most commonly used technologies. The battery pack serves as the primary energy storage unit, with lithium-ion batteries dominating due to their high energy density, long cycle life, and low self-discharge rate. Power electronics, including inverters and converters, manage the bidirectional conversion between the DC voltage of the battery and the AC voltage required by the motor, enabling precise control of speed and torque. An onboard charger allows the battery to be replenished from AC hands, while the battery management system (BMS) continuously monitors parameters such as voltage, temperature, state of charge (SoC), and state of health (SoH) to maintain safe and optimal battery operation. In addition, regenerative braking captures kinetic energy during deceleration and stores it back into the battery, enhancing overall efficiency. The operation principle of PEVs relies exclusively on the electric motor for propulsion: energy flows from the battery through the inverter to the motor to produce torque, and during braking or downhill motion, the motor acts as a generator that returns energy to the battery. The simplified drivetrain, free from multi-gear mechanical transmissions, reduces mechanical complexity and lowers maintenance needs. PEVs offer numerous advantages, including zero tailpipe emissions, high energy efficiency exceeding 90%, reduced operating and maintenance costs due to fewer moving parts, quiet operation, and enhanced driving performance thanks to instant torque delivery. However, they also face limitations such as restricted driving range due to current battery technologies, relatively long charging times compared to conventional refueling, gradual battery aging that reduces performance and capacity, and a high initial purchase cost largely driven by battery expenses. Today, PEVs are widely used in personal and commercial transportation, with adoption accelerated by incentives, stricter environmental regulations, and ongoing technological advancements. Current market trends emphasize the development of higher-energy-density batteries, advanced thermal management, and more efficient charging infrastructure to support the continued growth of pure electric vehicles(Covic et al. 2011).

Figure 5. Architecture and key components of Pure Electric Vehicle

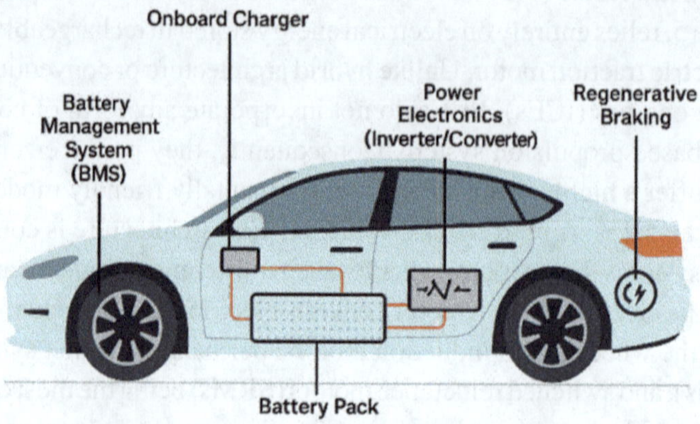

A primary classification is presented in table 2 according to the type of electrical energy source.

Table 2. Types of Pure Electric Vehicles (PEVs)

Type of Pure Electric Vehicle	Energy Source	Typical Powertrain	Key Characteristics	Applications
Battery Electric Vehicle (BEV)	Lithium-ion battery	Electric motor + inverter	Zero emissions, high efficiency, limited range depending on battery capacity	Passenger cars, buses, light-duty trucks
Solar Electric Vehicle (SEV)	Solar panels + battery storage	Electric motor + solar charging system	Uses solar energy; extended range in sunny regions; low power density	Research prototypes, low-speed vehicles, specialized applications
In-Wheel Motor EV	Lithium-ion battery	Individual motors inside each wheel	High traction control, simplified drivetrain, improved handling	Sports cars, high-performance EVs
Fuel-Cell-Derived Pure EV (using only fuel cell for power)	Hydrogen fuel cells	Electric motor + fuel cell stack	Zero tailpipe emissions, long range, fast refueling	Buses, commercial fleets
Light Electric Vehicle (LEV)	Small battery packs (Li-ion, NiMH)	Electric motor with simple controller	Lightweight, low power, short-range urban mobility	Scooters, e-bikes, micro-EVs
Autonomous EV (PEV with integrated sensing and AI)	High-capacity battery	Electric motor + AI systems + sensors	Autonomous driving capability; advanced electronics	Robotaxis, delivery vehicles

3. Other classification of EV

The rapid evolution of electric mobility has led to the emergence of several types of electric vehicles (EVs), each designed to meet specific performance, efficiency, and sustainability goals. These vehicle types differ primarily in their power source, energy storage system, and method of propulsion. Broadly, EVs can be categorized into Battery Electric Vehicles (BEVs), Plug-in Hybrid Electric Vehicles (PHEVs), and Fuel Cell Electric Vehicles (FCEVs).

Each category represents a distinct technological approach to reducing dependence on fossil fuels and mitigating greenhouse gas emissions. While BEVs rely entirely on battery-stored electrical energy, PHEVs combine electric and conventional fuel systems to extend driving range and flexibility. FCEVs, on the other hand, use hydrogen fuel cells to generate electricity through an electrochemical reaction, offering a promising long-term alternative for sustainable transport (Un-Noor et al. 2017). The following subsections provide a detailed overview of these three main EV types, highlighting their operational principles, advantages, and current limitations in the context of modern transportation systems (Grunditz and Thiringer 2016).

a. Battery Electric Vehicles (BEVs)

Battery Electric Vehicles (BEVs) are fully electric systems that operate exclusively on electrical energy stored in onboard batteries. Unlike conventional vehicles, BEVs are completely void of internal combustion engines (ICEs) and depend solely on external electricity sources for charging. The battery's energy capacity directly determines the vehicle's driving range and power performance. Modern BEVs typically achieve a range of approximately 100–400 km per full charge(Hwang et al. 2013). Some premium models offer even greater autonomy, though these often come at significantly higher costs due to larger battery capacities and advanced technologies. An important feature of BEVs is regenerative braking, which allows the vehicle to recover part of its kinetic energy during deceleration. In this process, the electric motor acts as a generator, converting mechanical energy into electrical energy that is stored back in the battery(Mohrehkesh and Nadeem 2011). BEVs offer several key advantages: they are mechanically simpler, easy to operate, and virtually silent in motion. Moreover, they produce zero tailpipe emissions, making them environmentally friendly and suitable for sustainable urban transport. However, their primary limitation remains the restricted driving range and relatively long charging times. Despite this, their clean operation and efficiency make BEVs particularly well-suited for city commuting and short-distance travel.

b. Plug-in Hybrid Electric Vehicles (PHEVs)

Plug-in Hybrid Electric Vehicles (PHEVs) were introduced as an advancement over conventional Hybrid Electric Vehicles (HEVs) to enhance overall efficiency and flexibility. PHEVs integrate both an internal combustion engine and an electric motor powered by a rechargeable battery. A defining feature of PHEVs is their ability to recharge the battery directly from the power grid through an external charging system, unlike traditional hybrids that rely solely on regenerative braking or the engine for battery charging. This capability allows PHEVs to operate in full-electric mode for short trips, reducing fuel consumption and emissions significantly. When the battery charge is depleted, the vehicle automatically switches to hybrid mode, using the ICE for propulsion or for recharging the battery during low-speed operation. Overall, PHEVs offer improved fuel economy and greater energy flexibility compared to HEVs, as they can draw energy from both electric and conventional fuel sources. This dual-mode operation positions them as an effective transitional technology toward full electrification(Lulhe and Date 2016).

c. Fuel Cell Electric Vehicles (FCEVs)

Fuel Cell Electric Vehicles (FCEVs) operate using a hydrogen-based electrochemical system rather than combustion or battery storage alone. These vehicles generate electricity onboard through a fuel cell stack, where hydrogen reacts with oxygen to produce electricity, water, and heat (Un-Noor et al. 2017). As long as a continuous hydrogen supply is available, the system can generate power indefinitely, similar in concept to a combustion engine that requires fuel to run. The most commonly employed technology in FCEVs is the Proton Exchange Membrane Fuel Cell (PEMFC), favored for its efficiency, compactness, and rapid start-up characteristics. One of the major advantages of FCEVs is their zero-emission output, as the only by-product of the electrochemical reaction is water vapor released through the tailpipe. However, despite their environmental promise, widespread adoption of FCEVs faces significant challenges. The lack of a well-developed hydrogen refueling infrastructure, the high cost of hydrogen production, and safety concerns since hydrogen is highly flammable and burns with an almost invisible flame limit their current practicality. Researchers are actively exploring methods to make hydrogen production more economical and sustainable, and advancements in storage and distribution technologies are expected to address these barriers in the coming years.

AI APPROACHES CLASSIFICATION IN EV

AI techniques applied to battery management can be grouped into two broad categories: computational intelligence (CI) and machine learning (ML). Figure 6 represent a general diagram of AI classification(Ahmed Baba, Naoui, and Cherkaoui 2023) (Mohamed, Aymen, Altamimi, et al. 2022).

Figure 6. AI classification applied on EV batteries

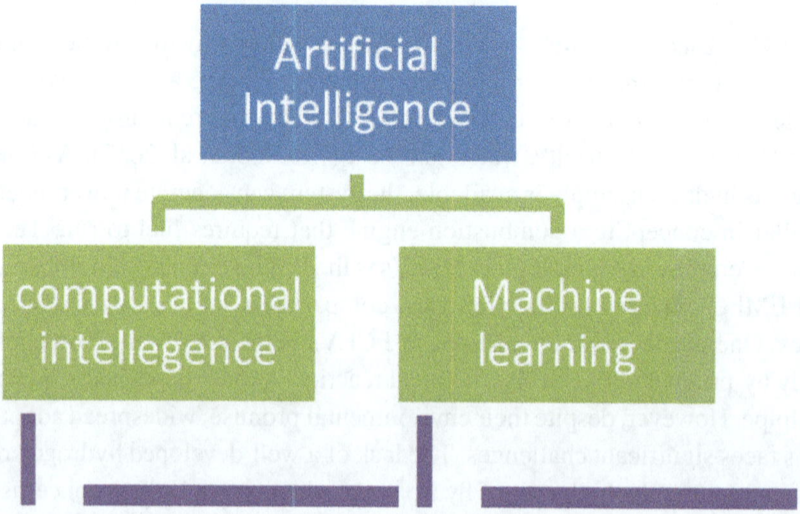

Computational intelligence is characterized by its ability to adapt to complex and uncertain environments. It relies on bio-inspired and heuristic approaches that enable the modeling of nonlinear systems and the provision of flexible solutions to energy management challenges and the optimization of charge and discharge flows. Machine learning, on the other hand, relies on statistical modeling and data analysis to accurately predict battery health and performance. It leverages the vast amounts of data generated by energy management systems to anticipate future behaviors and optimize decision-making.

1. Computational intelligence (CI)

Fuzzy logic provides a powerful approach for managing uncertainty and imprecision in estimating the state of charge (SOC) and state of health (SOH) of electric vehicle batteries. Unlike traditional methods, which rely on deterministic and linear models. By leveraging fuzzy rules and fuzzy sets, this approach can model the nonlinear relationships between input variables, such as voltage, current, and temperature, and battery parameters (Romdhane and Naoui 2023)(Baba, Naoui, and Cherkaoui 2023).

Evolutionary algorithms offer optimized solutions for battery charging and discharging strategies. These techniques, inspired by natural processes of evolution and adaptation, improve energy performance while maximizing battery lifespan. Among these algorithms: Genetic algorithms (GA) stand out for their ability to optimize charging and discharging schedules (Alqunun et al. 2025). By simulating selection, crossover, and mutation mechanisms, they explore a vast solution space to identify the most efficient configurations. Particle swarm optimization (PSO) is another widely used evolutionary approach to reduce energy consumption while extending battery lifespan. Inspired by the collective behavior of natural swarms, this method allows rapid convergence towards optimal solutions, even in complex and dynamic environments. Differential evolution (DE) also is particularly effective for adjusting charging parameters. Through iterative research based on the differences between individuals in a population, this technique allows for the refinement of charging strategies based on variations in operating conditions, thus ensuring better adaptation to operational requirements (Singh, Kour, and Singh 2017).

2. Machine Learning (ML) in Electric Vehicles

Machine Learning (ML) has become a central pillar in the advancement of intelligent energy management for electric vehicles (EVs). By leveraging large volumes of heterogeneous data collected from sensors, battery management systems (BMS), power converters, and charging infrastructures, ML enables predictive, adaptive, and automated decision-making. Unlike Computational Intelligence (CI), which relies on heuristic or bio-inspired principles, ML uses statistical learning, pattern recognition, and high-dimensional data modeling to provide accurate predictions and optimized control actions(Tarusan et al. 2020).

3. Role of ML in EV Battery State Estimation

EV battery state estimation such as State of Charge (SoC), State of Health (SoH), State of Function (SoF), and Remaining Useful Life (RUL) is crucial for ensuring safety, optimizing range, and prolonging battery lifespan. ML models excel at learning nonlinear relationships between voltage, current, temperature, degradation parameters, and dynamic operating conditions.(Naoui, Flah, and Sbita 2023)

Common ML techniques applied to state estimation include:

- Support Vector Regression (SVR): effective for SoC and SoH prediction under noisy conditions.
- Random Forests (RF): robust to measurement fluctuations and suitable for feature importance evaluation.
- Gaussian Process Regression (GPR): provides predictions with uncertainty quantification, valuable for safety margins.
- Deep Neural Networks (DNN): model complex degradation patterns over long-term cycles.

In scenarios where batteries operate under variable temperatures and charge–discharge cycles, ML-based estimators achieve higher accuracy than physics-based or equivalent-circuit models.

a. ML for Intelligent Charging Control

Machine learning (ML) enables the development of adaptive charging strategies that intelligently respond to real-time conditions, such as fluctuations in grid loads, changes in electricity prices, user preferences, and battery thermal constraints. By learning from continuous data streams, ML-based charge controllers dynamically optimize power allocation to enhance system efficiency and sustainability. One of the most widely used approaches is reinforcement learning (RL), which learns optimal charging procedures through interaction with the environment, making it highly effective in applications involving dynamic pricing and real-time scheduling. Deep reinforcement learning (DRL) extends the scope of reinforcement learning by incorporating deep neural networks, enabling highly scalable optimization suitable for managing large fleets of electric vehicles and operating under uncertain grid conditions. Additionally, multilayer perceptions (MLPs) are used to rapidly predict optimal charging setpoints, providing fast and reliable decision-making in complex power environments. In practical applications, reinforcement learning-based controllers not only reduce charging costs but also lower peak loads on the grid and extend battery life by preventing unnecessary fast-charging cycles. Together, these intelligent models contribute to building smarter, more flexible, and cost-effective electric vehicle charging infrastructures.

Figure 7 illustrate an example of ML-based adaptive EV charging control where it depends on grid load, electricity price, user schedule and battery state.

Figure 7. Concept of ML-based adaptive EV charging control

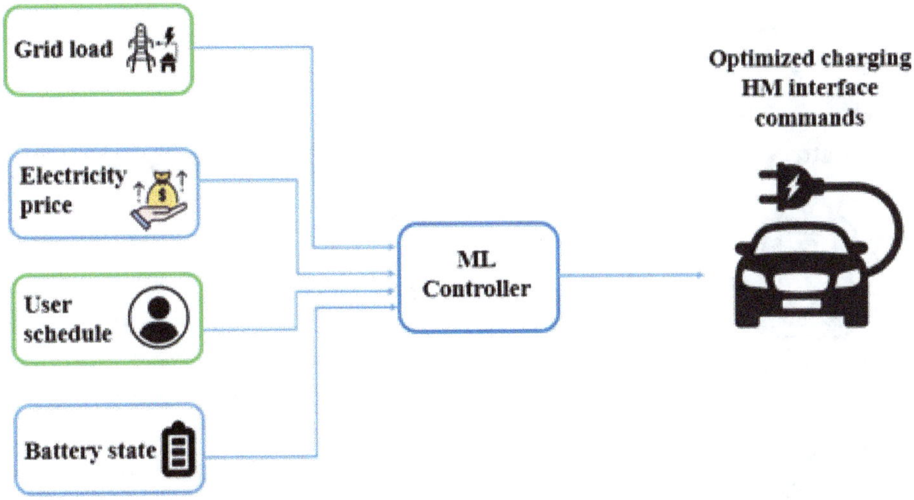

b. ML for Predictive Maintenance and Fault Diagnosis

Predictive maintenance has become a cornerstone of improving the reliability and performance of electric vehicle systems, particularly in large-scale applications such as vehicle fleets and public charging infrastructure. By leveraging machine learning techniques, predictive maintenance systems can analyze historical and real-time data to identify early signs of abnormal behavior, enabling proactive rather than reactive maintenance. Anomaly detection models monitor charging patterns and thermal behavior to identify irregularities, such as battery overheating or inefficient charging. Fault classification techniques help diagnose specific issues, including inverter failures, motor malfunctions, or sensor degradation, within the battery management system (BMS)(Huang et al. 2022). Furthermore, battery health prediction models forecast degradation trends over time, providing valuable insights into remaining battery life and optimal replacement schedules. Advanced machine learning architectures, such as convolutional neural networks (CNNs) and long-term memory networks (LSTMs), play a pivotal role in this process(Q. Zhang and Feng 2019). They can extract spatial and temporal dependencies from sensor data, enabling the detection of subtle degradation patterns that traditional monitoring systems might miss. Overall, predictive maintenance powered by machine learning enhances operational safety, reduces downtime, and significantly lowers maintenance costs across all electric vehicle systems (Ahmed et al. 2020).Figure 8 shown an overview of ML-based predictive maintenance in EV systems.

Figure 8. Overview of ML-based predictive maintenance in EV systems

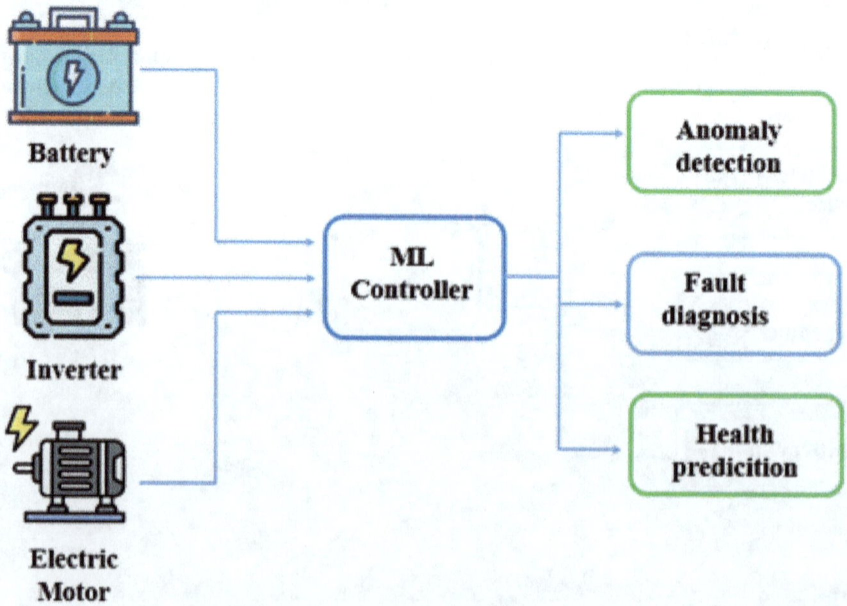

c. ML for Charging Infrastructure Planning

The optimal placement and sizing of electric vehicle (EV) charging stations play a crucial role in ensuring grid stability, reducing congestion, and maximizing user satisfaction. Machine learning (ML) provides powerful tools to support decision-making in this domain by analyzing complex and multidimensional datasets that include traffic flow patterns, population density, historical charging demand, energy consumption profiles, and renewable energy availability. By processing these data streams, ML models can predict future demand hotspots, identify under-served regions, and assist in designing charging infrastructures that balance accessibility with grid reliability(Naoui et al. 2023). Advanced ML techniques such as Multivariate Adaptive Regression Splines (MARS) are employed to model nonlinear relationships between geographic and energy variables, while clustering algorithms like K-means and DBSCAN are used to group potential locations based on demand intensity and accessibility. Moreover, hybrid ML-optimization frameworks such as HPSO-ML (Hybrid Particle Swarm Optimization with ML) and ETOPSIS-ML (Enhanced Technique for Order Preference by Similarity to Ideal Solution with ML) further refine location and capacity planning by minimizing grid stress and user waiting times. Through these intelligent approaches, ML contributes to building efficient, scalable, and sustainable EV charging infrastructures that adapt to real-world demand dynamics and renewable energy integration(Mohamed, Aymen, Alqarni, et al. 2022). An ML-based framework for optimal EV charging station placement is presented in Figure 9.

Figure 9. ML-based framework for optimal EV charging station placement

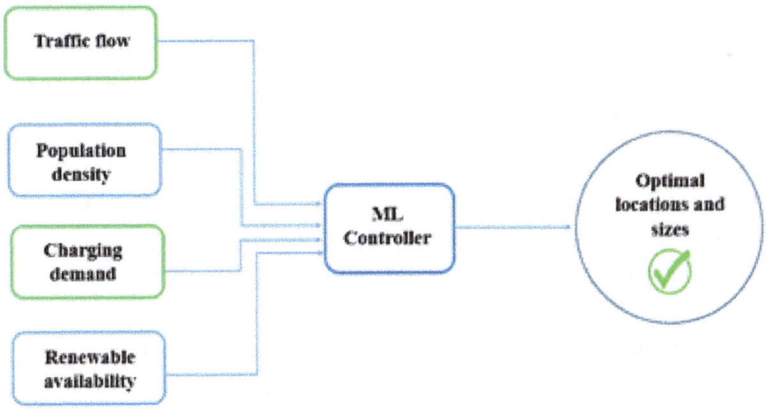

d. Cybersecurity and ML in EV Ecosystems

As EV charging infrastructures become increasingly interconnected, cybersecurity risks continue to grow, making machine learning (ML) an essential component in detecting cyberattacks, fraudulent activities, and unauthorized system access. ML-based intrusion detection systems (IDS) employ supervised and unsupervised learning to identify anomalies within vehicle-to-grid (V2G) communication channels, ensuring early detection of malicious behavior. In addition, Generative Adversarial Network (GAN)-based defense models strengthen system resilience by recognizing adversarial patterns and protecting ML-driven controllers from manipulation. Privacy-preserving ML techniques further enhance security by enabling federated learning across distributed EV fleets, allowing model training without exposing sensitive user or operational data. Together, these approaches significantly improve the trustworthiness, robustness, and resilience of AI-enabled EV charging ecosystems(M. H. Ali 2021).The relationship between cybersecurity and charging infrastructure for connected EV is shown in figure 10.

Figure 10. Cybersecurity in connected EV charging Infrastructures

e. Limitations and Challenges

The application of machine learning in electric vehicle (EV) systems offers significant potential but also faces several important limitations and challenges. ML based solutions are highly data dependent, requiring large, diverse, and high-quality datasets to ensure reliable performance. Even with sufficient data, generalization remains a major concern, as models trained under specific conditions may perform poorly when exposed to new environments or unseen operating scenarios. Additionally, many advanced ML techniques particularly deep learning demand substantial computational resources, making training and deployment costly, especially for real-time EV applications. Another key challenge is the lack of transparency and interpretability in deep learning models, which complicates their acceptance by regulators and limits trust in safety critical EV operations. These limitations and challenges are highlighted in Figure 11. Overcoming these challenges calls for hybrid approaches that integrate machine learning with computational intelligence methods, physics-based models, and high-performance computing to create more robust, explainable, and efficient EV management systems.

Figure 11. Methodological limitations

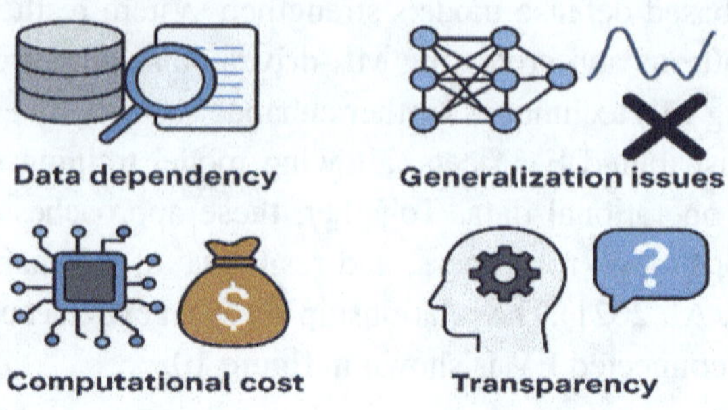

AI APPLIED ON EV BATTERY AND ENERGY MANAGEMENT

1. Implementation survey

In this section, we present and classify research related to the application of artificial intelligence (AI) in the management of electric vehicle (EV) battery charging, according to the main functions they address. These functions can be grouped into four main categories:

- Battery state estimation (State of Charge SoC, State of Health SoH, etc.)
- Energy production optimization and associated disruption management
- Prediction of optimal charging station locations
- Integration of charging systems into the smart grid

Table 3 is comparative table summarizes the AI approaches used in the first three functions. The fourth category, due to its complexity and strong interdependence with energy infrastructure, will be discussed in detail in the following section(Guner and Ozdemir 2017).

Table 3.AI application Survey

Function	AI Tool	Advantages	Reference
Battery State Estimation	FL+ANFIS	Optimizes torque for battery charging through regenerative braking based on (SOC) and engine speed	(M. U. Ali et al. 2019)
	TWO Layer Forwaerd NN	Minimize mean squared error (MSE) of State of Charge (SOC), State of Energy (SOE), and Power Loss (PL)	(Affanni et al. n.d.)
	GNN+ DL	Optimize battery data cleaning, restoring corrupted cloud data with only 1.3% error under varying temperatures.	(Y. Zhao and Behdad 2024)
Optimization of power generation and distribution	DL	Balance load during peak hours, improves accuracy and reliability of energy storage systems.	(Yang et al. 2023)
	Auto-regressive (AR) forecasting +SVR regression space vector LSTM	Solve the problem of unplanned EV charging	(Hussain et al. 2025)
Prediction of optimal charge station	MRAS, MARS, and COMPASS	Optimization of electric vehicle (EV) charging schedules by reducing the action space.	(Jiang, Jia, and Guan 2022)
	HPSO+ ETOPSIS	Maximizing captured charging demand and reducing electricity cost and user charging time and power grid load variance.	(Bai et al. 2022)

2. Integration with smart grid

The integration of smart grids as it is shown in Figure 12, represents a systemic modernization of traditional energy infrastructure through the synergistic deployment of advanced digital technologies, intelligent sensor networks, bidirectional communication systems, and predictive analytics tools. The incorporation of electric vehicles (EVs) via Vehicle-to-Grid (V2G) and its counterpart G2V (Grid-to-Vehicle) technology, combined with artificial intelligence (AI) algorithms, represents a disruptive innovation in electricity grid management (Naoui, Flah, and Sbita 2023).

Actually, The V2G-G2V designates two modes of bidirectional exchange of electrical energy between an electric vehicle (EV) and the electrical distribution network (grid). Energy is exchanged via an AC/DC (G2V) converter to charge the battery from the grid. And a DC/AC (V2G) converter to inject energy into the grid in synchronized AC form. Which necessitates a controller here becomes the role of AI tools. Figure 2 designs a simplified diagram of the V2G/G2V exchange (Mohamed, Flah, and Ben Hamed 2018).

Figure 12. Simplified V2G-G2V system

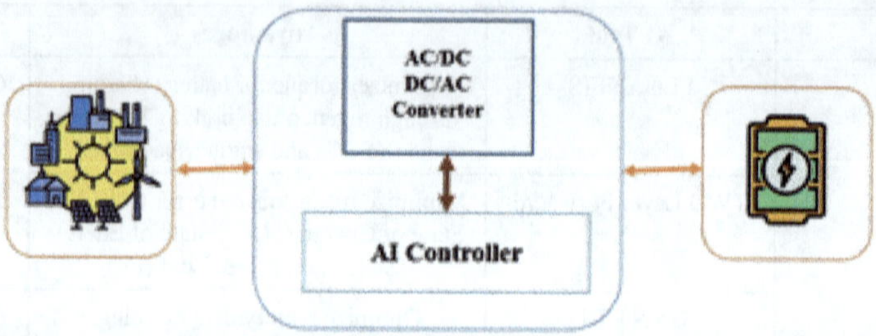

In this section, we are taking a deep close-up of the use of AI methods in bidirectional V2G/G2V charging. Table 4 summarizes the research of the past 5 years in terms of AI techniques used. We can conclude that the intelligent control depends on voltage, frequency, and phase synchronization with the grid. Beside the power factor and total harmonic distortion (THD) controls to comply with grid-code standards (IEEE 1547, EN 50549, etc.

Table 4. AI methods in bidirectional V2G/G2V charging.

Core aim	AI technique	Strengths	Drawbacks	Reference
Smoothing power in weak low voltage network	Fuzzy logic	Reduces variability and improves grid resilience.	Significant limitations at low battery SoC.	(Villa-ávila et al. 2025)
Manage and schedule the day-ahead discharging of Electric Vehicles (EVs) within a Vehicle-to-Grid (V2G) system.	Multi-Agent Reinforcement Learning (MARL)	Improved peak shaving decentralized decision making.	Requires large and diverse datasets.	(Fotopoulou and Rakopoulos 2021)
Optimize the charging (G2V) and discharging (V2G) profiles of EVs based on PV generation forecasts.	Improved Honey Badger Algorithm	Reduced power loss and THD.	Requires precise predictions for solar energy generation.	(Salvatti et al. 2020)
Protect V2M systems from inference and evasion attacks. (security risks associated with machine learning models) combines ML and GAN in a defensive architecture tailored to micro grid applications.	Generative Adversarial Network (GAN)+ML classifier	High performance on Adversarial Detection of 92.5%, significantly better than benchmark.	Absence of real time testing Focused only on two attack types inference and evasion attacks.	(Hakeem and Kim 2025)
Control and optimize the energy dispatch in a solar photovoltaic (SPV)-based micro grid.	Modified Grey Wolf Optimizer (GWO)	Cost reduction, efficient energy distribution, and grid interaction under variable solar conditions	Absence of real time validation or deployment. + limited renewable sources consideration.	(Kumar and Alaraj 2021)

COMPUTATIONAL INTELLIGENCE VS MACHINE LEARNING

Following a study of various artificial intelligence (AI) techniques and their applications, Table 5 presents a detailed comparison between Computational Intelligence (CI) and Machine Learning (ML). This comparison is based on several criteria, including the techniques used, accuracy, adaptability, and application optimization (Liao et al. 2021)(Cao et al. 2018).

Table 5. Computational Intelligent VS Machine Learning

Aspect	Computatinal Intelligent	Machine Learning
Framework	Metaheuristic, Evolutionary techniques	Analytics- based, Data-driven modeling
Precision	High (nonlinear systems)	High (big-data)
Adjustability	High	Moderate
Assessing uncertainty	Effective	Limited
Implementation & Adoption	Real-time control Adaptive charging Optimization	SOC,SOH

It emerges that ML techniques are particularly effective when sufficient labeled datasets are available for training. They stand out for their performance in prediction, pattern recognition, and anomaly detection. In contrast, CI techniques are more suited to contexts characterized by uncertainty, dynamic adaptation, and real-time control. They excel in the areas of optimization, adaptive control, and autonomous decision-making.

In practice, the hybrid integration of these two approaches is increasingly emerging as a preferred solution, enabling the development of more robust and adaptive battery management systems capable of combining predictive accuracy and decision-making flexibility.

HUMAN MACHINE INTERACTION AS AN ENABLER FOR INTELLIGENT ELECTRIC VEHICLE CHARGING

Human Machine Interaction (HMI) constitutes a critical functional layer in the architecture of intelligent electric vehicle (EV) charging systems, ensuring effective bidirectional communication between the user and the underlying control, estimation, and optimization modules. From a technical perspective, the HMI operates as the upper supervisory interface through which the smart charging controller communicates key operational variables, including battery State of Charge (SOC), terminal voltage, charging current, expected charging duration, and dynamic price signals obtained from the grid or energy management system. These data streams are visualized through smartphone applications, onboard vehicle displays, or charging station human machine interfaces, which expose system states and provide access to user-defined decision parameters such as time constraints, preferred charging profiles, or energy cost limitations.

In advanced implementations, the HMI layer also explicitly supports mode selection for various predefined operating scenarios. “Fast Charge” mode configures the EVSE (Electric Vehicle Supply Equipment) to apply maximum allowable power within battery safety constraints; "Scheduled Charge" triggers time-indexed charging sequences synchronized with user-selected intervals; and "Eco" or "Grid-Friendly" charging initiates a tariff- or load-dependent optimization process at aimed at minimizing peak load stress or operational cost. These selections are propagated to the lower control layers via communication protocols such as ISO 15118, OCPP, or proprietary CAN-based messages, ensuring that user commands are translated into valid set points for the power electronics interface.

Technically, HMI also incorporates real-time diagnostic and safety supervisory functions. Embedded sensors and Battery Management System (BMS) data are continuously parsed to detect deviations in temperature, impedance rise, overvoltage/under voltage events, or abnormal current ripple. When anomalies occur, the HMI instantly generates alarms, leveraging high-frequency monitoring streams to ensure compliance with IEC 61851 and ISO 6469 safety requirements. Such alerts are not merely informational; They are integrated into the feedback control loop, prompting the EVSE to reduce charging current, adjust duty cycles of the DC/DC converter, or terminate charging entirely when thresholds are violated.

An essential feature of the HMI stems from its integration with advanced Artificial Intelligence (AI) components, previously discussed in this work. Predictive models such as Long Short-Term Memory (LSTM) neural networks for SOC/SOH estimation, reinforcement-learning-based (RL) schedulers for tariff-driven optimization, and data-driven forecasting modules for load prediction feed processed information to the HMI layer. These models fuse historical driving patterns, energy price signals, environmental parameters, and battery aging models to compute optimal charging windows. The inferred control recommendations (e.g., optimal start time, power level trajectory, grid-aware constraint adjustments) are transmitted to the HMI, where they are synthesized into interpreted visual outputs and actionable user options.

Figure 12 illustrates the block diagram of the interaction between the Human-Machine interface HMI, Artificial Intelligence AI and the smart charging module. The HMI provides user inputs, status visualization, and diagnostic alerts, which are transmitted to the AI layer for predictive modeling, machine-learning-based decision-making, and charging optimization. The AI-generated control actions are then applied to the Smart Charging module, which regulates charging power and schedules according to operational constraints. Feedback from Smart Charging is returned to the HMI to ensure real-time user awareness and adaptive interaction.

Figure 13. AI- HM Interaction framework for EV smart charging.

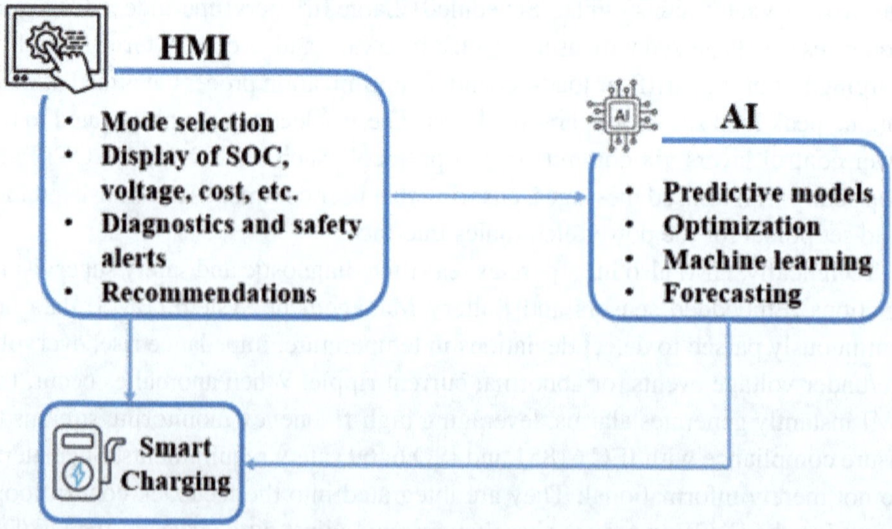

Thus, the HMI functions not only as a visualization tool but as a technical abstraction layer that encapsulates the complexity of AI-driven decision-making. By mapping high-dimensional optimization results into user-interpretable commands and by accepting user constraints that are reinjected into the optimization pipeline, the HMI effectively closes the loop between the AI computing core and the end user. This coupling ensures that smart charging decisions remain both computationally optimal and operationally aligned with user needs, ultimately positioning HMI as a structural enabler for intelligent EV charging and grid-interactive energy ecosystems.

CONCLUSION

The growing integration of electric vehicles into modern energy systems marks a defining step toward global decarbonization, yet it also introduces significant challenges related to battery management, charging efficiency, grid stability, and cybersecurity. This chapter has provided a comprehensive overview of both how Artificial Intelligence through Computational Intelligence (CI) and Machine Learning (ML) offers transformative solutions to these challenges by enabling predictive, adaptive, and intelligent control across the entire EV ecosystem. CI approaches such as fuzzy logic, evolutionary algorithms, and bio-inspired optimization have demonstrated strong capabilities in real-time decision making, uncertainty handling, and optimization of charging and discharging strategies. Meanwhile, ML models including deep learning, reinforcement learning, and hybrid ML optimization frameworks have proven highly effective for state estimation, predictive maintenance, infrastructure planning, and cybersecurity reinforcement. Through detailed classifications, comparative analyses, and recent implementation surveys, it becomes evident that no single AI technique is universally optimal; instead, each method excels under specific conditions and operational constraints. ML is particularly powerful when abundant data are available for accurate prediction and anomaly detection, whereas CI is better suited for dynamic, nonlinear, and uncertain environments requiring rapid adaptation. Increasingly, hybrid CI ML frameworks are emerging as the most promising path forward, combining predictive accuracy with flexible control to enhance battery longevity, improve charging performance, reduce grid stress, and strengthen system resilience.

The chapter also highlights the growing importance of bidirectional energy exchange (V2G/G2V), where AI plays a central role in ensuring synchronized, stable, and secure power flow between EVs and the grid. As future energy systems evolve toward smarter, more autonomous, and human-centric infrastructures, AI-enabled EV technologies will become indispensable components of next-generation mobility and energy management. Continued research is required to address persistent challenges, including data scarcity, model interpretability, cybersecurity threats, and real time implementation constraints. Nevertheless, the advancements reviewed in this chapter demonstrate that AI is not merely an add on technology but a foundational pillar driving the future of sustainable electric mobility and intelligent energy infrastructures.

REFERENCES

Abdel Hakeem, S. A., & Kim, H. (2025). Advancing intrusion detection in V2X networks: A comprehensive survey on machine learning, federated learning, and edge AI for V2X security. *IEEE Transactions on Intelligent Transportation Systems.*

Affanni, A.. (n.d.). EV battery state of charge: Neural network based estimation. *Proceedings of the International Electric Vehicle Conference.*

Ahmed, M., Abdelrahem, M., Harbi, I., & Kennel, R. (2020). An adaptive model-based MPPT technique with drift-avoidance for grid-connected PV systems. *Energies, 13*(24), 6656. Advance online publication. DOI: 10.3390/en13246656

Ahmed Baba, M., Naoui, M., & Cherkaoui, M. (2023). Fault-tolerant control strategy for Hall sensors in BLDC motor drive for electric vehicle applications. *Sustainability, 15*(13), 10430. Advance online publication. DOI: 10.3390/su151310430

Ali, M. H. (2021). Exploring cybersecurity issues in 5G enabled electric vehicle. *IET Generation, Transmission & Distribution, 15*(24), 3435–3449.

Ali, M. U., Zafar, A., Nengroo, S. H., Hussain, S., Alvi, M. J., & Kim, H.-J. (2019). Towards a smarter battery management system for electric vehicle applications: A critical review of lithium-ion battery state of charge estimation. *Energies, 12*(3), 446. Advance online publication. DOI: 10.3390/en12030446

Alqunun, K.. (2025). An efficient bearing fault detection strategy based on a hybrid machine learning technique. *Scientific Reports, 15*(1), 1–15. PMID: 40437009

Baba, M. A., Naoui, M., & Cherkaoui, M. (2023). Fault-tolerant control strategy for Hall sensors in BLDC motor drive for electric vehicle applications. *Sustainability, 15*(13), 10430. Advance online publication. DOI: 10.3390/su151310430

Bai, X.. (2022). Electric vehicle charging station planning with dynamic prediction of elastic charging demand: A hybrid particle swarm optimization algorithm. *Complex & Intelligent Systems, 8*(2), 1035–1046. DOI: 10.1007/s40747-021-00575-8

Bansal, P. (2015). Charging of electric vehicles: Technology and policy implications. *Journal of Science Policy & Governance, 6*(1), 1–38.

Bechiri, M. B.. (2024). Effective diagnosis approach for broken rotor bar fault using Bayesian-based optimization of machine learning hyperparameters. *Journal of Electrical Engineering, 12*(October), 139923–139936.

Cao, P., ... (2018). An IPT system with constant current and constant voltage output features for EV charging. *Proceedings: IECON 2018 - 44th Annual Conference of the IEEE Industrial Electronics Society, 1*, 4775–4780.

Cheng, M., Sun, L., Buja, G., & Song, L. (2015). Advanced electrical machines and machine-based systems for electric and hybrid vehicles. *Energies*, *8*(9), 9541–9564.

Covic, G. A., Boys, J. T., Budhia, M., & Huang, C. Y. (2011). Electric vehicles—Personal transportation for the future. *World Electric Vehicle Journal*, *4*(1), 693–704.

Fabbri, G.. (2014). Impact of V2G/G2V technologies on distributed generation systems. *IEEE International Symposium on Industrial Electronics*, 1677–1682.

Faria, J., Pombo, J., Calado, M. R., & Mariano, S. (2019). Power management control strategy based on artificial neural networks for standalone PV applications with a hybrid energy storage system. *Energies*, *12*(5), 902. Advance online publication. DOI: 10.3390/en12050902

Fotopoulou, M., & Rakopoulos, D. (2021). Day ahead optimal dispatch schedule in a smart grid containing distributed energy resources and electric vehicles. *Sensors (Basel)*, *21*(21), 7295. Advance online publication. DOI: 10.3390/s21217295 PMID: 34770613

Grunditz, E. A., & Thiringer, T. (2016). Performance analysis of current BEVs based on a comprehensive review of specifications. *IEEE Transactions on Transportation Electrification*, *2*(3), 270–289.

Guner, S., & Ozdemir, A. (2017). Stochastic energy storage capacity model of EV parking lots. *IET Generation, Transmission & Distribution*, *11*(7), 1754–1761.

Huang, W.. (2022). A hybrid model-based diagnosis approach for open-switch faults in PMSM drives. *IEEE Transactions on Power Electronics*, *37*(4), 3728–3732.

Hussain, A., Eswarakrishnan, V., Aslam, A., & Tripura, S. (2025). Charging stations demand forecasting using LSTM based hybrid transformer model (pp. 1–27). *Scientific Reports*, *15*(1), 36639. Advance online publication. DOI: 10.1038/s41598-025-20421-y PMID: 41120429

Hwang, J. J.. (2013). Lifecycle performance assessment of fuel cell/battery electric vehicles. *International Journal of Hydrogen Energy*, *38*(8), 3433–3446. DOI: 10.1016/j.ijhydene.2012.12.148

Jiang, Z., Jia, Q.-S., & Guan, X. (2022). On large action space in EV charging scheduling optimization. *Science China. Information Sciences*, *65*(2), 122201. Advance online publication. DOI: 10.1007/s11432-020-3106-7

Kumar, A., & Alaraj, M. (2021). Novel AI based energy management system for smart grid with RES integration. *IEEE Access: Practical Innovations, Open Solutions, 9*, 162530–162542.

Lammouchi, Z.. (2025). Model-free predictive control based on the execution time of active vectors using ultra-local model. *Arabian Journal for Science and Engineering, 50*(8), 5885–5898. DOI: 10.1007/s13369-024-09629-8

Li, W.. (2016). Comparison study on SS and double-sided LCC compensation topologies for EV/PHEV wireless chargers. *IEEE Transactions on Vehicular Technology, 65*(6), 4429–4439.

Liao, J.-T., Huang, H.-W., Yang, H.-T., & Li, D. (2021). Decentralized V2G/G2V scheduling of EV charging stations by considering the conversion efficiency of bidirectional chargers. *Energies, 14*(4), 962. Advance online publication. DOI: 10.3390/en14040962

Lulhe, A. M., & Date, T. N. (2016). A technology review paper for drives used in electrical vehicle (EV) & hybrid electrical vehicles (HEV). *2015 International Conference on Control Instrumentation Communication and Computational Technologies (ICCICCT 2015)*, 632–636.

Mohrehkesh, S., & Nadeem, T. (2011). Toward a wireless charging for battery electric vehicles at traffic intersections. *IEEE Conference on Intelligent Transportation Systems, Proceedings, ITSC*, 113–118.

Naoui, M.. (2023). Intelligent control system for hybrid electric vehicle with autonomous charging. In Azar, A. T., & Koubaa, A. (Eds.), *Artificial intelligence for robotics and autonomous systems applications* (pp. 405–437). Springer., DOI: 10.1007/978-3-031-28715-2_13

Naoui, M., Flah, A., & Alqarni, M.. (2022). A new wireless charging system for electric vehicles using two receiver coils. *Ain Shams Engineering Journal, 13*(2). Advance online publication. DOI: 10.1016/j.asej.2021.08.012

Naoui, M., Flah, A., Altamimi, A., & Khan, Z. A. (2022). Power management and control of a hybrid electric vehicle based on photovoltaic, fuel cells, and battery energy sources. *Sustainability, 14*(5).

Naoui, M., Flah, A., & Ben Hamed, M. (2018). Wireless charging system for a mobile hybrid electric vehicle. *International Symposium on Advanced Electrical and Communication Technologies (ISAECT 2018)*, 1–5.

Naoui, M., Flah, A., Hamed, M. B., & Sbita, L. (2017). Review on autonomous charger for EV and HEV. *International Conference on Green Energy Conversion Systems (GECS)*, 1–6.

Naoui, M., Flah, A., & Sbita, L. (2023). Efficiency of intelligent control design for dual-stator machine on the EV traction system. *Arabian Journal for Science and Engineering*. Advance online publication. DOI: 10.1007/s13369-023-08070-7

Romdhane, M., & Naoui, M. (2023). PMSM inter-turn short circuit fault detection using the fuzzy-extended Kalman filter in electric vehicles. *Electronics (Basel)*, *12*(18), 3758. Advance online publication. DOI: 10.3390/electronics12183758

Sallán, J., Villa, J. L., Llombart, A., & Sanz, J. F. (2009). Optimal design of ICPT systems applied to electric vehicle battery charge. *IEEE Transactions on Industrial Electronics*, *56*(6), 2140–2149.

Salvatti, G. A., Carati, E. G., Cardoso, R., da Costa, J. P., & Stein, C. M. O. (2020). Electric vehicles energy management with V2G/G2V multifactor optimization of smart grids. *Energies*, *13*(5), 1191. Advance online publication. DOI: 10.3390/en13051191

Singh, G., Kour, V., & Singh, L. (2017). Design and performance analysis of PID controller for automatic generation control of an autonomous power system. *International Journal of Engineering Research & Technology (Ahmedabad)*.

Tarusan, S. A. A.. (2020). A review of direct torque control development in various multilevel inverter applications. *International Journal of Power Electronics and Drive Systems*, *11*(3), 1675–1688.

Un-Noor, F.. (2017). A comprehensive study of key electric vehicle (EV) components, technologies, challenges, impacts, and future direction of development. *Energies*, *10*(8).

Villa-Ávila, E.. (2025). Improving V2G systems performance with low-pass filter and fuzzy logic for PV power smoothing in weak low-voltage networks. *Applied Sciences (Basel, Switzerland)*, *15*(4), 1952.

Wang, J.. (2020). Review on multi-power sources dynamic coordinated control of hybrid electric vehicle during driving mode transition process. *International Journal of Energy Research*, *44*(8), 6128–6148.

Yang, Z.. (2023). Review on optimal planning of new power systems with distributed generations and electric vehicles. *Energy Reports*, *9*, 501–509. DOI: 10.1016/j.egyr.2022.11.168

Youssef, R.. (2024). Comparative study of the MPPT control for the photovoltaic water pumping system between FSS-P&O and VSS-P&O. *WSEAS Transactions on Power Systems*, *19*, 229–241.

Zhang, H.. (2018). Six-plate capacitive coupler to reduce electric field emission in large air-gap capacitive power transfer. *IEEE Transactions on Power Electronics*, *33*(1), 665–675.

Zhang, Q., & Feng, M. (2019). Fast fault diagnosis method for Hall sensors in brushless DC motor drives. *IEEE Transactions on Power Electronics*, *34*(3), 2585–2596.

Zhao, C.. (2020). Design and analysis of an engine-start control strategy for a single-shaft parallel hybrid electric vehicle. *Energy*, *202*, 117621.

Zhao, Y., & Behdad, S. (2024). State-of-health estimation for sustainable electric vehicle batteries using temporal-enhanced self-attention graph neural networks. *Journal of Energy Storage*, *146*(June), 1–9.

Chapter 10
Underwater Robotic Tracking of Human-Defined Targets in Offshore Energy Applications

Wided Chlibi
ENIG, Tunisia

ABSTRACT

The development of underwater vehicles is a vital tool for exploring, monitoring, and responding to marine environments that are typically inaccessible or frightening to human beings. The intervention of underwater robots makes cross-oceans more manageable. As part of this work, the chapter focus on the fundamental aspects of underwater robot modeling and control. The second part of this chapter deals with the stabilization problem of the robot. The use of a control system is essential for underwater robots to ensure precise trajectory tracking, position stability, and attitude stabilization. The author offer in this section an explicit comparison between PID and FOPID Controllers. Finally, the study emphasizes the importance of Human-Machine Interaction (HMI) in robot control. An intuitive Dashboard (Simulink) was developed for real-time trajectory adjustment. Results validate the robot's ability to follow complex 3D paths and the HMI's effectiveness, although further controller optimization is required to improve performance during dynamic movements.

DOI: 10.4018/979-8-3373-5621-1.ch010

INTRODUCTION AND MOTIVATION

1. Historical Overview

The exploration of the underwater environment has indeed been a human concern for decades, and even centuries, although the methods and objectives have changed considerably. The first attempts at underwater exploration date back to ancient times, with the use of diving bells to recover objects or work on shipwrecks, as described by philosophers like Aristotle. The first diving bell, also known as the 'Barrel of Alexander,' was allegedly used by Alexander the Great to observe the marine life. The contraption consisted of a large glass barrel open at its base. The bell was lowered vertically into the water using a rope, descending to depths of up to 20 meters. It stands as one of the earliest pieces of equipment that facilitated human underwater exploration (Stoneman, 2008)

In 1690, two submarine models were created by the Frenchman Denis Papin in Germany (Lee-Pearce, 1998) (Blair, 1975). It was a prototype of a submersible in the shape of a reinforced and sealed iron parallelepiped as shown in Figure 1. Papin used a pump to compress the air inside and a barometer to measure the pressure, as well as a large syringe to take in or expel water. However, this machine was damaged during its first test attempt.

Figure 1. First submarine model

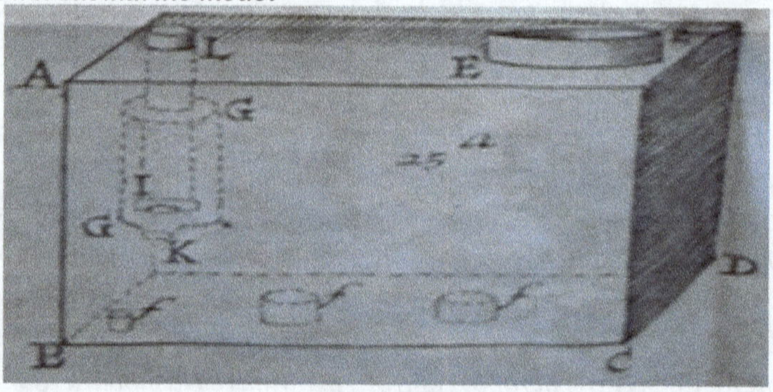

Undaunted, Papin proceeded to design a second model with military objectives in 1692, which incorporated several improvements. Thanks to this prototype, he could stand in the horizontal cylinder and extend an arm through the hole, once this second cylinder was pressurized with compressed air using the pump.

The American David Bushnell created also his Turtle in 1775, built entirely of wood (Figure 2). To move forward, the single pilot on board turned a hand crank that operated a propeller. To dive, he opened valves to fill the ballast tanks; to resurface, he expelled the water using a pump.

Figure 2. The Turtle in wood

The evolution of submarines during this period was continuous, progressing from their first rudimentary designs to the development of a truly operational submarine in 1887 (Figure 3) by the French engineers Henri Dupuy de Lôme and Gustave Zédé. This vessel was propelled by a 50-horsepower electric motor, achieved 8 knots on the surface and 4 while submerged. It was manned by a crew of five, armed with two torpedoes, and had a range of 65 miles (on the surface, but only a third while submerged).

Figure 3. Gymnote, launched in 1888 in France

During the period of 1914 to 1918, a significant number of Diesel-electric propelled submersibles were extensively deployed in wartime. These vessels relied on an accumulator battery to power their electric propulsion motor, with the batteries being recharged on the surface by a diesel engine-driven generator (Morgan, 2011). By 1944, the Germans enhanced the snorkel (originally a Dutch invention). This air tube enabled U-Boats to operate their diesel engines at shallow depths, thereby eliminating the need to surface, a position where they were highly vulnerable.

Then, The Nautilus's design phase commenced in March 1950 with the initiation of Project SCB 64, which sought to develop the world's first nuclear-powered submarine (Figure 4). Captain Hyman Rickover oversaw the entirety of this groundbreaking project.

Figure 4. The Nautilus

These represent some instances of humanity's initial endeavors in constructing underwater robots for the purpose of exploring marine life. Efforts and improvements continue to this day, driven by technological advancements.

2. FIELDS AND EXAMPLES OF UNDERWATER USE

Nowadays, we have access to a wide variety of underwater vehicles capable of performing different tasks (such as in demining, dealing with pollution, or working in great depths) and sometimes even reduces operational costs by automating certain tasks, thanks to research conducted over many decades. Thus, underwater robots are very important in offshore energy infrastructure, where subsea inspections of pipelines, power cables, and platforms are required for efficient operation and maintenance.

Numerous nations have been engaged in sustained research efforts concerning underwater robots used in several fields. Such us, the defense sector, in which the underwater robot is used to detect and destroy the underwater mines. Neutralizing these objects is a hazardous and highly specialized task, for which robotic systems can locate and disarm these threats, thereby eliminating the need to expose human personnel. Concurrently, port surveillance is an increasingly vital activity. To prevent the infiltration of hostile divers or underwater vehicles, autonomous robots can patrol and grid areas near port entrances. Underwater vehicles play also a crucial role in the scientific research and Oceanography, enabling direct exploration of the ocean's most challenging environments. They are indispensable for deep-sea exploration, allowing scientists to study remote areas such as ocean trenches, active underwater volcanoes, and unique extreme ecosystems that thrive without sunlight. Furthermore, these versatile vessels facilitate detailed marine biology observations, providing unparalleled insights into underwater flora and fauna in their natural habitats. In marine geology and geophysics, submarines are used to map seabeds, investigate tectonic plates, and examine mid-ocean ridges and faults. Finally, they are vital tools for climatology, collecting essential data on ocean currents, temperature, salinity, and acidification, all of which are critical for understanding and monitoring global climate change.

The U.S.military as an example, designed the “Bluefin” autonomous underwater vehicle (AUV), which is capable of autonomous underwater navigation and object detection and played a significant role in the search for the missing Malaysia Airlines MH370 data in 2014.

Russia developed the unique 'Peace 1' and 'Peace 2' manned submersibles, which are the only pair globally capable of collaborative underwater exploration (Wu, 2024). Germany contributed to underwater robotics with 'Deep C', an AUV designed for operations at depths of up to 4000 meters and boasting a 60-hour endurance in the deep sea. France's 'VICTOR 6000', depicted in Figure 5 (a), is a cable-operated underwater robot known for its ability to capture high-quality optical images underwater (Schillai, 2019). Meanwhile, Britain introduced the fully autonomous 'Autosub6000' submarine, shown in Figure 5 (b), which is equipped with batteries and sensors for independent navigation (Barker et al., 2020).

Figure 5. The 'VICTOR 6000' and 'Autosub6000' underwater robot

3. CLASSIFICATION OF UNDERWATER ROBOTS

Underwater vehicles can be classified basically in two groups, Remotely Operated Underwater Vehicles, commonly known as ROVs and another is Autonomous Underwater Vehicles, generally known as AUVs. Remotely operated vehicles (ROVs) rely on human operators and a physical tether for real-time control from the surface. In contrast, autonomous underwater vehicles (AUVs) doesn't involve human intervention, executing tasks independently underwater. For complex inspection missions, constant human oversight is often necessary, making full autonomy impractical. This has given rise to human-in-the-loop systems, where a human operator initiates the mission by defining a target or trajectory, and the robot takes over from there.

Underwater vehicles have undergone significant technological advancements, allowing for their classification based on various criteria or specifications. This section presents a classification of underwater vehicles according to distinct characteristics. Additionally, an alternative classification groups these vehicles into four primary categories: manned vehicles, unmanned vehicles (Feng, 2025), underwater gliders, and biomimetic systems.

In conclusion, the evolution of underwater robots, from early concepts to sophisticated modern systems, marks a profound technological leap. This chapter successfully outlined a comprehensive framework for modeling and controlling underwater robots, specifically addressing the critical human-machine interaction (HMI) challenge within energy infrastructure operations.

We present in this chapter a detailed 6-DOF kinematic and dynamic modeling framework, which prove effective in representing the complex behavior of underwater robotic systems. Our examination of various control strategies, including both classical (PID) and advanced (MPC, robust) methods for target tracking in uncertain underwater environments, demonstrated their respective strengths and limitations.

Crucially, the role of human-machine interaction was effectively demonstrated through a use-case scenario where a human operator's target selection was autonomously tracked by the robotic system in a simulated offshore energy infrastructure environment (pipeline or cable). The simulations provided valuable insights into robot behavior, highlighting the importance of model fidelity, sensor uncertainty, and control robustness for reliable real-world deployment in energy-related underwater operations.

These findings underscore the potential for leveraging human cognitive strengths in complex underwater tasks while relying on robotic autonomy for precise execution. Future work will focus on validating these control strategies in real-world conditions and further refining the HMI interface for enhanced operational efficiency and safety.

UNDERWATER ROBOT MODELING

1. Kinematic Modeling

Modeling an underwater robot's motion begins by defining reference frames. To describe the robot's 6-degrees-of-freedom motion in Euclidean space, two frames are used (see *Figure 6*). The first is the global inertial reference frame, denoted as $R_0 = (O, X_0, Y_0, Z_0)$.It is earth-fixed.

A second frame, the local body-fixed frame $R_v=(C, X_v, Y_v, Z_v)$, has its origin C at the robot's center of buoyancy. The vehicle's motion is described by two key vectors.

The state vector η (a 6×1 vector) defines the vehicle's position and orientation in the global frame:

- **Linear Position (x,y,z)**: The coordinates of the vehicle's center of gravity (or a reference point) in the Earth-fixed frame, representing East, North, and depth (Down) positions.
- **Angular Orientation (ϕ,θ,ψ)**: The Euler angles that describe the orientation of the vehicle's body-fixed frame relative to the Earth-fixed frame, corresponding to Roll, Pitch, and Yaw rotations.

The **velocity vector ν** is a 6×1 vector containing the linear and angular velocities in the body-fixed frame R_b.

- **Linear Velocities (u,v,w)**: Expressed in m/s, these are the velocities along the vehicle's X_v, Y_v, and Z_v axes, known as surge, sway, and heave, respectively.
- **Angular Velocities (p,q,r)**: Expressed in rad/s, these are the rotational velocities around the X_v, Y_v, and Z_v axes, known as the roll, pitch, and yaw rates.

Figure 6. Reference systems

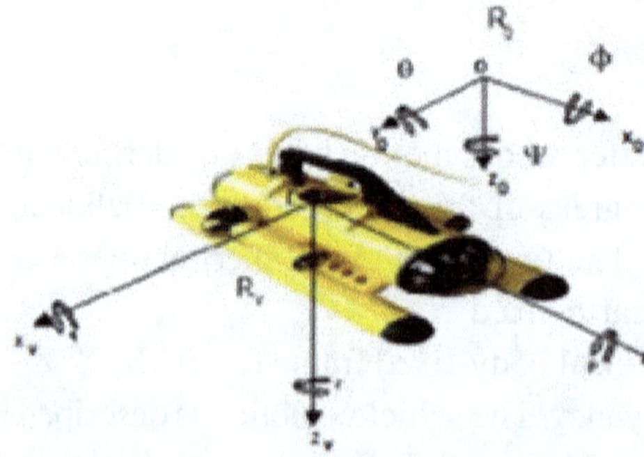

The vehicle's configuration is determined by three orientation angles: yaw, pitch, and roll. The rotation matrix for the first of these, a yaw angle rotation, is given by equation (1) and denoted as H_Ψ

$$H_\Psi = \begin{bmatrix} \cos(\Psi) & -\sin(\Psi) & 0 \\ \sin(\Psi) & \cos(\Psi) & 0 \\ 0 & 0 & 1 \end{bmatrix} \quad (1)$$

It refers to a movement around the yaw axis of a rigid body that changes the direction it is pointing, to the left or right of its direction of motion. It is considered positive in the counter-clockwise direction relative to the Z_0 axis (see *Figure 7*)

Figure 7. Yaw rotation

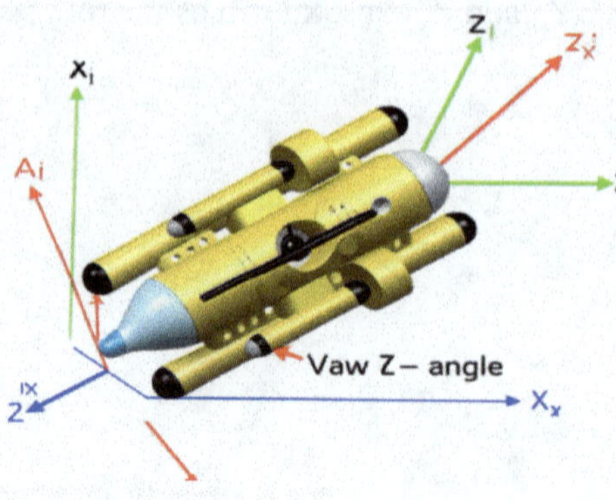

The second matrix describes the pitch angle rotation around Y1-axis (Figure 8), yields the transformation matrix H_θ is given by the equation (2)

$$H_{\theta} = \begin{bmatrix} \cos(\theta) & 0 & \sin(\theta) \\ 0 & 1 & 0 \\ -\sin(\theta) & 0 & \cos(\theta) \end{bmatrix} \quad (2)$$

Figure 8. Pitch rotation

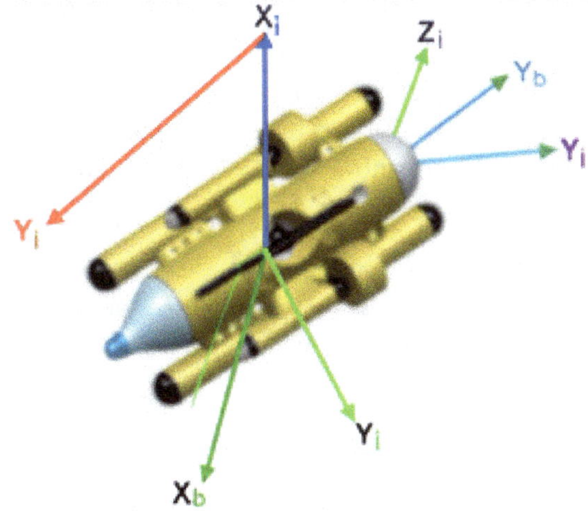

The last rotation shown in equation (3), a roll angle rotation, occurs around the X_v-axis (see *Figure 9*).

H=

$$H_{\Phi} = \begin{bmatrix} 1 & 0 & 0 \\ 0 & cos(\Phi) & -sin(\Phi) \\ 0 & sin(\Phi) & cos(\Phi) \end{bmatrix} \quad (3)$$

Figure 9. Roll rotation

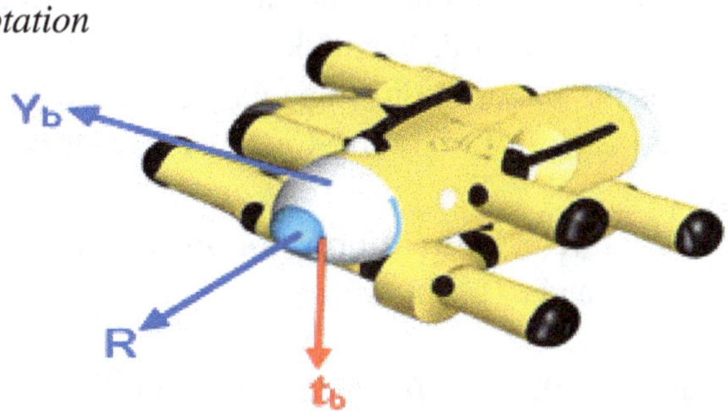

To describe the transformation from the global frame R_0 to the local frame R_V, we have to calculate the rotation matrix J_1. Its formal is given by the equation (4)

$$J_1 = J_1(\eta_2) = H_\psi H_\theta H \qquad (4)$$

$\eta_2 = [\phi, \theta, \psi]$ represents the vector of Euler angles (attitude) relative to the fixed frame R_0.

$$J_1 = \begin{bmatrix} \cos(\Psi) & -\sin(\Psi) & 0 \\ \sin(\Psi) & \cos(\Psi) & 0 \\ 0 & 0 & 1 \end{bmatrix} \begin{bmatrix} \cos(\theta) & 0 & \sin(\theta) \\ 0 & 1 & 0 \\ -\sin(\theta) & 0 & \cos(\theta) \end{bmatrix} \begin{bmatrix} 1 & 0 & 0 \\ 0 & \cos(\Phi) & -\sin(\Phi) \\ 0 & \sin(\Phi) & \cos(\Phi) \end{bmatrix} \qquad (5)$$

$$= \begin{bmatrix} \cos(\psi)\cos(\theta) & -\sin(\psi)c(\varphi) + \sin(\varphi)\cos(\psi)\sin(\theta) & \sin(\varphi)\sin(\psi) + \sin(\theta)\cos(\psi)\cos(\varphi) \\ \cos(\theta)\sin(\psi) & \cos(\psi)\cos(\varphi) + \sin(\theta)\sin(\psi)\sin(\varphi) & -\cos(\psi)\sin(\varphi) + \cos(\varphi)\sin(\theta)\sin(\psi) \\ -\sin(\theta) & \cos(\theta)\sin(\varphi) & \cos(\theta)\cos(\varphi) \end{bmatrix} \qquad (6)$$

To express the transformation of the translational velocity components with respect to R_0, the following equations (7) and (8) can be used:

$$\eta^{\cdot}_1 = J_1(\eta_2)\, \nu_1 \qquad (7)$$

$$\eta^{\cdot}_2 = J_2(\eta_2)\nu_2 \qquad (8)$$

Where:

- $\nu_1 = [u, v, w]^T$ The local translational velocity vector with respect to R_0, expressed in the local frame R_v
- $\eta^{\cdot}_1 = [x^{\cdot}, y^{\cdot}, z^{\cdot}]^T$ describes the linear velocity of the robot's center of gravity (or reference point) in the Earth-fixed (global inertial) frame
- $\eta^{\cdot}_2 = [\phi^{\cdot}, \theta^{\cdot}, \psi^{\cdot}]^T$ is the vector of Euler angular velocities in the fixed frame.
- $v_2 = [p,q,r]^T$ is the vector of body-fixed angular velocities expressed in the local frame R_v
- $J_2(\eta_2)$ is the angular velocity transformation matrix which depends on the Euler angles ϕ (roll) and θ (pitch).

Based on these equations (9) and (10), which describe the local rotational velocity:

$$\nu_2 = \dot{\Phi} Z_0 + \dot{\theta} Y_1 + \dot{\Psi} X_v \tag{9}$$

$$= \begin{pmatrix} \dot{\Phi} \\ 0 \\ 0 \end{pmatrix} + \begin{pmatrix} 0 \\ \dot{\theta} \\ 0 \end{pmatrix} H_\Phi + \begin{pmatrix} 0 \\ 0 \\ \dot{\psi} \end{pmatrix} H_\theta H_\Phi \tag{10}$$

The equations (11), (12) and (13) describe a detailed calculation of the matrix

$$= \begin{pmatrix} \dot{\Phi} \\ 0 \\ 0 \end{pmatrix} + \begin{pmatrix} 0 \\ \dot{\theta} \\ 0 \end{pmatrix} \begin{bmatrix} 1 & 0 & 0 \\ 0 & \cos(\Phi) & -\sin(\Phi) \\ 0 & \sin(\Phi) & \cos(\Phi) \end{bmatrix} + \begin{pmatrix} 0 \\ 0 \\ \dot{\psi} \end{pmatrix} \begin{bmatrix} 1 & 0 & 0 \\ 0 & \cos(\Phi) & -\sin(\Phi) \\ 0 & \sin(\Phi) & \cos(\Phi) \end{bmatrix}$$

$$\begin{bmatrix} \cos(\theta) & 0 & \sin\theta \\ 0 & 1 & 0 \\ -\sin(\theta) & 0 & \cos\theta \end{bmatrix} \tag{11}$$

$$= \begin{pmatrix} \dot{\varphi} \\ \dot{\theta}\ \cos\varphi \\ -\dot{\theta}\sin\varphi \end{pmatrix} + \begin{pmatrix} -\dot{\Phi}\sin\theta \\ 0 \\ \dot{\Phi}\cos\theta \end{pmatrix} \cdot \begin{bmatrix} 1 & 0 & 0 \\ 0 & \cos(\Phi) & -\sin(\Phi) \\ 0 & \sin(\Phi) & \cos(\Phi) \end{bmatrix} \tag{12}$$

$$= \begin{pmatrix} \dot{\Phi} \\ \dot{\theta}\ \cos\Phi \\ -\dot{\theta}\sin\Phi \end{pmatrix} + \begin{pmatrix} -\dot{\Phi}\sin\theta \\ \dot{\Phi}\cos\theta\sin\Phi \\ \dot{\Phi}\cos\theta\cos\Phi \end{pmatrix} = \begin{pmatrix} \dot{\Phi} - \dot{\Phi}\sin\theta \\ \dot{\theta}\cos\Phi + \dot{\Phi}\cos\theta\sin\Phi \\ -\dot{\theta}\sin\Phi + \dot{\Phi}\cos\theta\cos\Phi \end{pmatrix} \tag{13}$$

Consequently, we get the matrix J_2 (η_2) in the equation (14)

$$= \begin{pmatrix} \dot{\Phi} \\ \dot{\theta} \\ \dot{\Psi} \end{pmatrix} \times \begin{bmatrix} 1 & 0 & -\sin(\theta) \\ 0 & \cos\Phi & \cos\theta\sin\Phi \\ 0 & -\sin\Phi & \cos\theta\cos\Phi \end{bmatrix} = J_2(\eta_2) \tag{14}$$

To obtain the angular velocity vector ν_2, it is necessary to calculate the inverse of the $J_2(\eta_2)$ matrix. It is defined in equations (15) and (16)

$$\nu_2 = \begin{bmatrix} 1 & 0 & -\sin(\theta) \\ 0 & \cos\Phi & \cos\theta\sin\Phi \\ 0 & -\sin\Phi & \cos\theta\cos\Phi \end{bmatrix}^{-1} \tag{15}$$

$$= \begin{bmatrix} 1 & \sin\Phi\tan\theta & \cos\Phi\tan\theta \\ 0 & \cos\Phi & -\sin\Phi \\ 0 & \sin\Phi/\cos\theta & \cos\Phi/\cos\theta \end{bmatrix} \tag{16}$$

The equation representing the kinematic transformation relating the body-fixed velocities to the rates of change of the global position and orientation for the underwater robot is given by the equation (17).

$$\begin{pmatrix} \dot{\eta}_1 \\ \dot{\eta}_2 \end{pmatrix} = \begin{pmatrix} J_1(\eta_2) & 0_{3*3} \\ 0_{3*3} & J_2(\eta_2) \end{pmatrix} \tag{17}$$

It should be noted that a singularity occurs in the pitch parameterization at θ=2π +kπ. Nevertheless, this singular point is inaccessible in practice because hydrostatic and hydrodynamic effects, coupled with the robot's propulsion capabilities, preclude it from ever attaining a 90° pitch angle

2. Dynamic Modelling

a. Parametrization

Dynamic modelling of an underwater robot is a crucial step that builds upon kinematic modelling. It explains the robot's movement by analyzing the forces and moments applied to it and their impact on its accelerations.

To achieve this model, we use a complete Newton–Euler approach to drive the equations of motion. This model depends also on a set of physical parameters which are obtained from experimentation.

Determining the vehicle's position and orientation necessitates knowledge of its linear and angular accelerations. These accelerations, of course, result from forces acting upon the vehicle. The specific forces and torques influencing the vehicle will be detailed in the subsequent section. The combined effect of these forces and torques produces a resultant force and torque. This section's primary objective is to calculate the acceleration caused by this resultant force and torque.

Based on the Newton-Euler formalism, we will determine the relationships between generalized forces and generalized position, velocity, and acceleration vectors.

We note $\Gamma_1 = [X, Y, Z]^T$ and $\Gamma_2 = [K, M, N]^T$ respectively the forces and moments applied to the rigid body at point G, the robot's center of gravity.

Firstly, we will express the three translations with respect to a fixed frame.

The figure 10 shows the relation between the fixed frame and the rigid-body frame.

Figure 10. the relationship between the coordination systems

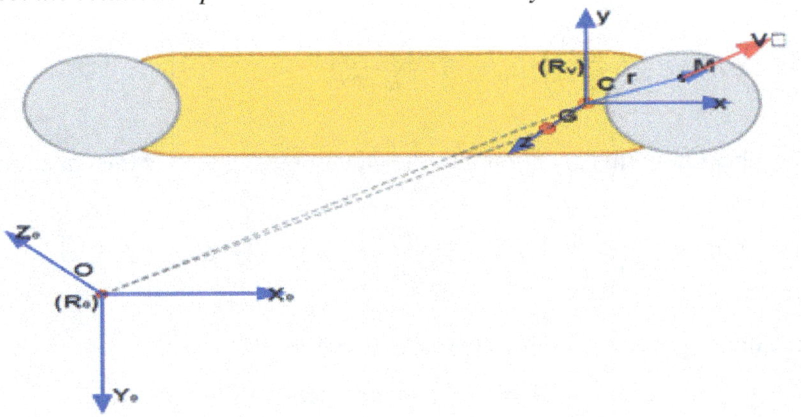

We can express the vector $\overrightarrow{OG}$ by the equation (18) given as follow

$$\overrightarrow{OG} = \overrightarrow{OC} + \overrightarrow{CG} \quad (18)$$

To obtain the velocity of the center of gravity in the terrestrial frame, we differentiates the previous equation with respect to time. So, we get the equation (19)

$$\frac{d\overrightarrow{OG}}{dt}/R_0 = \frac{d\overrightarrow{OC}}{dt}/R_0 + \frac{d\overrightarrow{CG}}{dt}/R_0 \quad (19)$$

Or we can simplify: $\frac{d\overrightarrow{CG}}{dt}/R_0$ we obtain the equation (20)

$$\frac{d\overrightarrow{CG}}{dt}/R_0 = \frac{d\overrightarrow{CG}}{dt}/R_V + \Omega \wedge \overrightarrow{CG}/R_0 = \Omega \wedge \overrightarrow{CG}/R_0 \quad (20)$$

$\frac{d\overrightarrow{CG}}{dt}/R_v = 0$ because the body is undeformable. With Ω is the rotational velocity of the R_v

frame with respect to the fixed frame R_0. The equation become (21):

$$\frac{d\overrightarrow{OG}}{dt}/R_0 = \frac{d\overrightarrow{OC}}{dt}/R_0 + \Omega \wedge \overrightarrow{CG}/R_0 \quad (21)$$

Now, we derivate the equation (21) with respect to time. We obtain the equation (22)

$$\frac{d}{dt}\left(\frac{d\overrightarrow{OG}}{dt}\right)/R_0 = \frac{d}{dt}\left(\frac{d\overrightarrow{OC}}{dt}\right)/R_0 + \Omega \wedge \left(\frac{d\overrightarrow{CG}}{dt}\right)/R_0 + \left(\frac{d\vec{\Omega}}{dt}\right) \wedge \overrightarrow{CG} \quad (22)$$

This term can be computed as the equation (23) below

$$\frac{d}{dt}\left(\frac{d\overrightarrow{OC}}{dt}\right)/R_0 = \frac{d}{dt}\left(\frac{d\overrightarrow{OC}}{dt}\right)/R_V + \Omega\wedge\frac{d\overrightarrow{OC}}{dt}/R_0 \quad (23)$$

By substituting this equation into the equation (22), we get the equation (24)

$$\frac{d}{dt}\left(\frac{d\overrightarrow{OG}}{dt}\right)/R_0 = \frac{d}{dt}\left(\frac{d\overrightarrow{OC}}{dt}\right)/R_V + \Omega\wedge\frac{d\overrightarrow{OC}}{dt}/R_0 + \Omega\wedge\,(\Omega\wedge\overrightarrow{CG}\,)+\dot{\vec{\Omega}}\wedge\overrightarrow{CG} \quad (24)$$

By adopting the conventions (SNAME, 1950)

$\left(\frac{d\overrightarrow{OC}}{dt}\right) = v_1,\ \Omega = v_2,\ \overrightarrow{CG} = r_g$ (the gravity center of the robot)

So, we have a new equation (25):

$$\frac{d}{dt}\left(\frac{d\overrightarrow{OG}}{dt}\right)/R_0 = \dot{v}_1 + v_2\wedge v_1 + v_2\wedge(v_2\wedge r_g) + \dot{v}_2\wedge r_g \quad (25)$$

By applying the fundamental principle of dynamics (Newton's second law relate the force exerted on a body to the lineal acceleration): $\sum\vec{F} = m\vec{a}$, with m the mass of the robot, $\sum\vec{F}$ represents the vector sum of all external forces acting on the body (the net or resultant force), expressed in Newtons (N) and $\vec{a}$ is the acceleration of the body . The previous equation becomes (26)

$$m\ [\dot{v}_1 + v_2\wedge v_1 + v_2\wedge(v_2\wedge r_g)+\dot{v}_2\wedge r_g] = \Gamma_1 \quad (26)$$

Now, we move to the second part which is the expression of Γ_2, the vector that generates the moments. To compute this expression, we can use the derivation of the angular momentum.

In the first part, we will focus on calculating angular momentum based on the following expression (27):

$$\vec{h} = \eth(\vec{r}\wedge\rho\vec{v})\,dv \quad (27)$$

Where dv is an element of volume of the solid and ρ its density. The calculation of angular momentum is quite complex, we cannot do it directly. But, we can use the velocity composition method.so, we use the equation (28)

$$\vec{v} = \frac{d\overrightarrow{OC}}{dt}/R_v + \frac{d\vec{r}}{dt} = \frac{d\overrightarrow{OC}}{dt}/R_V + \Omega\wedge\vec{r} \quad (28)$$

The equation (27) becomes

$$\vec{h} = \int \vec{r}\rho dv \wedge \frac{d\overrightarrow{OC}}{dt} + \int \rho \vec{r} \wedge \vec{\Omega} \wedge \vec{r} dv$$

We can also introduce the notion of mass based on this expression (29)

$$\vec{r}_g = \frac{1}{m}\int \rho \vec{r} dv \qquad (29)$$

With $\vec{r}_g = [x_g, y_g, z_g]^T$ the gravity center's vector

We replace $\int \rho \vec{r} \wedge \vec{\Omega} \wedge \vec{r}$dv with $I_b\vec{\Omega}$ where I_b is the inertial matrix. In consequence the new equation (30) describe the angular momentum

$$\vec{h} = m\vec{r}_g \wedge \frac{d\overrightarrow{OC}}{dt} + I_b\vec{\Omega} \qquad (30)$$

Now, to compute the dynamic moment, we have to derivate the previous equation (30)

$$\frac{d\vec{h}}{dt} = m\frac{d\vec{r}_g}{dt} \wedge \frac{d\overrightarrow{OC}}{dt} + m\vec{r}_g \wedge \frac{d}{dt}\frac{d\overrightarrow{OC}}{dt} + I_b\frac{d\vec{\Omega}}{dt} + \vec{\Omega} \wedge I_b\vec{\Omega}$$

By simplification, we obtain the relation (31)

$$\dot{\vec{h}} = (m\vec{\Omega} \wedge \vec{r}_g) \wedge \frac{d\overrightarrow{OC}}{dt} + m\vec{r}_g \wedge \frac{d}{dt}\frac{d\overrightarrow{OC}}{dt} + \vec{\Omega} \wedge I_b\vec{\Omega} + I_b\dot{\vec{\Omega}} + \vec{\Omega} \wedge \frac{d\overrightarrow{OC}}{dt} \qquad (31)$$

If we return to the expression for kinetic moment and do the derivation, we obtain the equation (32)

$$\dot{\vec{h}} = \int \dot{\vec{r}} \wedge \rho \vec{v})dv + \int (\vec{r} \wedge \rho \dot{\vec{v}})\, dv \qquad (32)$$

Where $\int (\vec{r} \wedge \dot{\vec{v}})$ dv $= \vec{M}$ (the dynamic moment applied to a undeformable solid), we obtain the equation (33)

$$\dot{\vec{h}} = \int \dot{\vec{r}} \wedge \rho \vec{v})dv + \vec{M} = \int \dot{\vec{r}}\rho dv \wedge \frac{d\overrightarrow{OC}}{dt} + \int \rho \dot{\vec{r}} \wedge \dot{\vec{r}} dv + \vec{M} \qquad (33)$$

By equalization of equations (31) and (32) we obtain the expression (34)

$$\vec{M} = m\vec{r}_g \wedge \frac{d}{dt}\frac{d\overrightarrow{OC}}{dt} + \vec{\Omega} \wedge I_b\vec{\Omega} + I_b\dot{\vec{\Omega}} + \vec{\Omega} \wedge \frac{d\overrightarrow{OC}}{dt} \quad (34)$$

By adopting the conventions [3]: $\left(\frac{d\overrightarrow{OC}}{dt}\right) = v_1,\ \Omega = v_2$
So the final equation (35) generates the moments that apply to the robot:

$$\Gamma_2 = m r_g \wedge \dot{v}_1 + v_2 \wedge I_b v_2 + v_2 \wedge v_1 + I_b \dot{v}_2 \quad (35)$$

Based on the equations obtained in the previous part, we obtain these equations (36):

$$m[\dot{u} - vr + wq + zg\,(pr + \dot{q})] = X$$

$$m[\dot{v} - wp + ur + zg\,(qr + \dot{p})] = Y$$

$$m[\dot{w} - uq + vp - zg\,(q^2 + p^2)] = Z \quad (36)$$

$$Ixx\dot{p} + (Izz - Iyy)qr - m[zg(\dot{v} - wp + ur)] = K$$

$$Iyy\dot{q} + (Ixx - Izz)rp + m[zg(\dot{u} - vr + wq)] = M$$

$$Izz\dot{r} + (Izz - Iyy)pq = N$$

b. Forces

The formulation of Fossen (Fossen, 2011), equation (37), is introduced.

$$M\dot{\nu} + C(\nu)\nu + D(\nu)\nu + g(\eta) = \tau \quad (37)$$

This general form involves the Rigid-body dynamics (mass, inertia), Hydrodynamic effects (added mass, drag), Buoyancy and gravity forces, also Thruster forces and control allocation.

i. The Rigid-body dynamics

M is the Mass Matrix which represents the total inertia of the vehicle. It can be written as the equation (38) bellow:

$$M = M_A + M_{RB} \quad (38)$$

Where:

M_A: Added mass matrix (due to the fluid moving with the body) is given by the equation(39). This is frequency-dependent in real scenarios but often approximated as constant for simplicity.

$$M_A = \begin{bmatrix} M_A^{11} & M_A^{12} \\ M_A^{21} & M_A^{22} \end{bmatrix} \tag{39}$$

Encoding the antisymmetric nature of the cross-product, this matrix is crucial for dynamic modeling. The mass, Coriolis, and centripetal matrices, essential for defining system behavior, can be derived from Kirchhoff's equations. This methodology finds its formal basis in Sagatun and Fossen's 1991 theorem on the Lagrangian formulation of vehicle dynamics.

The corresponding added mass matrices can be expressed as functions of the hydrodynamic derivatives, derived using a Lagrangian formulation based on Kirchhoff's equations. We obtain the equation (40)

$$M_A = \begin{bmatrix} X_{\dot{u}} & X_{\dot{v}} & X_{\dot{w}} & X_{\dot{p}} & X_{\dot{q}} & X_{\dot{r}} \\ Y_{\dot{u}} & Y_{\dot{v}} & Y_{\dot{w}} & Y_{\dot{p}} & Y_{\dot{q}} & Y_{\dot{r}} \\ Z_{\dot{u}} & Z_{\dot{v}} & Z_{\dot{w}} & Z_{\dot{p}} & Z_{\dot{q}} & Z_{\dot{r}} \\ K_{\dot{u}} & K_{\dot{v}} & K_{\dot{w}} & K_{\dot{p}} & K_{\dot{q}} & K_{\dot{r}} \\ M_{\dot{u}} & M_{\dot{v}} & M_{\dot{w}} & M_{\dot{p}} & M_{\dot{q}} & M_{\dot{r}} \\ N_{\dot{u}} & N_{\dot{v}} & N_{\dot{w}} & N_{\dot{p}} & N_{\dot{q}} & N_{\dot{r}} \end{bmatrix} \tag{40}$$

The Added mass matrix can be simplified and written as equation (41)

$$M_A = \begin{bmatrix} X_{\dot{u}} & 0 & 0 & 0 & 0 & 0 \\ 0 & Y_{\dot{v}} & 0 & 0 & 0 & 0 \\ 0 & 0 & Z_{\dot{w}} & 0 & 0 & 0 \\ 0 & 0 & 0 & K_{\dot{p}} & 0 & 0 \\ 0 & 0 & 0 & 0 & M_{\dot{q}} & 0 \\ 0 & 0 & 0 & 0 & 0 & N_{\dot{r}} \end{bmatrix} \tag{41}$$

In addition, simplifying the added mass matrix M_A to a diagonal form is a common and often justifiable approximation for underwater robots that reduces the complexity of the model especially in calculations, facilitate control design, and leverage the inherent symmetries of typical underwater vehicles, even if it's not perfectly precise for highly asymmetric bodies or complex maneuvers where coupling effects become prominent.

M_{RB} is the Rigid-body mass matrix (constant). This matrix (42) includes the actual mass and moments of inertia of the vehicle and it has this form:

$$M_{RB} = \begin{bmatrix} m & 0 & 0 & 0 & mz_g & -my_g \\ 0 & m & 0 & -mz_g & 0 & mx_g \\ 0 & 0 & m & my_g & -mx_g & 0 \\ 0 & -mz_g & my_g & I_{xx} & -I_{xy} & -I_{xz} \\ mz_g & 0 & -mx_g & -I_{xy} & I_{yy} & -I_{yz} \\ -my_g & mx_g & 0 & -I_{xz} & -I_{zy} & I_{zz} \end{bmatrix} \qquad (42)$$

In most cases, for an underwater robot designed with a certain symmetry (e.g., cylindrical or with a plane of symmetry), and if the local frame R_v is chosen to coincide with the principal axes of inertia and its origin is placed at the center of gravity, the M_{RB} matrix (43) simplifies considerably.

If the local frame R_v has its origin at the center of gravity ($x_g=y_g=z_g=0$) and its axes aligned with the principal axes of inertia (which makes the products of inertia zero, $I_{xy}=I_{xz}=I_{yz}=0$), then the M_{RB} matrix becomes diagonal:

$$M_{RB} = \begin{bmatrix} m & 0 & 0 & 0 & 0 & 0 \\ 0 & m & 0 & 0 & 0 & 0 \\ 0 & 0 & m & 0 & 0 & 0 \\ 0 & 0 & 0 & I_{xx} & 0 & 0 \\ 0 & 0 & 0 & 0 & I_{yy} & 0 \\ 0 & 0 & 0 & 0 & 0 & I_{zz} \end{bmatrix} \qquad (43)$$

Generally, the mass matrix of the underwater robot is expressed as the equation (44)

$$M = diag\{m \quad m \quad m \quad I_{xx} \quad I_{yy} \quad I_{zz}\} + diag\{X_{\dot{u}} \quad Y_{\dot{v}} \quad Z_{\dot{w}} \quad K_{\dot{p}} \quad M_{\dot{q}} \quad N_{\dot{r}}\} \qquad (44)$$

As well as the mass matrix, the Coriolis matrix expression is given by the equation (45)

$$C(v) = C_{RB}(v) + C_A(v) \qquad (45)$$

It consist of contributions from both the rigid-body dynamics of the vehicle and the hydrodynamic effects due to interaction with the surrounding fluid.

The Coriolis and centripetal matrix, due to hydrodynamic added mass, according to Fossen is expressed by the equations (46) and (47)

$$C_A(v) = \begin{bmatrix} 0_{3*3} & -S(M_A^{11}v_1 + M_A^{12}v_2) \\ -S(M_A^{11}v_1 + M_A^{12}v_2) & -S(M_A^{21}v_1 + M_A^{22}v_2) \end{bmatrix} \quad (46)$$

With

- The terms M_A^{11}, M_A^{12}, M_A^{21}, M_A^{22} refer to sub-blocks of the added mass matrix.
- S is the skew-symmetric matrix that transforms a cross-product operation into a matrix multiplication.

$$= \begin{bmatrix} 0 & 0 & 0 & 0 & -Z_{\dot{w}}w & Y_{\dot{v}}v \\ 0 & 0 & 0 & Z_{\dot{w}}w & 0 & -X_{\dot{u}}u \\ 0 & 0 & 0 & -Y_{\dot{v}}v & X_{\dot{u}}u & 0 \\ 0 & -Z_{\dot{w}}w & Y_{\dot{v}}v & 0 & -N_{\dot{r}}r & M_{\dot{q}}q \\ Z_{\dot{w}}w & 0 & -X_{\dot{u}}u & N_{\dot{r}}r & 0 & -K_{\dot{p}}p \\ -Y_{\dot{v}}v & X_{\dot{u}}u & 0 & -M_{\dot{q}}q & K_{\dot{p}}p & 0 \end{bmatrix} \quad (47)$$

Where:

- u,v,w are the body-fixed linear velocities.
- p,q,r are the body-fixed angular velocities.
- $X_{\dot{u}}$, $Y_{\dot{v}}$, $Z_{\dot{w}}$ are the added mass coefficients for translation.
- $K_{\dot{p}}$, $M_{\dot{q}}$, $N_{\dot{r}}$ are the added mass coefficients for rotation.

$$C_{RB}(v) = \begin{bmatrix} 0 & 0 & 0 & 0 & -mw & mv \\ 0 & 0 & 0 & mw & 0 & -mu \\ 0 & 0 & 0 & -mv & mu & 0 \\ 0 & -mw & mv & 0 & -I_{zz}r & I_{yy}q \\ mw & 0 & -mu & I_{zz}r & 0 & -I_{xx}p \\ mv & mu & 0 & -I_{yy}q & I_{xx}p & 0 \end{bmatrix} \quad (48)$$

For Equation (48) to be valid, it is assumed that the local frame's origin coincides with the center of gravity (xg=yg=zg=0). Furthermore, the frame's axes must be aligned with the robot's principal axes of inertia, which sets the products of inertia (Ixy, Ixz, Iyz) to zero.

ii. Hydrodynamic forces

The hydrodynamic damping is donated by D(v). While 6-DOF damping in underwater vehicles is highly nonlinear and coupled, especially at high speeds, Fossen (1995) points out two principal hydrodynamic damping forces that must be considered: quadratic drag and lift forces, and linear skin friction. This can be expressed by the equation (49):

$$D(v) = D_{linear}(v) + D_{nonlinear}(v) \qquad (49)$$

Consequently, most authors suggest simplifying assumptions for practical modelling. This common simplification for damping forces models the linear and quadratic drag along a given axis 'i' as solely dependent on its corresponding velocity component. This effectively assumes uncoupled vehicle motion and the presence of three (near) planes of symmetry.

Opposing the relative motion of a solid through a fluid, linear damping forces are exerted in the direction contrary to the object's relative velocity. The matrix representation for these forces (50) is given by:

$$D_{linear} = -\begin{bmatrix} X_u & 0 & 0 & 0 & 0 & 0 \\ 0 & Y_v & 0 & 0 & 0 & 0 \\ 0 & 0 & Z_w & 0 & 0 & 0 \\ 0 & 0 & 0 & K_p & 0 & 0 \\ 0 & 0 & 0 & 0 & M_q & 0 \\ 0 & 0 & 0 & 0 & 0 & N_r \end{bmatrix} \qquad (50)$$

While The Nonlinear Damping Matrix represents the part of the water's resistance (damping) that does not increase simply in proportion to speed. The general form of the diagonal nonlinear damping matrix is given by the equation (51):

$$
D_{nonlinear} = \begin{bmatrix} -(\mathbf{X}_{|u|u} \mid \mathbf{u} \mid) & 0 & 0 & 0 & 0 & 0 \\ 0 & -(\mathbf{Y}_{|v|v} \mid \mathbf{v} \mid) & 0 & 0 & 0 & 0 \\ 0 & 0 & -(\mathbf{Z}_{|w|w} \mid \mathbf{w} \mid) & 0 & 0 & 0 \\ 0 & 0 & 0 & -(\mathbf{K}_{|p|p} \mid \mathbf{p} \mid) & 0 & 0 \\ 0 & 0 & 0 & 0 & -(\mathbf{M}_{|q|q} \mid \mathbf{q} \mid) & 0 \\ 0 & 0 & 0 & 0 & 0 & -(\mathbf{N}_{|r|r} \mid \mathrm{r} \mid) \end{bmatrix} \tag{51}
$$

So, the damping force D(v) is expressed by the equations (52)

$$
F_D = -\begin{pmatrix} X_u \\ Y_v \\ Z_w \\ K_p \\ M_q \\ N_r \end{pmatrix} - \begin{pmatrix} (\mathbf{X}_{|u|u} \mid \mathbf{u} \mid) \\ (\mathbf{Y}_{|v|v} \mid \mathbf{v} \mid) \\ (\mathbf{Z}_{|w|w} \mid \mathbf{w} \mid) \\ (\mathbf{K}_{|p|p} \mid \mathbf{p} \mid) \\ (\mathbf{M}_{|q|q} \mid \mathbf{q} \mid) \\ (\mathbf{N}_{|r|r} \mid \mathrm{r} \mid) \end{pmatrix} \tag{52}
$$

Where X_u and $(\mathbf{X}_{|u|u}$ are the lineal viscous coefficient and the quadratic coefficient respectively in the x direction.

In the field of underwater vehicle modelling, most researchers and engineers (G. Conte, 1997)(McMillan, 1995) consider the lift force to be of a magnitude so much smaller than the drag force that it can be neglected or considered insignificant in the dynamic model.

The Drag force which is opposite to the robot's movement through water is always defined whenever there is relative motion between the robot and the water. While, the lift force which is perpendicular to the direction of motion, generated by fluid flow around an asymmetrical shape is insignificant because of the robot's geometry (cylindrical or symmetric).

So, we can consider this equation (53)

$$
\mathbf{X}_{|u|u} \mid \mathbf{u} \mid = \frac{1}{2}\rho C_{D_x} A_x |u|u \tag{53}
$$

Where r is the fluid density, C_{Dx} is the drag coefficient in the x direction and A_x is the area of the section of the body in the x direction, so equation (54) define the new form of the matrix (52)

$$F_D = \begin{pmatrix} X_u \\ Y_v \\ Z_w \\ K_p \\ M_q \\ N_r \end{pmatrix} - \begin{pmatrix} \frac{1}{2}\rho C_{D_x} A_x |u| u \\ \frac{1}{2}\rho C_{D_y} A_y |v| v \\ \frac{1}{2}\rho C_{D_z} A_z |w| w \\ \frac{1}{2}\rho C_{D_p} A_p |p| p \\ \frac{1}{2}\rho C_{D_q} A_q |q| q \\ \frac{1}{2}\rho C_{D_r} A_r |r| r \end{pmatrix} \tag{54}$$

Or it can be written as a matrix (55)

$$F_D(\nu) = -\begin{bmatrix} X_u + (X_{|u|u} \mid u \mid) & 0 & 0 & 0 & 0 & 0 \\ 0 & Y_v + (\mathbf{Y}_{|v|v} \mid \mathbf{v} \mid) & 0 & 0 & 0 & 0 \\ 0 & 0 & Z_w + (\mathbf{Z}_{|w|w} \mid \mathbf{w} \mid) & 0 & 0 & 0 \\ 0 & 0 & 0 & K_p + (\mathbf{K}_{|p|p} \mid \mathbf{p} \mid) & 0 & 0 \\ 0 & 0 & 0 & 0 & M_q + (\mathbf{M}_{|q|q} \mid \mathbf{q} \mid) & 0 \\ 0 & 0 & 0 & 0 & 0 & N_r + (\mathbf{N}_{|r|r} \mid \mathbf{r} \mid) \end{bmatrix} \tag{55}$$

iii. Buoyancy and gravity forces

Based on the Archimedes's principle 'Any body immersed in a liquid experiences an upward vertical buoyant force from that liquid, equal to the weight of the volume of liquid displaced.'

The form of the buoyancy force is given by the equation (56)

$$F_B = \rho \nabla g \tag{56}$$

Where ρ is the fluid density, ∇ the body volume and g the gravity acceleration. Or it can be described in the earth fixed frame R_0 as the equation (57)

$$f_B = \begin{pmatrix} 0 \\ 0 \\ -\rho \ \nabla \ g \end{pmatrix} \tag{57}$$

The buoyant force is exclusively vertical. The Earth-fixed frame's Z0 axis is defined as positive downwards. That's why the buoyant force is expressed as a negative component along the Z-axis.

To calculate the buoyancy torque (when the body's gravity centre and the body's buoyancy centre do not coincide) which is crucial for the stability of the robot, we use the equation (58) bellow:

$$\mathbf{T_B} = r_B \times f_B \tag{58}$$

Where r_B represents the position vector of the center of buoyancy (CB) with respect to the origin of the global frame.

To express the buoyant force in the body frame, the buoyant force in the Earth-fixed frame is transformed using the rotation matrix, we obtain the equation (59)

$$
\begin{aligned}
{}^{G}f_B &= \begin{pmatrix} c\Psi c\theta & s\Psi c\theta & -s\theta \\ -s\Psi c\Phi + c\Psi s\theta s\Phi & c\Psi c\Phi + s\Phi s\theta s\Psi & c\theta s\Phi \\ s\Psi s\Phi & -c\Psi s\Phi + c\Phi s\theta s\Psi & c\theta c\Phi \end{pmatrix} \cdot \begin{pmatrix} 0 \\ 0 \\ -\rho\ \nabla\ g \end{pmatrix} \\
&= \begin{pmatrix} s\theta B \\ -c\theta s\Phi B \\ -c\theta c\Phi B \end{pmatrix}
\end{aligned}
\tag{59}
$$

With $B = \rho \nabla g$

So, through a rigorous calculation, we get these two equations (60)

$$
{}^{G}\mathbf{T_B} = \begin{pmatrix} x_b \\ y_b \\ z_b \end{pmatrix} \times \begin{pmatrix} s\theta B \\ -c\theta s\Phi B \\ -c\theta c\Phi B \end{pmatrix} = \begin{pmatrix} -y_b c\theta c\Phi B + z_b c\theta s\Phi B \\ z_b s\theta B + x_b c\theta s\Phi B \\ -x_b c\theta c\Phi B - y_b s\theta B \end{pmatrix}
\tag{60}
$$

So, we obtain this vector (61) which accounts for the buoyancy's contribution to the restoring forces and moments ($g(\eta)$) in the body-fixed frame.

$$
{}^{G}F_B = \begin{pmatrix} s\theta B \\ -c\theta s\Phi B \\ -c\theta c\Phi B \\ -y_b c\theta c\Phi B + z_b c\theta s\Phi B \\ z_b s\theta B + x_b c\theta s\Phi B \\ -x_b c\theta c\Phi B - y_b s\theta B \end{pmatrix}
\tag{61}
$$

To obtain the complete $g(\eta)$ term, one would also need to add the forces and moments due to gravity acting at the CG. First, we focus on the gravity's force which is easy to express in the fixed-body frame. It is given by the equation (62)

$$f_G = \begin{pmatrix} 0 \\ 0 \\ mg \end{pmatrix} \qquad (62)$$

With m the mass of the vehicle. To calculate the gravity torque we need this equation (63)

$$\mathbf{T}_G = r_C \times f_G \qquad (63)$$

The vector (64) represents the gravity force (W=mg represents net weight) expressed in the body-fixed frame, after being transformed from the Earth-fixed frame. This force is vertical and opposite of the buoyancy force.

$${}^{G}f_G = \begin{pmatrix} -s\theta W \\ c\theta s\Phi W \\ c\theta c\Phi BW \end{pmatrix} \qquad (64)$$

The same calculation as previously is performed to derive the gravitational torque vector (65)

$$\mathbf{T}_G = \begin{pmatrix} x_C \\ y_C \\ z_C \end{pmatrix} \times \begin{pmatrix} -s\theta W \\ c\theta s\Phi W \\ c\theta c\Phi W \end{pmatrix} = \begin{pmatrix} y_C c\theta c\Phi W - z_C c\theta s\Phi W \\ -z_C s\theta W - x_C c\theta c\Phi W \\ x_C c\theta s\Phi W + y_C s\theta W \end{pmatrix} \qquad (65)$$

We get consequently the matrix (66)

$${}^{G}F_G = \begin{pmatrix} -s\theta W \\ c\theta s\Phi W \\ c\theta c\Phi BW \\ y_C c\theta c\Phi W - z_C c\theta s\Phi W \\ -z_C s\theta W - x_C c\theta c\Phi W \\ -z_C s\theta W - x_C c\theta c\Phi W \end{pmatrix} \qquad (66)$$

Equations (61) and (66) give the complete Restoring Forces and Moments Vector as shown in the equation (67)

$$\mathbf{g}\,(\boldsymbol{\eta}) = {}^{G}F_G + {}^{G}F_B = \begin{pmatrix} s\theta B \\ -c\theta s\Phi B \\ -c\theta c\Phi B \\ -y_b c\theta c\Phi B + z_b c\theta s\Phi B \\ z_b s\theta B + x_b c\theta s\Phi B \\ -x_b c\theta c\Phi B - y_b s\theta B \end{pmatrix} + \begin{pmatrix} -s\theta W \\ c\theta s\Phi W \\ c\theta c\Phi BW \\ y_C c\theta c\Phi W - z_C c\theta s\Phi W \\ -z_C s\theta W - x_C c\theta c\Phi W \\ -y_c s\theta W - x_C c\theta c\Phi W \end{pmatrix}$$

$$
= \begin{pmatrix} s\theta B - s\theta W \\ -c\theta s\Phi B + c\theta s\Phi W \\ -c\theta c\Phi B + c\theta c\Phi W \\ (-\mathbf{y}_b c\theta c\Phi B + \mathbf{z}_b c\theta s\Phi B) + (\mathbf{y}_C c\theta c\Phi W - \mathbf{z}_C c\theta s\Phi W) \\ (\mathbf{z}_b s\theta B + \mathbf{x}_b c\theta s\Phi B) + (-\mathbf{z}_C s\theta W - \mathbf{x}_C c\theta c\Phi W) \\ (-\mathbf{x}_b c\theta c\Phi B - \mathbf{y}_b s\theta B) + (-\mathbf{z}_C s\theta W - \mathbf{x}_C c\theta c\Phi W) \end{pmatrix} \tag{67}
$$

By simplifying the last equation, we obtain the equation (68)

$$
\mathbf{g}\,(\boldsymbol{\eta}) = \begin{pmatrix} (W - B)s(\theta) \\ -(W - B)c(\theta)s(\phi) \\ -(W - B)c(\theta)c(\phi) \\ -(y_c W - y_b B)c(\theta)c(\phi) + (z_c W - z_b B)c(\theta)s(\phi) \\ (z_c W - z_b B)s\theta + (x_c W - x_b B)c\theta c\Phi \\ -(x_c W + x_b B)c\theta s\Phi + (y_c W - y_b B)s\theta \end{pmatrix} \tag{68}
$$

It is important to note that when the center of gravity coincides with the center of buoyancy, the moments (or torques) caused by gravity and buoyancy vanish. This is because, as the force of gravity and the buoyant force act along the same vertical line (passing through the common CG/CB point), they are unable to create a net torque that would induce rotation in the robot

In that case, g (η) assumes this expression (69)

$$
g\,(\eta) = \begin{pmatrix} F_x \\ F_y \\ F_z \\ 0 \\ 0 \\ 0 \end{pmatrix} \tag{69}
$$

Where $F_{x,}$ Fy and F_z are the are the components of the net gravitational/buoyant force, expressed in the body-fixed frame.

iv. Thruster forces and control allocation

The propulsion units, or thrusters, are responsible for generating the force required to propel and steer the underwater robot. Their operation typically relies on the principle of propeller-driven propulsion, where the spinning of a propeller displaces water rearward, thereby creating a reactive force that pushes the robot forward (thrust).

The force (F) generated by a thruster depends on several factors, including its rotational speed (RPM), the voltage or current applied to the motor, the geometry of the propeller, and the properties of the fluid (water density).

Thruster configuration is a critical design choice for underwater vehicles (see fig.11), affecting their maneuverability, ability to withstand faults, and energy use. Traditional 6-DOF designs, often with orthogonal thrusters, provide fundamental motion control but offer limited resilience to thruster malfunctions (Vega, 2016) (Sarkar, 2002) (Kalikatzarakis, 2021). Research by Kadiyam et al. (Kadiyam, 2021) (KS, Kadiyam, & Mohan, 2023)compared fixed versus vector thruster schemes, highlighting their differing trajectory tracking capabilities. For a more general approach, Xu et al. (Xu, 2024) proposed a configuration solution that minimizes energy consumption while respecting thruster saturation, solving this non-convex optimization problem via quadratic programming. Another innovative approach by Tanakitkorn et al. (Tanakitkorn, 2022)presented a convertible thruster setup, demonstrating strong performance in both path tracking and dynamic hovering.

Figure 11. Underwater Thruster Operation

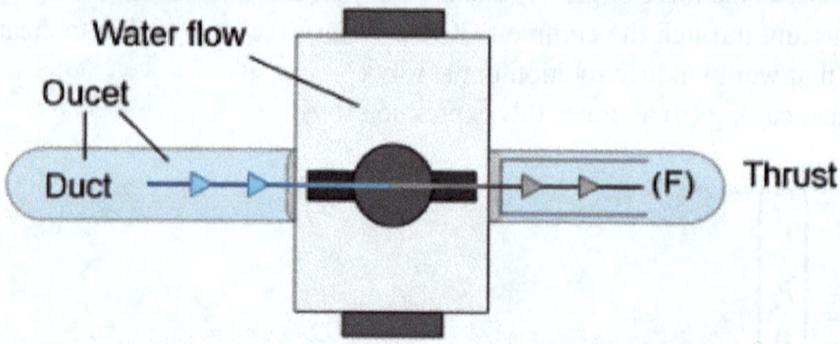

The function of thrust allocation is to translate the 6-DOF control forces and torques, commanded by the vehicle's higher-level controller, into specific outputs for each thruster within the propulsion system. Consequently, the thrusters produce thrust based on the optimal solution derived from the allocation optimization.

Redundant thruster systems, while beneficial, introduce significant complexity into control allocation. Traditional techniques, such as pseudo-inverse-based least squares, typically focus solely on minimizing thrust errors, often neglecting crucial factors like energy efficiency and thruster saturation, which can result in suboptimal performance or instability (Baldini, et al., 2018). Consequently, researchers have developed more advanced methods, including weighted pseudoinverse, task-priority approaches, and nonlinear optimization, to address these

limitations (Podder, 2001) (Wang, 2022) (Wang, Huang, Li, & Ren, 2024). Among these, optimization-based thrust allocation strategies, particularly those leveraging quadratic programming (QP), have gained prominence. QP frameworks excel at simultaneously minimizing energy consumption and adhering to physical constraints like thruster saturation. A notable application is Rethfeldt et al.'s work (Rethfeldt, 2024), where QP was used to optimize actuator utilization for an underwater vehicle equipped with variable buoyancy systems.

CONTROL ARCHITECTURE

While the kinematic and dynamic modeling of underwater vehicles primarily relies on Fossen's framework, the reliance on constant hydrodynamic coefficients imposes crucial limitations in operational contexts. These nominal models fail to accurately capture complex non-linear effects in the marine environment, most notably the progressive changes induced by biofouling and vehicle aging (S. Wang, 2021), thereby justifying the essential need for robust control strategies.

A controller is crucial for an underwater vehicle as it guarantees stability, manoeuvrability, and accurate performance in the often unpredictable and challenging underwater environment. It is the brain of the underwater robot. Without a robust control system, vehicles like Autonomous Underwater Vehicles (AUVs) and Remotely Operated Vehicles (ROVs) would struggle to navigate, complete tasks, and maintain their desired position.

There is a wide range of underwater robot. Control methods range from the basic classical Proportional-Integral-Derivative (PID) controllers, which are effective and simple for many tasks, to more complex approaches such as Model Predictive Control (MPC) (Lammouchi, 2025) (Youssef, 2023), adaptive control, and robust control, which deal with

nonlinearities and external perturbations like ocean currents.We have for exemple the Kalman Filter which is a recursive algorithm essential in robotics, serving to estimate a dynamic systems internal state from noisy mesurements (Romdhane, 2023). The effectiveness of these controllers is strongly dependent on reliable real-time

state information that is frequently obtained via noisy sensor data.

In this section we will focus on the FOPID/ PID controllers.

1. PID Controller

A PID controller is a closed-loop regulation instrument that processes sensor input to determine the difference between the actual and desired values (Figure 12). It then adjusts outputs to manage variables like temperature, flow, speed, pressure, or voltage. This is achieved through three mechanisms: proportional control (responding to current error), integral control (correcting past accumulated errors), and derivative control (anticipating future errors). The sum of these three components calculates the final output, enabling efficient process control and system stability.

Figure 12. Block Diagram for PID control

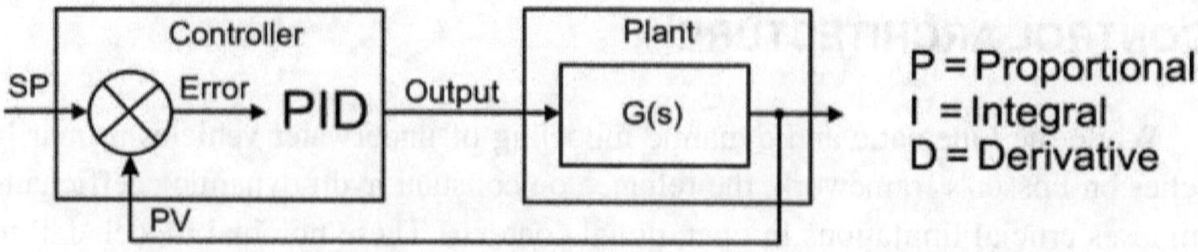

The transfer function of the PID is given by the expression (70)

$$C(S) = K_p + \frac{K_I}{S} + K_D S \qquad (70)$$

In this underwater robot project, the PID (Proportional-Integral-Derivative) controller is a fundamental and widely used control strategy for managing the state output, which represents the robot's position and orientation in the terrestrial frame (or inertial frame).

A PID controller operates by calculating an error between the desired value (setpoint) and the measured output value. The form of expression is given by the equation (71)

$$Error = state\_desired - state \qquad (71)$$

The figure 13 describes the integration of PID controller in the whole model

Figure 13. Control schema of the underwater robot

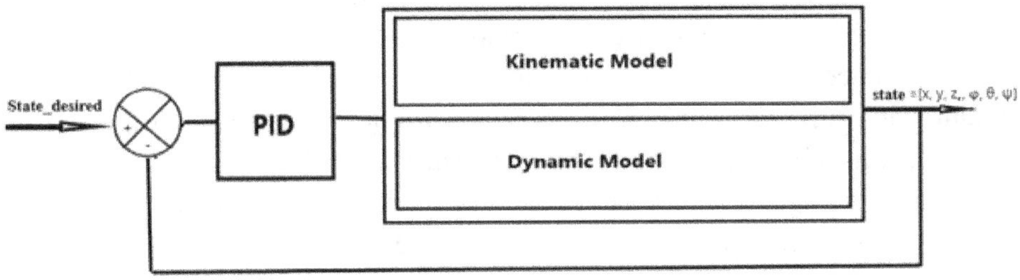

Now we focus on the three terms (P-I-D):

Proportional Term (P): It is proportional to the current error. The larger the error, the stronger the correction. This term helps to quickly reduce the error.

If the robot is far from its target position or orientation, the P term will generate a command proportional to this distance/angular difference.

Integral Term (I): It is proportional to the integral of the error over time. This term helps to eliminate static errors (steady-state errors) that might persist with the P term alone, often due to constant disturbances (like a weak current or a buoyancy imbalance).

If the robot slowly drifts from its target, the I term will accumulate this small error and increase the command to fully correct it.

Derivative Term (D): It is proportional to the rate of change of the error. This term anticipates future errors and dampens oscillations. It helps to stabilize the system and prevent overshoots. If the robot approaches its target too quickly or oscillates around it, the D term will generate a command to slow down the movement and stabilize the response.

However, the problem that arises is how to adjust the PID gains. Because, tuning the terms of a PID controller (KP,KI,KD) for an underwater robot which is a nonlinear system is a significant challenge, as the optimal performance of the gains can vary depending on the robot's state (speed, orientation) and the environment (currents, depth). These Challenges related to underwater robot nonlinearity arise from several factors. Firstly, variable dynamics, including Coriolis terms, quadratic damping, and restoring forces (gravity/buoyancy), render the robot's dynamics highly nonlinear, meaning that PID gains effective at one speed or orientation may become suboptimal or unstable at others. Secondly, movements in 6-DOF are intrinsically coupled, which significantly complicates the independent tuning of each axis. Lastly, uncertainties are prevalent, as model parameters (such as added masses and damping coefficients) are often poorly known or subject to variation, and the environment itself introduces unpredictable disturbances like currents and waves.

PID gain tuning for nonlinear underwater robots can be approached through several methods, each with its own advantages and complexities.

- **Empirical (Trial-and-Error) methods** offer a simple starting point, involving manual adjustment of gains (K_P,K_I,K_D) by observing system response, though their effectiveness is limited in highly nonlinear systems where optimal gains vary with operating conditions.
- **Ziegler-Nichols Rule-Based Methods** provide initial gain calculations from system responses, but similarly require linearization around specific operating points for nonlinear applications. Model-Based Tuning leverages an accurate dynamic model for controller design, using techniques like pole placement or H-infinity synthesis, and can be applied to nonlinear systems through linearization or direct nonlinear control methods.
- **Optimization Techniques**, such as genetic algorithms, are highly effective for nonlinear systems, finding optimal PID gains by minimizing a cost function during simulations. Gain Scheduling Control is a common and highly relevant approach for underwater robots, as it involves defining multiple sets of optimal PID gains for different operating regions (e.g., speed or depth ranges) and dynamically switching between them based on the robot's current state.
- **Adaptive PID Control** represents an advanced solution where the controller adjusts its gains in real-time to manage nonlinearities and uncertainties, though it demands more complex implementation and stability guarantees.

Table (1) describes the effect of the parameter of each action on time response and stability of the system (Vukic Z. &., 2002)

Table 1. Effects of increasing the value of each parameter of conventional PID controller

Parameter	Rise time	Overshoot	Setting time	State error	Stability
K_P	Decrease	Increase	Small change	Decrease	Degrade
K_i	Decrease	Increase	Increase	Eliminate	Degrage
K_d	Minor change	Decrease	Decrease	No effect	Improve if K_d small

For an underwater robot project with nonlinear dynamics, we start with empirical tuning to find basic gains for a "nominal" operating point. Then, we define multiple sets of gains for different speed or orientation ranges. After that, we simulate the robot's behavior with the nonlinear model and we test different gain sets and scheduling strategies to see how the robot responds to various setpoints and disturbances. The results are reasonable.

2. Fopid

Initially introduced by Podlubny in 1999, the FOPID controller represents an advancement of the traditional PID controller, incorporating principles from fractional calculus. (Joshu & Talange, 2013)

It is same as conventional PID controller when λ=1 and δ=1, it is PI controller when δ=0, and it is PD controller when λ=0. The figure 14 illustrates the block diagram of a single input - single output closed loop control system with fractional controller (FOPID) (Vukic, 2002) (Meng, 2009) (Cao et al., 2005)

Figure 14. Block diagram of FOPID controller

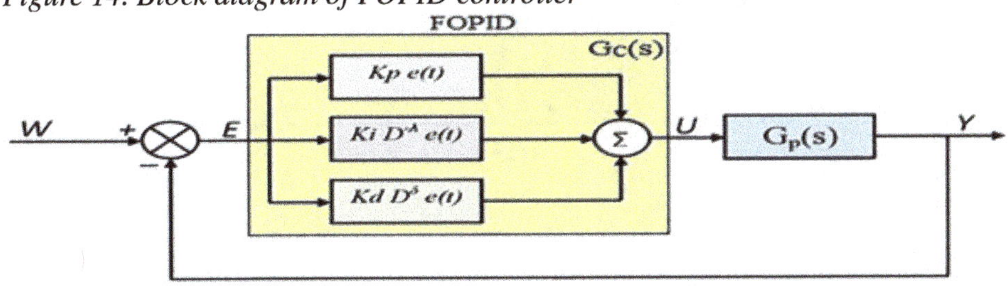

Fractional Order PID (FOPID) controllers offer significant advantages over their conventional counterparts, primarily due to their increased flexibility. With five tuning parameters (K_P,K_I,K_D,λ,μ) instead of three, FOPIDs allow for a finer controller design capable of achieving superior performance and improved robustness against model uncertainties and disturbances. This flexibility is particularly beneficial for managing complex and nonlinear systems, such as underwater robots, where dynamics vary with the environment, ensuring more stable performance over a broader operating range. Furthermore, adjusting the fractional orders λ and μ enables more precise shaping of the transient response (rise time, overshoot) and optimized elimination of steady-state error. Lastly, fractional operators imbue FOPIDs with "memory" properties that integrate the signal's past behavior over a longer period, which can be advantageous for certain systems.

The differential equation of the FOPID controller is described by the equation (72):

$U(t) = K_p e(t) + K_i D - \lambda () + K_d D \delta e(t)$ (72)

Where, λ and δ are fractional orders. By taking Laplace transform for the above equation, the controller equation becomes:

$U(s) = K_p e(s) + K_i s^{-\lambda} e(s) + K_d s^{\delta} e(s)$ (73)

The PID controller is a special case of FOPID controller, where $\lambda = \delta = 1$.The conventional PID controller is represented by fourpoints (P, PI, PD, PID), while fractional FOPID controller is expanded to plane. This expansion adds more flexibility and accuracy, Figure 15 displays this concept (Tehrani, 2011)

Figure 15. Generalization of FOPID Controller: From point to plan

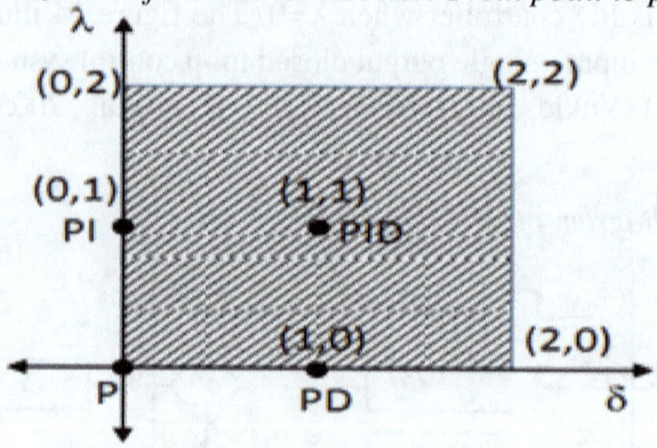

In the context of the underwater robot project, the FOPID could potentially offer superior performance to the classical PID since it is more advanced, particularly for navigation in the presence of unpredictable currents, precise position keeping or trajectory tracking despite the robot's dynamic nonlinearities, and improved robustness to variations in robot parameters (for example, due to changes in payload). However, this would involve a greater design and tuning effort. Because, FOPID controllers present two major challenges. Firstly, their tuning complexity is increased, as they require the adjustment of five parameters (K_P, K_I, K_D, λ, μ) compared to the three of conventional PID, often necessitating optimization methods like genetic algorithms or PSO to find optimal gains. Secondly, their practical implementation is more complex, as it requires the use of numerical approximations (such as Oustaloup or Grünwald-Letnikov filters) to handle fractional orders, which can add significant computational overhead.

First of all, we must know how to implement and tune this controller. Fractional-order derivative and integral operators (s^{μ} and $s^{-\lambda}$) cannot be directly implemented in a numerical (discrete) system. They must be approximated by integer-order filters. As well as the PID controller, there are multiple approaches that we can use such us:

- **Oustaloup Approximation:** This is one of the most popular methods. It approximates the fractional operator by an integer-order rational transfer function within a specified frequency band. The choice of filter order (number of poles and zeros) and the approximation frequency band is crucial for accuracy. The higher the order, the more precise the approximation, but the higher the computational complexity.
- **Grünwald-Letnikov Approximation (or Riemann-Liouville):** This method uses a discrete definition of the fractional derivative/integral based on a weighted sum of past signal values. It is more intuitive for direct implementation but may require significant computational "memory" (error history) for good accuracy.
- **CFE (Continued Fraction Expansion) Approximation:** Other methods exist, offering different properties in terms of precision and complexity.

Tuning the five parameters of the FOPID is an optimization problem. Unlike the classical PID where empirical or rule-based methods (like Ziegler-Nichols) can yield good results, the FOPID generally requires more sophisticated approaches:

- **Numerical optimization methods** represent the most common approach for tuning the five parameters of a FOPID controller. This process begins with defining a cost function (or objective function) that quantifies the desired system performance, using metrics such as ISE, ITAE, IAE, or ISTE, which penalize tracking or position-keeping errors and can include terms for overshoot, response time, or control effort. Subsequently, metaheuristic optimization algorithms like Genetic Algorithms (GA), Particle Swarm Optimization (PSO), Simulated Annealing (SA), or Ant Colony Optimization (ACO) are employed to explore the five-parameter space and identify the combination that minimizes this cost function. This exploration involves numerous simulations of the underwater robot with various FOPID parameter sets, where the cost function is calculated for each simulation, allowing the algorithm to iteratively adjust parameters towards an optimal solution. Finally, once optimal parameters are determined through simulation, rigorous validation is essential, first in simulated environments with realistic disturbances, and then, if feasible, on a physical prototype.
- **Methods based on Robustness and Frequency Domain Specification:** Some approaches aim to guarantee system robustness by specifying properties in the frequency domain (e.g., specific phase and gain margins). These

methods can be used to derive FOPID parameters, but they are more mathematically complex.

- **Gain Scheduling with FOPID:** For a highly nonlinear underwater robot, even a FOPID with fixed gains may not be sufficient across the entire operating range. The FOPID can then be combined with a gain scheduling strategy, where different sets of five FOPID parameters are used depending on the robot's state (e.g., speed, depth).

3. Comparison between PID and FOPID

The process of selecting an appropriate controller for an underwater robot typically involves choosing between the well-established classical PID controller and its sophisticated counterpart, the FOPID (Fractional Order PID) (Mohamed, 2014). While both serve as foundational regulation instruments, their differing characteristics dictate their suitability for particular applications, notably in the context of nonlinear and complex underwater vehicle systems.

So that we can make the choice of the adequate controller, we rely on this table which presents the advantages and disadvantages of each type of control:

Table 2. Comparison between PID and FOPID

	Advantages	Disadvantages
PID	• Simplicity (only three terms to tune) • easy to implement • Wide of use (it is the most widely used controller industry with various well-established methods of tuning) • Efficiency specially for linear system	• Limited flexibility(PID uses only 3 terms so it can be difficult to optimize all aspects of performance simultaneously) • Suboptimal performance in nonlinear (For underwater robots with highly nonlinear and variable dynamics, PID cannot give the expected results)
FOPID	• Increased flexibility(the intervention of the two additional parameters λ and μ) • Superior performance(FOPID offer better response) • improved robustness • Better handling for non-linearities	• Adjustments complexity, • Implementation complexity (Fractional orders cannot be directly implemented in digital systems. • Their integration requires numerical approximations (e.g., Oustaloup or Grünwald-Letnikov filters), which adds computational and design complexity)

Now we will compare the two controllers PID and FOPID based on the simulation of errors on Matlab/Simulink. As we noted before, the expression of the error given in equation (71) is

Error = state_desired – state

Where *state_desired* is the setpoint and *state* is measured output value.

The Fig.12 illustrates the error's simulation of both the controllers PID and FOPID. Scope blocks are used to display the time-domain plots of the error signals from both the PID and FOPID loops. By connecting them to the outputs of the error "Sum" blocks, we can visually compare how quickly and accurately each controller reduces the error to zero.

This Simulink setup allows to run simulations and directly observe the differences in error response between a classical PID and a FOPID controller for the same commanded input and robot dynamics.

Figure 16. The error responses of PID and FOPID controllers

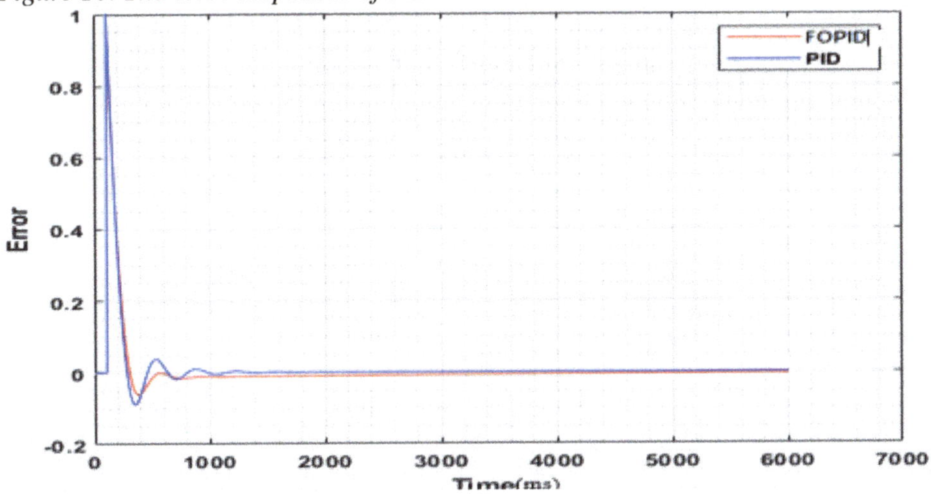

This figure describes the evolution of the tracking error over time for a system controlled by a conventional PID controller (in blue) and a FOPID controller (in red). The error represents the difference between the desired setpoint and the actual output of the system. The objective of a good controller is to reduce this error to zero as quickly as possible and to keep it there.

Analyzing the error figure, it is observed that, from the beginning of the simulation (at t=0), the initially high error is rapidly reduced by both controllers. However, a slight undershoot (the error becomes negative, indicating that the output overshoots the setpoint before returning) is visible for both systems, with the PID (Blue) showing a more pronounced initial undershoot (reaching approximately -0.05) than the FOPID (Red). After this undershoot, both controllers exhibit oscillations around zero. The PID (Blue) displays oscillations with a slightly larger amplitude that damp out more slowly, requiring several cycles before complete stabilization. The FOPID (Red), on the other hand, shows oscillations of smaller amplitude and faster damping, converging to zero more smoothly and directly. In terms of Error Settling Time, the FOPID reaches an error close to zero and maintains it stably more quickly than the PID, being practically zero after approximately 1000-1500 mseconds (depending on the time scale). The residual oscillations of the PID are visible for longer. Finally, regarding Steady-State Error, both controllers successfully reduce it to zero once the transients have faded, which is an essential characteristic. In conclusion, the FOPID demonstrates superior performance in reducing transient error, with a less pronounced undershoot, faster damping, and a more prompt convergence to zero error.We also tried to

simulate the output signal for both FOPID and PID controllers. The results are shown in Figure 17:

Figure 17. Output responses of PID and FOPD controllers

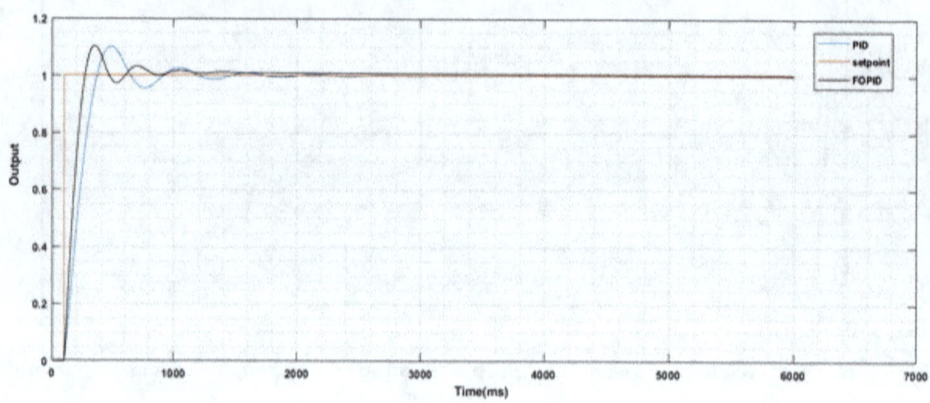

By analyzing the output responses of PID and FOPID controllers reveals several key observations. Regarding Rise Time, both controllers demonstrate a fast initial response time, reaching the setpoint of 1 in less than 0.5 seconds (about 250-300 ms for the first PID overshoot, and a little slower for the FOPID), which indicates a rapid establishment dynamic for the system. However, notable differences appear in terms of Overshoot: the PID controller (represented in blue) exhibits a significant overshoot, reaching a value of approximately 1.12 before starting to oscillate. This pronounced overshoot can be undesirable in applications requiring high precision and an absence of over-reaction (for instance, to avoid collisions or abrupt robot movements). In contrast, the FOPID controller (in black) shows a much lower overshoot, reaching about 1.08, indicating better precision and less aggressiveness during the transient phase. As for Oscillations and Settling Time, the PID's response, after its initial overshoot, is characterized by damped oscillations that take approximately 1.5 to 2 seconds to settle around the setpoint. The FOPID, on the other hand, also exhibits oscillations and appears to settle at a similar rate to the PID, or perhaps slightly slower, also around 1.5 to 2 seconds. Finally, concerning Steady-State Error, both controllers effectively manage to eliminate this error, with the output stabilizing stably at the setpoint of 1 in the long term (after 2-3 seconds). Overall, the FOPID demonstrates superior performance in terms of overshoot reduction, as its peak is lower, offering a smoother and more precise response.

To conclude, we tried to implement both PID and FOPID controllers. We find that for an underwater robot, the FOPID has the potential to offer superior control performance compared to the classical PID, particularly for precise navigation, position keeping in the presence of disturbances, and the management of nonlinear dynamics. However, this increased performance comes at the cost of significantly higher design, tuning, and implementation complexity. The choice between the two will depend on the specific requirements of the project, the resources available for design and tuning, and the desired level of performance and robustness.

SIMULATION SETUP AND RESULTS

The results are operated on the virtual platform Matlab/Simulink. Simulink/MATLAB allows you to simulate human interactions using blocks that represent human inputs (instructions, adjustments) and visualization tools that simulate feedback to the operator. For more complex interactions, Simulink can be integrated with MATLAB graphical interfaces or even external platforms.

Figure 18. The six sub-plots

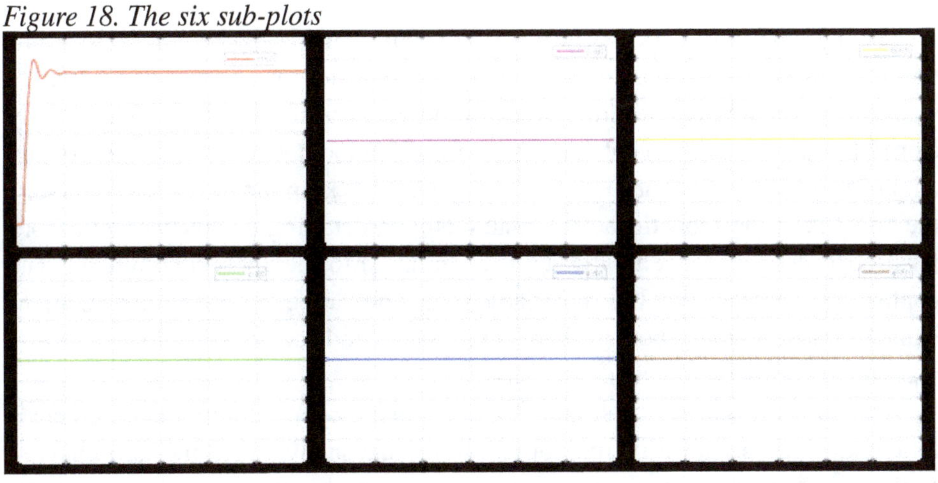

As depicted in figure 18 the robot swiftly moves from its starting point at x=0 to the desired position of x=1 in just a few seconds. This rapid response is accompanied by an overshoot, which is attributed to the PID controller's characteristics. Nevertheless, the other results confirm that the robot maintains stable control across its depth, lateral position, and all rotational axes (roll, pitch, and yaw).

Figure 19. The setpoint/actual state comparison plot

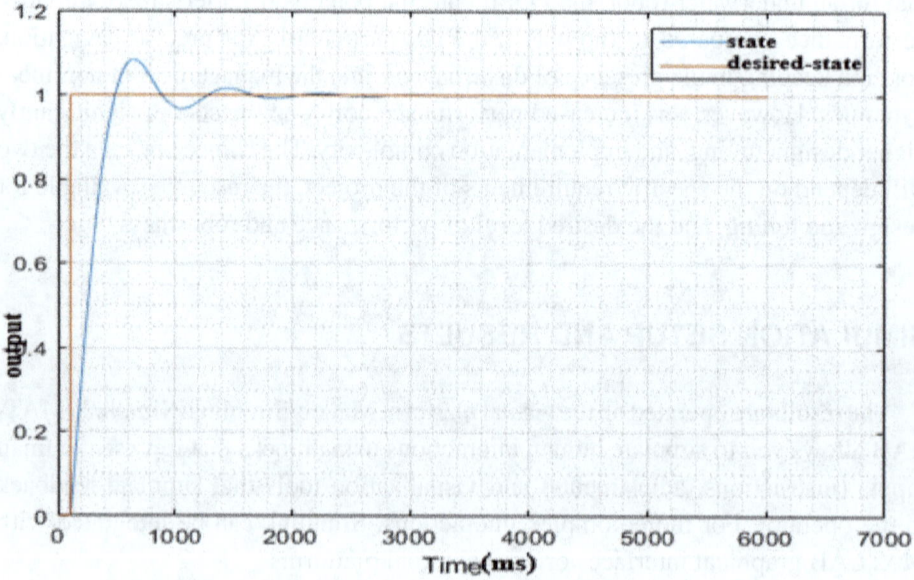

The figure 19 presents a time-domain plot comparing the desired input (setpoint) with the actual system output (η). The red trace signifies the commanded reference signal, implemented as a step input where the target value transitions instantaneously from 0 to 1 at t=0 and is maintained thereafter. This methodology is conventionally employed to evaluate a control system's command-tracking capabilities. Conversely, the blue trace represents the actual system output, detailing the underwater robot's positional response (specifically its 'x' coordinate) to the applied command. The control system governing the 'η' variable (presumed to be the 'x' position) demonstrates robust design for both rapid responsiveness and accuracy, exhibiting a swift reaction to abrupt command alterations.

Now we will focus on 3D simulation of an underwater robot in MATLAB environment, which is a powerful tool for visualizing and analyzing its motion behavior, testing controllers, and validating mission strategies before physical deployment. It combines Simulink's dynamic modeling with MATLAB's 3D visualization capabilities.

This 3D simulation is given by the Figure 20 below:

Figure 20. A linear trajectory of the robot

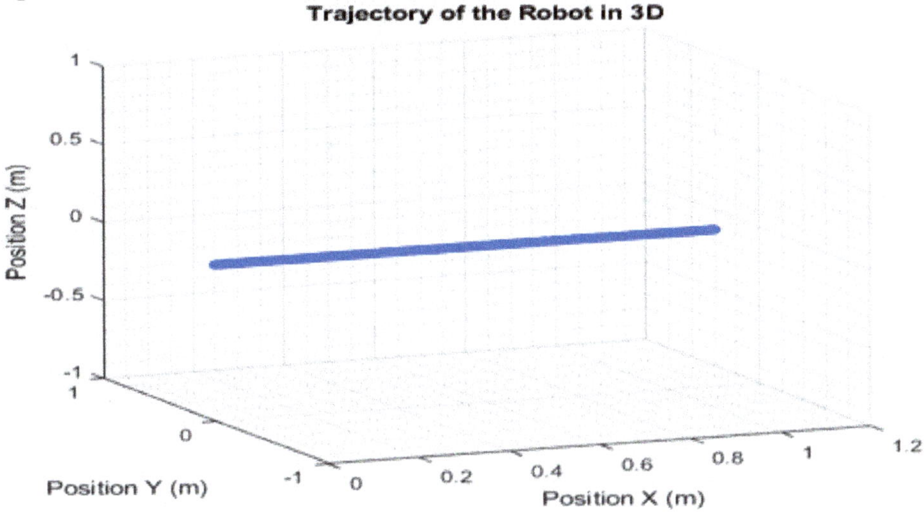

The figure 20 shows that the robot follows a linear path moving mainly in the positive direction of the X axis while maintaining relatively stable Y and Z coordinates.

If we change the robot's setpoint positions or commands radically altered its trajectory. This new figure 21 shows a robot performing a more elaborate and potentially complex movement: it initially ascends and moves laterally, then descends almost vertically. This indicates that the robot is capable of sophisticated movements and is not limited to a simple straight line.

Figure 21. 3D Movement of the Robot

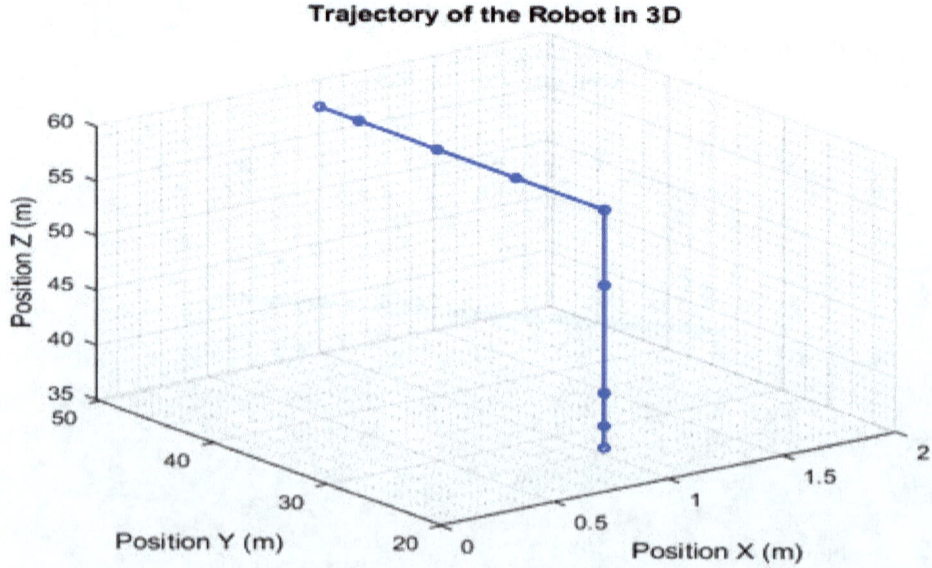

In conclusion, we can say that the comparative analysis of the two 3D trajectory figures highlights the flexibility and the capacity of the robot to execute varied and complex movements in the space. This significant difference between the two trajectories is a direct illustration of the crucial importance of human-machine interaction (HMI) in robotic control.

It ensures efficiency and precision of commands which will be transmitted via a well-designed human-machine interface (HMI).

It is the human operator who, by controlling the set points, transcribes the complexity and nature of the path that the robot will follow.

In the next section, we will focus on the crucial role of human intervention in the control of underwater robots.

HUMAN-IN-THE-LOOP MISSION PLANNING

Human interaction in underwater robots is crucial. It presents an active participant who can make critical decisions, adjust plans in real time, or intervene in unforeseen circumstances, while allowing the robot autonomy to perform low-level tasks. This crucial importance is seen mainly in the complex and uncertain nature of the underwater environment. (Wu X. X.-i.-t.-l.-3., 2022)

Human interaction in the control of underwater robots is not essentially limited to following a well-defined trajectory, but extends to more dynamic and critical roles.

This unpredictable environment, characterized by variable currents, reduced visibility, unmapped obstacles, and equipment failure risks, makes human interpretation and high-level decision-making indispensable, whereas even advanced artificial intelligence capabilities cannot always replicate them.

Furthermore, for complex or high-risk missions, such as the inspection of critical infrastructures (oil, cables), delicate scientific research, rescue operations, or mine clearance, human involvement is essential to guarantee safety, compliance, and efficiency. The human operator also plays a crucial role in the robot's learning and adaptation (Alqunun et al., 2025), by providing real-time feedback that allows it to adjust to new or unforeseen situations. In the event of an autonomous system failure or unexpected conditions, humans ensure the management of uncertainties and breakdowns, being able to take over, diagnose the problem, and reorient the mission. Finally, precision and efficiency are considerably improved by the human's ability to refine trajectories, optimize points of interest for data collection, or adjust control parameters for optimal performance in specific scenarios. (Wagner, 2025)

Human interaction in the control of underwater robots occurs in several key ways, allowing the combination of human decision-making capabilities with robotic autonomy.

This interaction concretely manifests through high-level planning and supervision, where the human operator defines the overall mission objectives, operational areas, safety constraints, and crucial waypoints. The robot's execution is then supervised via sophisticated human-machine interfaces (HMIs), which display real-time sensor data (video, sonar, position and orientation information) and 3D visualizations. This interaction also allows for real-time adjustment and reorientation in the face of unforeseen situations, such as the detection of unmapped obstacles, the presence of strong currents, or changes in mission objectives. In such cases, humans can intervene to modify the robot's trajectory or setpoint, change its control mode (for example, switch to teleoperation for delicate tasks), or validate plan proposals generated by the robotic system.

In this project, we used Dashboard from Simulink/Matlab which is the central interface for controlling and monitoring the vehicle's systems and operations. It allows us to know at any time the position, depth, orientation of the vehicle, as well as the status of its sensors and other subsystems. We send the commands to the robot such us changing its trajectory or adjust the speed. The Figure 22 illustrates the overall control system architecture for an underwater robot, specifically highlighting the interaction between human input, the controller, the robot model, and a 3D simulation.

Figure 22. Control system architecture for the robot

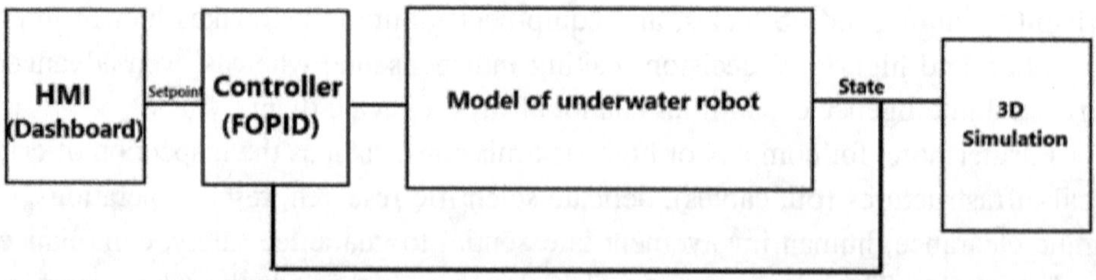

These three subgraphs show the evolution of the individual positions (X, Y, Z) of the robot with time:

Figure 23. Variation of positions (X, Y, Z) over time

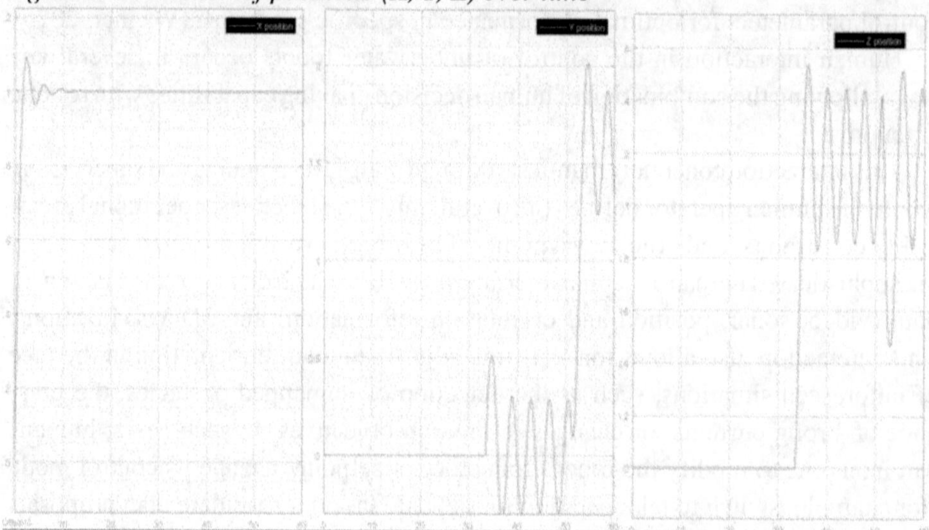

The robot starts at X=0 and quickly reaches a stable position of about 0.9 m, with a slight overshoot and some oscillations at the beginning (around t=5s to 10s). Then, the X position remains stable until the end of the simulation. For position Y it remains at 0 until approximately t=35s. Around t=35s, there is a one-step change to a new setpoint, which causes oscillations. The Y position attempts to reach the setpoint value, but appears to oscillate around 0.2-0.3m.

Then, at t=45s, there is another change, this time the Y position increases to approximately 1.8-2m, with notable oscillations around this value. Which is a bit similar to Z position.

The Figure 24 combines information from all three positions (X, Y, Z) to show the actual path traveled by the robot in 3D space.

Figure 24. Trajectory travelled by the robot

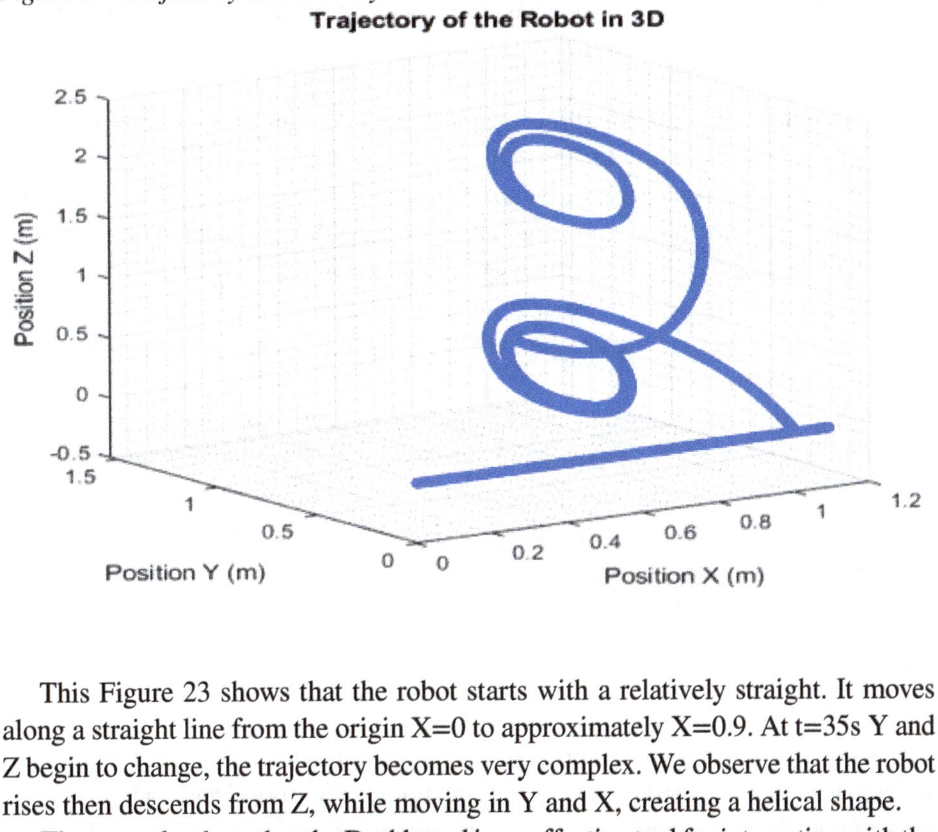

This Figure 23 shows that the robot starts with a relatively straight. It moves along a straight line from the origin X=0 to approximately X=0.9. At t=35s Y and Z begin to change, the trajectory becomes very complex. We observe that the robot rises then descends from Z, while moving in Y and X, creating a helical shape.

These results show that the Dashboard is an effective tool for interacting with the system and generating movement instructions for an underwater robot. However, these same results also reveal the need for significant improvements at the controller level.

EMERGING CHALLENGES OF MODELING AND AUTONOMY

The Fossen's framework is limited because the growing demands for performance, endurance, and autonomy in offshore applications raise two major challenges that exceed the scope of the nominal dynamic model: the precise integration of the electrical model and cognitive autonomy through AI.

The electrical non-linearities involves electric actuators: Motors and thrusters (Ahmed Baba et al., 2023) (Allal et al., 2024).

For long-duration missions, the controller's performance directly impacts endurance. Modeling must therefore extend to predicting the State of Charge the batteries. Control precision fundamentally becomes a system-wide energy optimization problem. Artificial Intelligence is required for high-level challenges: perception and decision-making under uncertainty:

- **Detection and Classification in Complex Environments**: Computer vision based on Deep Learning (DL) (Moniruzzaman, 2017) must compensate for poor visibility and underwater artifacts to detect and classify inspection targets (e.g., identifying a hairline crack on a pipeline despite turbidity). The challenge lies in the robustness of AI models against varying lighting conditions and limited training data in the marine environment. DL also is necessary to overcome the unique optical challenges of the underwater environment. (Ilioudi, 2025)
- **Dynamic Mission Planning**: AI must enable reactive planning based on perceived data. If it detects any anomaly, the system must autonomously decide to slow down, change its depth, or its trajectory to minimize risk (McMahon, 2016) or optimize data collection, thereby moving beyond the scope of traditional control laws.

CONCLUSION

This project addressed the crucial issue of human-defined target tracking by underwater robots for offshore energy applications, a rapidly expanding field requiring autonomous and precise inspection and maintenance solutions. We developed a comprehensive modeling of the underwater robot, encompassing its kinematics and dynamics, which is essential for an in-depth understanding of its behavior in complex aquatic environments.

The study then implemented and compared two control approaches: the classical Proportional-Integral-Derivative (PID) controller and the Fractional-Order PID (FOPID) controller. Detailed 3D simulations allowed us to visualize and evaluate the trajectory tracking performance of these controllers when faced with moving targets. Our results showed that while the standard PID controller offers commendable robustness and good tracking capability, the FOPID, thanks to its increased flexibility provided by fractional orders, demonstrated a significant reduction in overshoot. This ability to minimize oscillations and overshoots is particularly critical in offshore environments, where precision is paramount to avoid collisions with delicate infrastructure and to ensure operational safety. Furthermore, the 3D simulation not only validated the kinematic and dynamic models but also provided an immersive visual environment for evaluating the control algorithms, highlighting the technical feasibility of the tracking task. Although the settling time was comparable between the two approaches in our current configurations, the FOPID's advantage in terms of the smoothness of the transient response positions it as a promising candidate for demanding applications requiring precision and safety in underwater environments. To enrich and extend this promising work, several avenues for research and development can be considered. These include Advanced Controller Optimization, where more sophisticated algorithms such as genetic algorithms, particle swarm optimization, or ant colony optimization could be explored for tuning PID and FOPID parameters to achieve optimal trade-offs in performance, and adaptive or learning-based controllers like Reinforcement Learning could be investigated to enable dynamic parameter adjustments based on environmental disturbances and target characteristics. A critical next step is the Integration of the Robot's Electrical Model, which involves developing and implementing a comprehensive electrical model encompassing power consumption, battery discharge, and thruster electrical dynamics, to gain a more realistic understanding of the robot's endurance, power limitations, and the impact of electrical constraints on control performance and mission duration, ultimately allowing for the evaluation of energy-efficient control strategies and mission planning. Further efforts should focus on the Consideration of Real-World Uncertainties and Disturbances, by integrating more realistic environmental models (variable ocean currents, waves, turbulence) and accounting for uncertainties in robot model parameters (mass, inertia, hydrodynamic coefficients) to develop robust controllers. Perception and Sensor Fusion for Target Tracking is another key area, extending beyond defined targets to explore detection and tracking algorithms based on actual sensors (cameras, sonars, DVL), including computer vision techniques and data fusion filters (Extended Kalman Filter, Particle Filter) for accurate target state estimation in noisy, low-visibility environments, and developing strategies to handle target occlusion. Trajectory Planning and Obstacle Avoidance will involve integrating algorithms for generating optimal paths to reach targets while avoiding obstacles (underwater infrastructure, other ROVs) and respecting robot constraints, alongside developing real-time replanning strategies for dynamic targets or unforeseen obstacles. Crucially, Experimental Validation of the models and control algorithms on a physical underwater robot in test tanks and real environments will be necessary to confront simulation results with practical challenges such as latencies, sensor noise, and unmodeled dynamics. Finally, the Development

REFERENCES

Ahmed Baba, M., Naoui, M., & Cherkaoui, M. (2023). Fault-Tolerant Control Strategy for Hall Sensors in BLDC Motor Drive for Electric Vehicle Applications. *Sustainability*, *15*(13), 10430.

Allal, A., Lammouchi, Z., Khechekhouche, A., Mohamed, N., Alqunun, K., Alghadhban, A., & Guesmi, T. (2024). Advanced Residual Optimal Mapping Approach for Precise Detection of Stator Faults in Induction Motors. *IEEE Access: Practical Innovations, Open Solutions*, *12*, 143515–143530.

Alqunun, K., Bechiri, M. B., Naoui, M., Khechekhouche, A., Marouani, I., Guesmi, T., & Allal, A. (2025). An efficient bearing fault detection strategy based on a hybrid machine learning technique. *Scientific Reports*, *15*(1), 18739. PMID: 40437009

Baldini, A., Ciabattoni, L., Felicetti, R., Ferracuti, F., Freddi, A., & Monteriù, A. (2018). Dynamic surface fault tolerant control for underwater remotely operated vehicles. *ISA Transactions*, *78*, 10–20. PMID: 29503009

Barker, L. D., Jakuba, M. V., Bowen, A. D., German, C. R., Maksym, T., Mayer, L., & Whitcomb, L. L. (2020). Scientific challenges and present capabilities in underwater robotic vehicle design and navigation for oceanographic exploration under-ice. *Remote Sensing*, *12*(16), 2588.

Blair, C. (1975). *Silent victory: the US submarine war against Japan*. No Title.

Cao, J. Y., Liang, J., & Cao, B. G. (2005, August). *Optimization of fractional order PID controllers based on genetic algorithms. In 2005 international conference on machine learning and cybernetics* (Vol. 9). IEEE.

Fossen, T. I. (2011). *Handbook of marine craft hydrodynamics and motion control*. John Wiley & Sons.

Ilioudi, A. (2025). Deep-learning-based computer vision for underwater environments.

Joshu, S. D., & Talange, B. D. (2013, July). Fractional order modeling and control of depth system of AUV a case study. In *2nd International Conference on Advances in Computer Science and Engineering (CSE 2013)* (pp. 164-168). Atlantis Press.

Kadiyam, J., Mohan, S., Deshmukh, D., & Seo, T. (2021). Simulation-based semi-empirical comparative study of fixed and vectored thruster configurations for an underwater vehicle. *Ocean Engineering*, *234*, 109231.

Kalikatzarakis, M., Coraddu, A., Oneto, L., & Anguita, D. (2021). Optimizing fuel consumption in thrust allocation for marine dynamic positioning systems. *IEEE Transactions on Automation Science and Engineering*, *19*(1), 122–142.

KS. (2023). A.K.; Kadiyam, J.; Mohan, S. Comparative performance investigations of the intervention-class underwater vehicle with different possible thruster configurations using eight identical thrusters. *Ocean Engineering*, *288*, 116147.

Lammouchi, Z., Naoui, M., Bekakra, Y., Khechekhouche, A., & Sbita, L. (2025). Model-free predictive control based on the execution time of active vectors using ultra-local model. *Arabian Journal for Science and Engineering*, *50*(8), 5885–5898.

Lee-Pearce, M. L., Plowman, T. S., & Touchstone, D. (1998). Starbase-Atlantis, a school without walls: A comparative study of an innovative science program for at-risk urban elementary students. *Journal of Education for Students Placed at Risk*, *3*(3), 223–235.

Liu, F., Xu, W., Feng, Z., Yu, C., Liang, X., Su, Q., & Gao, J. (2025). Task Allocation and Path Planning Method for Unmanned Underwater Vehicles. *Drones (Basel)*, *9*(6), 411.

McMahon, J., & Plaku, E. (2016). Mission and motion planning for autonomous underwater vehicles operating in spatially and temporally complex environments. *IEEE Journal of Oceanic Engineering*, *41*(4), 893–912.

Meng, L., & Xue, D. (2009, June). Design of an optimal fractional-order PID controller using multi-objective GA optimization. In *2009 Chinese Control and Decision Conference* (pp. 3849-3853). IEEE.

Mohamed, M. J., & Khashan, A. (2014, February). Comparison between PID and FOPID controllers based on particle swarm optimization. In *The Second Engineering Conference of Control, Computers and Mechatronics Engineering (ECCCM2, 2014)*.

Moniruzzaman, M., Islam, S. M. S., Bennamoun, M., & Lavery, P. (2017, September). Deep learning on underwater marine object detection: A survey. In *International Conference on Advanced Concepts for Intelligent Vision Systems* (pp. 150-160). Cham: Springer International Publishing.

Morgan, D., & Taylor, B. (2011). *U-Boat Attack Logs: A Complete Record of Warship Sinkings from Original Sources 1939-1945*. Seaforth Publishing.

Podder, T. K., & Sarkar, N. (2001). Fault-tolerant control of an autonomous underwater vehicle under thruster redundancy. *Robotics and Autonomous Systems*, *34*(1), 39–52.

Rethfeldt, C., & Jeinsch, T. (2024, September). Combined Actuator Allocation for Underwater Vehicles with Variable Buoyancy Systems Using QP-Based Optimization. In *OCEANS 2024-Halifax* (pp. 1–7). IEEE.

Romdhane, M., Naoui, M., & Mansouri, A. (2023). PMSM Inter-Turn Short Circuit Fault Detection Using the Fuzzy-Extended Kalman Filter in Electric Vehicles. *Electronics (Basel)*, *12*(18), 3758.

Sarkar, N., Podder, T. K., & Antonelli, G. (2002). Fault-accommodating thruster force allocation of an AUV considering thruster redundancy and saturation. *IEEE Transactions on Robotics and Automation*, *18*(2), 223–233.

Schillai, S. M., Turnock, S. R., Rogers, E., & Phillips, A. B. (2019). Experimental analysis of low-altitude terrain following for hover-capable flight-style autonomous underwater vehicles. *Journal of Field Robotics*, *36*(8), 1399–1421.

SNAME, T. (1950). Nomenclature for treating the motion of a submerged body through a fluid. *The Society of Naval Architects and Marine Engineers, Technical and Research Bulletin*, (1950), 1-5.

Stoneman, R. (2008). *Alexander the Great: A life in legend.* Yale University Press.

Tanakitkorn, K., & Phoemsapthawee, S. (2022). Impacts of thruster configurations on the task performance of an unmanned surface vehicle. *Ocean Engineering*, *256*, 111465.

Tehrani, K. A., Capitaine, T., Barrandon, L., Hamzaoui, M., Rafiei, S. M. R., & Lebrun, A. (2011, August). Current control design with a fractional-order PID for a three-level inverter. In *Proceedings of the 2011 14th European Conference on Power Electronics and Applications* (pp. 1-7). IEEE.

Vega, E. P., Chocron, O., & Benbouzid, M. (2016). A flat design and a validated model for an AUV reconfigurable magnetic coupling thruster. *IEEE/ASME Transactions on Mechatronics*, *21*(6), 2892–2901.

Vukic, Z., & Kuljaca, O. (2002). *Lectures on PID controllers*. Faculty of Electrical Engineering and Computing University of Zagreb.

Wagner, B., Kuebler, J., & Zalnieriute, M. (2025). Humans in the loop: Exploring the challenges of human participation in automated decision-making systems. *Frontiers in Political Science*, *7*, 1611563.

Wang, C., Mei, D., Wang, Y., Yu, X., Sun, W., Wang, D., & Chen, J. (2022). Task allocation for Multi-AUV system: A review. *Ocean Engineering*, *266*, 112911.

Wang, S., Yang, M., Wang, Y., Yang, S., Lan, S., & Zhang, X. (2021, July). Optimization of Flight Parameters for Petrel-L Underwater Glider. *IEEE Journal of Oceanic Engineering*, *46*(3), 817–828.

Wang, Y., Huang, Y., Li, Y., & Ren, C. (2024). Optimization of thruster configuration and control allocation for a spherical magnetic coupling thrusters actuated underwater robot. *Ocean Engineering*, *309*, 118445.

Wu, H., Chen, Y., Yang, Q., Yan, B., & Yang, X. (2024). A review of underwater Robot localization in confined spaces. *Journal of Marine Science and Engineering*, *12*(3), 428.

Wu, X., Xiao, L., Sun, Y., Zhang, J., Ma, T., & He, L. (2022). A survey of human-in-the-loop for machine learning. *Future Generation Computer Systems*, *135*, 364–381.

Xu, D., Yan, Z., Xie, H., Lu, L., Xiao, Z., & Han, B. (2024). Demand-driven optimization method for thruster configuration of underwater vehicles considering capability boundary and system efficiency. *Ocean Engineering*, *310*, 118749.

Youssef, R., Hamida, M. A., Mohamed, N., Lassaad, S., Djamel, T., & Yasmine, A. (2023). Single-Stage Standalone Photovoltaic Water Pumping System Using Predictive Torque Control (PTC) of Induction Machine. *Electric Power Components and Systems*, •••, 1–17.

Chapter 11
Modeling and Control of Wind Energy Conversion Systems Based on Permanent Magnet Synchronous Generators

Laila Bouaziz
https://orcid.org/0009-0002-1890-3358
National Engineering School of Gabes, University of Gabes, Tunisia

Mehdi Dhaoui
National Engineering School of Gabes, University of Gabes, Tunisia

Mouna Ben Hamed
https://orcid.org/0000-0002-8331-3728
National Engineering School of Gabes, University of Gabes, Tunisia

ABSTRACT

This chapter presents a comprehensive modeling and control framework for a grid-connected wind energy conversion system (WECS) based on a Permanent Magnet Synchronous Generator (PMSG). The proposed system integrates aerodynamic and electromechanical models, including wind profile, turbine, shaft dynamics, PMSG, and full-scale back-to-back converters. A Tip Speed Ratio (TSR)-based Maximum Power Point Tracking (MPPT) method is implemented to ensure optimal energy extraction, while pitch angle control maintains rated power under high wind conditions. Vector control is applied to both machine-side and grid-side converters to regulate torque, flux, DC-link voltage, and manage active/reactive power injection into the grid. A

DOI: 10.4018/979-8-3373-5621-1.ch011

Phase-Locked Loop (PLL) ensures synchronization. MATLAB/Simulink simulations validate the system's performance, demonstrating efficient tracking of the optimal operating point, stable dynamics, and enhanced energy efficiency. This work offers a solid basis for intelligent control in modern wind turbine systems.

INTRODUCTION

Describe the general perspective of the chapter. End by specifically stating the objectives of the chapter.

A wind turbine is a device designed to convert the kinetic energy of wind into mechanical energy and subsequently into electrical energy through a generator (Abogrean et al., 2022). Modeling a wind energy conversion chain involves several critical parameters, such as wind profile characteristics, the aerodynamic behavior of the blades, as well as the models of power converters and the associated control systems.

This chapter focuses on the modeling of a wind energy system based on a Permanent Magnet Synchronous Generator (PMSG). The discussion successively addresses the aerodynamic, electrical, and electronic components, highlighting their interactions to achieve optimal wind energy conversion.

The selection of the generator plays a crucial role in the overall efficiency of the system. The PMSG offers multiple advantages, including improved energy efficiency and reduced maintenance costs due to the absence of brushes and slip rings (Ailliot & Monbet, 2012). Additionally, the management of the generated electrical energy relies on the quality of the power converters and the control system, which ensures the wind turbine operates effectively under varying wind conditions and grid requirements.

Following the detailed modeling of each component of the conversion chain, the second part of this chapter focuses on wind turbine control. Several Maximum Power Point Tracking (MPPT) techniques have been widely studied in the literature, such as the Perturb and Observe (P&O) method, fuzzy logic control, and intelligent algorithms (Alqunun et al., 2025). In this work, the Tip Speed Ratio (TSR)-based method has been adopted, as it demonstrates superior performance for the studied system, ensuring efficient and stable power extraction. Furthermore, to control the Permanent Magnet Synchronous Generator, a vector control strategy has been implemented, enabling precise torque and flux regulation, thereby ensuring optimal dynamic performance of the overall system.

1. MODELING OF THE WIND ENERGY CONVERSION CHAIN

The wind energy conversion chain enables the transformation of wind kinetic energy into usable electrical energy. It consists of multiple interconnected subsystems, ranging from wind energy capture to power delivery into the grid (Allal et al., 2024).

In this study, the system under consideration comprises a three-bladed wind turbine driving a Permanent Magnet Synchronous Generator (PMSG), a three-phase IGBT rectifier, a Boost PFC (Power Factor Correction) converter, a DC-link, and a three-phase inverter. This configuration ensures the conversion of mechanical energy into electrical energy, its conditioning, and its integration into the grid through a step-up transformer, while meeting power quality and stability requirements. Figure 1 illustrates the overall architecture of the conversion chain.

The system analyzed in this work corresponds to a 1.5 MW wind turbine generator connected to a 30 kV distribution network. Energy produced by the turbine is transmitted through a step-up transformer and power transmission lines.

In the configuration shown in Figure 1, the PWM rectifier is placed between the generator and the DC bus, while the connection to the grid is achieved through the PWM inverter. The advantage of this configuration lies in the use of vector control, which allows the generator to operate near its optimal point. However, the performance of this control strongly depends on accurate knowledge of the generator's parameters.

Figure 1. Architecture of the wind energy conversion chain

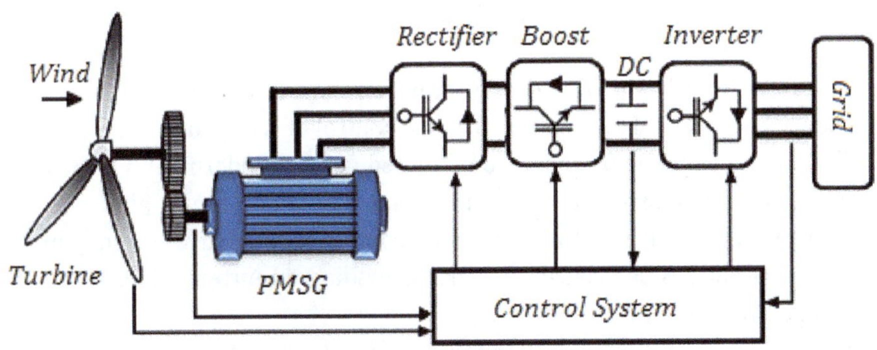

a. Wind Modeling

Wind modeling is an essential step in the analysis and design of wind energy systems. Due to its spatial and temporal variability, wind is commonly represented using mathematical models that capture its natural fluctuations and allow accurate simulation of real operating conditions (Allal et al., 2024).

i. Wind Characteristics

Wind is a complex meteorological phenomenon that results from the uneven heating of the Earth's surface by the sun, combined with the rotation of the planet and geographical features. Its behavior is influenced by a variety of factors, including local topography, temperature gradients, atmospheric pressure variations, and surface roughness. Understanding these characteristics is essential for the design, optimization, and operation of wind energy conversion systems (Aoumri et al., 2024).

Several key aspects define the nature of wind and its impact on wind turbine performance:

- **Wind Speed:** Wind speed is one of the most critical parameters affecting wind energy generation. It varies both temporally and spatially, influenced by daily and seasonal weather patterns. Generally, wind speed increases with altitude due to reduced surface friction. This vertical profile of wind speed is often modeled using empirical laws such as the logarithmic or power law profiles. Precise measurement and modeling of wind speed distribution are crucial for predicting the available wind energy and for sizing wind turbines appropriately.
- **Wind Direction:** The direction from which the wind blows significantly influences the optimal orientation and yaw control of wind turbines. Since wind direction can change frequently and unpredictably, modern wind turbines are equipped with sensors and yaw mechanisms to continuously align the rotor with the incoming wind to maximize energy capture. Moreover, wind direction variations can cause asymmetric loading on turbine blades, affecting fatigue life and structural integrity.
- **Turbulence:** Turbulence refers to the rapid and irregular fluctuations in wind speed and direction, often caused by atmospheric instabilities, obstacles such as buildings or trees, and terrain features. Turbulence intensity is a critical factor in wind turbine design as it affects mechanical loads and stress on turbine components. High turbulence levels can lead to increased wear and tear, reducing the lifespan of turbines. Accurate characterization of turbulence

helps in developing robust control strategies and ensuring the reliability of wind turbines.

- **Other Influential Factors:** Additional meteorological factors such as air density, temperature, and humidity also play a role in wind energy conversion efficiency. Air density decreases with altitude and temperature, impacting the amount of kinetic energy available in the wind. Seasonal and diurnal temperature variations can lead to changes in atmospheric stability, further influencing wind characteristics.

ii. Mathematical Modeling of Wind

Several models are used to represent wind speed:

- **Statistical models**: Based on probability distributions, such as the Weibull distribution, widely used to describe the wind speed distribution at a given site (Baba et al., 2023).
- **Stochastic models**: These account for random wind fluctuations using stochastic processes such as the Markov model or time series approaches (Benbouhenni & Bizon, 2021).
- **Harmonic models (Fourier)**: These models represent wind speed as a sum of sinusoidal components, allowing periodic variations in wind speed to be described (Chen et al., 2023).

In this study, we adopted a general representation of wind speed expressed as:

$$\mathbf{V_w(t) = V_n + \sum_{k=1}^{i} a_k \sin(\omega_k t)} \quad (1)$$

Where $V_n(t)$ is the average wind speed, a_k is the amplitude of the harmonic components, and ω_k their respective frequencies.

We retained an empirical model based on a simplified Fourier series:

$$V_w(t) = 12.1 + \sin(x) - 0.87\sin(3x) + 0.75\sin(5x) - 0.625\sin(10x)$$
$$+0.5\sin(30x) + 0.25\sin(50x) + 0.125\sin(100x) \quad (2)$$

With:

$$x = \omega t = \frac{2\pi}{10}t$$

This model captures the temporal variations of wind while simplifying its implementation in simulations.

The wind speed profile used in our analysis is illustrated in Figure 2.

Figure 2. Temporal profile of wind speed based on a harmonic model

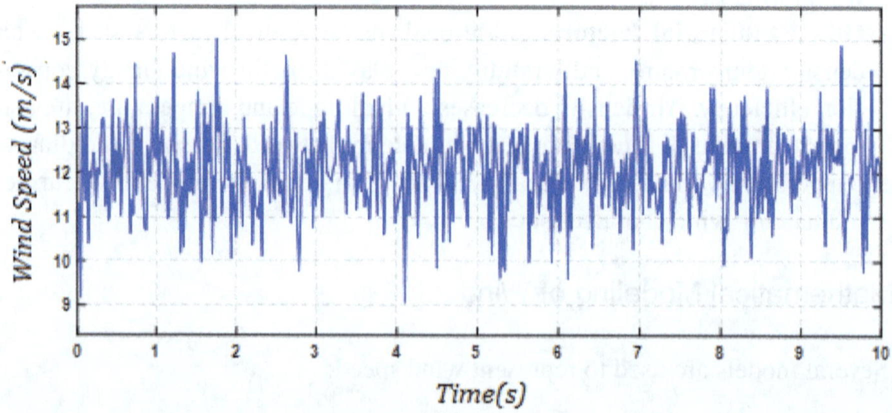

b. Turbine Model

The wind turbine is the central element of the energy conversion chain. It captures the kinetic energy of the wind and converts it into mechanical energy, which is then transformed into electrical energy by the Permanent Magnet Synchronous Generator (PMSG). The turbine model is based on aerodynamic and mechanical relationships that allow predicting its behavior under different environmental conditions, such as wind speed.

iii. Aerodynamic Model of the Turbine

The aerodynamic model describes the conversion of wind kinetic energy into mechanical energy by the turbine blades. The power captured by the turbine depends on the wind speed and the turbine's efficiency in extracting energy, which is represented by the power coefficient Cp.

The mechanical power Pt captured by the turbine is expressed as:

$$P_t = \frac{1}{2}.\rho.\pi.R^2.C_p(\lambda,\beta).V_w^3 \tag{3}$$

Where:

ρ: Air density (kg/m3).

$S=\pi R^2$: Swept area of the blades, with R being the blade radius (m).

Vw: Wind speed (m/s).

Cp: Power coefficient, depending on tip speed ratio λ and blade pitch angle β.

The coefficient Cp represents the turbine efficiency in converting wind energy into mechanical energy. According to Betz's law, the maximum power that can be extracted from wind is limited to 59% of the available kinetic energy, Cpmax=0.59 (Chen et al., 2009).

Several studies express the variation of Cp as:

$$C_p(\lambda,\beta) = (0.44 - 0.0167\beta).\sin\left(\pi\frac{\lambda-3}{15} - 0.3\beta\right)-0.00184.(\lambda-3)\beta \quad (4)$$

Figure 3 illustrates the variation of Cp as a function of the tip speed ratio λ. The coefficient reaches its maximum value Cpmax=0.44 when the tip speed ratio is optimal λ=10.5, indicating the ideal conditions for extracting maximum energy from the wind.

Figure 3. Evolution of the coefficient Cp as a function of λ

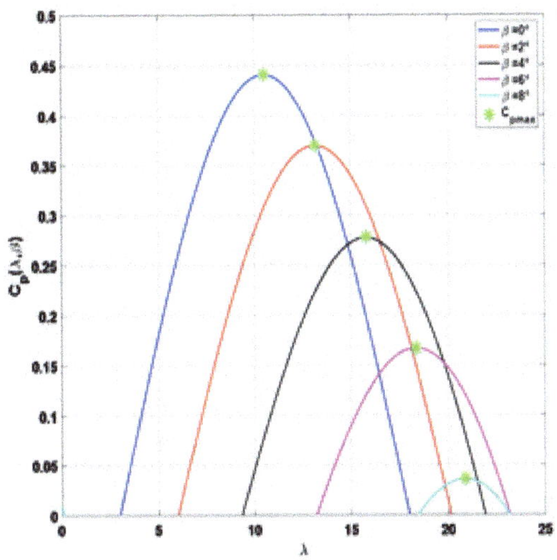

Using Equation (3), the mechanical power characteristics of the turbine for different wind speeds can be obtained, as shown in figure 4. These characteristics correspond to a turbine rated at 1.5 MW with a blade radius of 33.05 m. For a nominal wind speed of 12.1 m/s, the turbine provides a mechanical power of 1.6384 MW at a rotational speed of 3.8442 rad/s.

Figure 4. Variation of mechanical power P_t as a function of turbine speed Ω_t for different wind speeds

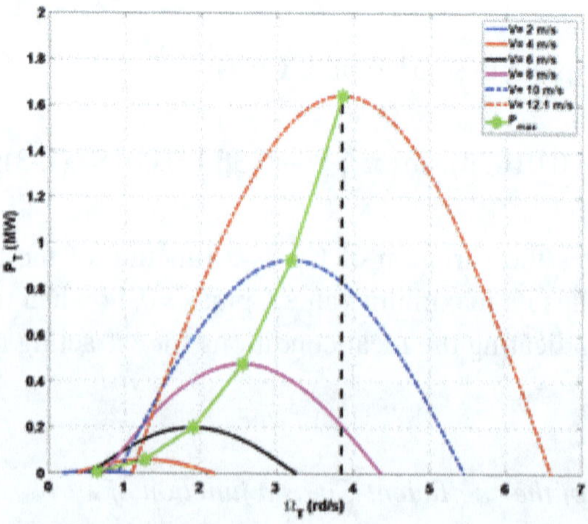

iv. Dynamic Model of the Turbine

This model describes the interaction between the aerodynamic forces applied on the blades and the rotational dynamics of the rotor. The motion equation is given by:

$$J_t.\frac{d\Omega_t}{dt} = T_t - T_g - f.\Omega_t \tag{5}$$

Where:

Jt: Rotor inertia ($Kg.m^2$)

Ωt: Rtor angular speed (rad/s)

Tt: Aérodynamique torque (N.m)

Tg: Mecanical torque on the generator shaft (N·m)

f.Ωt: Friction torque (N.m)

The aerodynamic torque Tt is related to the mechanical power:

$$T_t = \frac{P_t}{\Omega_t} = \frac{1}{2}.\rho.\pi.R^3.V_w^2.C_p(\lambda,\beta) \tag{6}$$

The tip speed ratio λ, a key parameter for maximizing energy extraction, is defined as:

$$\lambda = \frac{R \cdot \Omega_t}{V_w} \tag{7}$$

An optimal operation (with Cpmax) is achieved when λremains close to λopt.

v. Turbine Control

The main objective of turbine control is to maximize the extracted power while maintaining system stability and safety. It is primarily based on (Chowdary et al., 2018):

- Adjusting the blade pitch angle (β) according to wind speed.
- Maintaining the tip speed ratio (λ) close to its optimal value (λopt).
- Limiting the rotational speed during high wind conditions to avoid mechanical or electrical overload.
- This control strategy ensures efficient energy conversion while preventing excessive mechanical stress or potential system damage.

c. Drive Train Model

The drive train is a critical component of the wind energy conversion chain, ensuring the mechanical connection between the wind turbine rotor and the Permanent Magnet Synchronous Generator (PMSG). Its main function is to transmit the aerodynamic torque generated by the turbine to the generator through a gearbox, enabling optimal conversion of wind kinetic energy into electrical energy (Collier & Heldwein, 2011).

i. Mechanical Dynamics of the Drive Train

The gearbox increases the turbine's low rotational speed to a level suitable for the generator's operating requirements. The mechanical model of the drive train typically considers a two-mass system, distinguishing the low-speed shaft (connected to the turbine) and the high-speed shaft (connected to the generator), coupled through the gearbox. This representation accounts for:

- The inertia of each shaft.
- Torsional stiffness (kdt).
- Torsional damping (fdt).
- Viscous friction on both shafts (ft,fg).

This modeling approach is essential for analyzing torsional oscillations, resonance phenomena, and mechanical constraints caused by rapid torque variations.

ii. State-Space Representation

The drive train system is represented in the form of a state-space model to establish the dynamic relationships between key mechanical variables such as rotational speed, angular displacement, and torsional forces. This model provides a solid foundation for stability analysis, control design, and transient studies (Dani et al., 2024).

The state-space model of the two-mass drive train system is given by the following equations:

$$\frac{d}{dt}\begin{pmatrix} \Omega_r(t) \\ \Omega_g(t) \\ \theta_\Delta(t) \end{pmatrix} = A_{dt}\begin{pmatrix} \Omega_r(t) \\ \Omega_g(t) \\ \theta_\Delta(t) \end{pmatrix} + B_{dt}\begin{pmatrix} T_t(t) \\ T_g(t) \\ 0 \end{pmatrix} y = C_{dt}\begin{pmatrix} \Omega_r(t) \\ \Omega_g(t) \\ \theta_\Delta(t) \end{pmatrix} + D_{dt}\begin{pmatrix} T_t(t) \\ T_g(t) \\ 0 \end{pmatrix} \tag{8}$$

Where: Adt, Bdt, Cdt and Ddt are the system matrices, defined as follows:

$$A_{dt} = \begin{pmatrix} -\frac{(f_{dt}+f_t)}{J_t} & \frac{f_{dt}}{G.J_g} & -\frac{k_{dt}}{J_t} \\ \frac{\eta_{dt}.f_{dt}}{G.J_g} & -\frac{\left(\frac{\eta_{dt}.f_{dt}}{G^2}+f_g\right)}{J_g} & \frac{\eta_{dt}.k_{dt}}{G.J_g} \\ 1 & -\frac{1}{G} & 0 \end{pmatrix}, B_{dt} = \begin{pmatrix} \frac{1}{J_t} & 0 & 0 \\ 0 & -\frac{1}{J_g} & 0 \\ 0 & 0 & 0 \end{pmatrix}, C_{dt} = \begin{pmatrix} 1 & 0 & 0 \\ 0 & 1 & 0 \end{pmatrix} \text{ and}$$

$$D_{dt} = 0$$

The physical quantities associated with this model are:

θΔ(t): Relative angular displacement due to torsion

Ωr(t) et Ωg(t): Angular speeds of the turbine rotor and generator, respectively

Jt et Jg: Moments of inertia of the low-speed and high-speed shafts

ft et fg: Viscous friction coefficients of the low-speed and high-speed shafts

fdt: Damping coefficient due to torsion

G: Gearbox ratio

ηdt: Overall efficiency of the mechanical drive train

Figure 5 illustrates the kinematic and dynamic model of the two-mass drive train, highlighting the mechanical interactions between its components, including the effect of the gearbox on torque transmission to the generator. The numerical specifications of the parameters used are presented in the appendix.

Figure 5. Kinematic model of the two-mass drive train

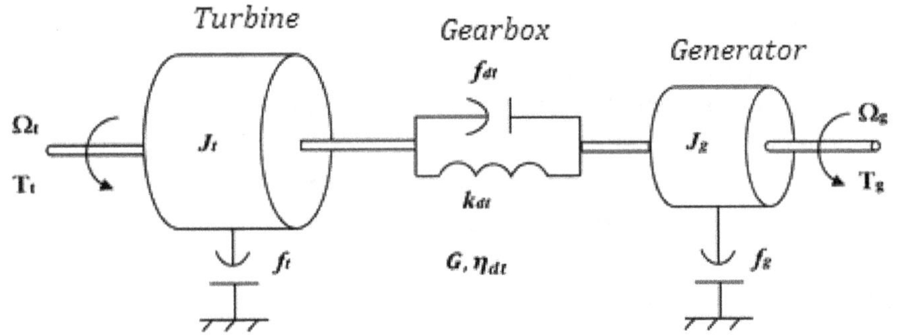

d. Modeling of the Permanent Magnet Synchronous Generator (PMSG)

The Permanent Magnet Synchronous Generator (PMSG) is widely used in wind energy systems due to its high performance, robustness, and the absence of an external excitation system. Unlike induction machines, the PMSG does not require an additional power source for rotor excitation, which enhances its energy efficiency (Collier & Heldwein, 2011).

i. Role of PMSG in the Wind Energy Conversion Chain

In a wind energy conversion system, the PMSG is the key component responsible for converting mechanical energy from the turbine rotor into electrical energy. It is associated with a control system, generally based on Maximum Power Point Tracking (MPPT) algorithms, to optimize the extraction of wind energy (Dannehl et al., 2011).

The choice of PMSG in this study is motivated by its superior performance and compatibility with advanced control techniques used for regulating generated power.

The modeling of the PMSG aims to mathematically represent the electromagnetic and mechanical phenomena governing its operation, in order to predict its behavior under various operating conditions, in both dynamic and steady-state modes. This model is also essential for numerical simulations.

Advantages:

- High efficiency due to the absence of excitation losses.
- Higher power density and compact size.

- Fast dynamic response, well-suited for wind speed variations.
- Ease of control using vector control techniques.

Disadvantages:

- High cost of permanent magnets.
- Thermal sensitivity (magnetic properties of the magnets can deteriorate at high temperatures).
- Increased control complexity in the absence of position sensors or efficient observer algorithms (Dannehl et al., 2011).

ii. Simplifying Assumptions

The modeling of PMSG is based on the following assumptions, commonly adopted in the literature (Darabian et al., 2023):

- No magnetic circuit saturation.
- Sinusoidal distribution of the magnetomotive force (MMF) produced by the stator windings.
- Neglect of hysteresis, eddy currents, and skin effect.
- Slotting effect considered negligible.
- Stator winding resistance assumed constant and independent of temperature.

iii. Mathematical Formulation in the Three-Phase Reference Frame

✓ Electrical Equations

The PMSG consists of a stator with three-phase windings connected in a star configuration with an isolated neutral point. Consequently, the instantaneous sum of the stator currents is zero. Rotor excitation is provided by permanent magnets.

In the three-phase reference frame (a, b, c), the stator voltage equations are expressed as:

$$[\mathbf{V_s}] = \mathbf{R_s} \cdot [\mathbf{I_s}] + \frac{\mathbf{d}}{\mathbf{dt}}[\boldsymbol{\varphi}_s] \quad (9)$$

Where:

$$V_s = \begin{bmatrix} V_a \\ V_b \\ V_c \end{bmatrix} i_s = \begin{bmatrix} i_a \\ i_b \\ i_c \end{bmatrix} \varphi_s = \begin{bmatrix} \varphi_a \\ \varphi_b \\ \varphi_c \end{bmatrix}$$

[Vs]: Stator voltage vector, [I_S]: Stator current vector, [ϕ_S]: Flux linkage vector of stator windings and [Rs]: Stator phase resistance.

✓ Magnetic Equations

The stator flux linkages consist of two components: the self-inductance flux due to the stator windings and the flux produced by the permanent magnets (Collier & Heldwein, 2011).

$$[\varphi_S] = [L_S][I_S] + [\varphi_f] \quad (10)$$

Where: [ϕf]: Flux vector created by the permanent magnets across the stator windings.

The inductance matrix [Ls] contains constant terms expressed by [Ls0] and variable terms depending on the rotor position (θr) expressed by [Ls1 (θr)].

$$[\mathbf{L}_S] = [\mathbf{L}_{s0}] + [\mathbf{L}_{s1}(\boldsymbol{\theta}_r)] \quad (11)$$

Where: $$[L_{s0}] = \begin{bmatrix} L_{s0} & M_{s0} & M_{s0} \\ M_{s0} & L_{s0} & M_{s0} \\ M_{s0} & M_{s0} & L_{s0} \end{bmatrix}$$

$$[L_{S1}(\theta_r)] = \begin{bmatrix} \cos(2\theta_r) & \cos2\left(\theta_r - \frac{2\pi}{3}\right) & \cos2\left(\theta_r - \frac{4\pi}{3}\right) \\ \cos2\left(\theta_r - \frac{2\pi}{3}\right) & \cos2\left(\theta_r - \frac{4\pi}{3}\right) & \cos(2\theta_r) \\ \cos2\left(\theta_r - \frac{4\pi}{3}\right) & \cos(2\theta_r) & \cos2\left(\theta_r - \frac{2\pi}{3}\right) \end{bmatrix} . L_{S1}$$

Where:

Ms: Mutual inductance between two stator Phases, Ls: Self-inductance of a stator phase, θr: Rotor angular position relative to the stator.

The system in (9) can thus be rewritten as:

$$[V_s]_{abc} = [R_s] \cdot [i_s]_{abc} + \frac{d}{dt}\left\{ [L_{ss}]\,[i_s]_{abc} + [\phi_f]_{abc} \right\} \quad (12)$$

✓ Mechanical Equations

Considering that the mechanical shaft consists of an inertia Jt on the turbine side and an inertia Jg on the generator side, the total inertia J is given by:

$$J = G^2.J_t + J_g \quad (13)$$

The dynamic behavior of the machine is then described by the following mechanical equation:

$$J.\frac{d\Omega_g}{dt} = T_g - T_{em} - f.\Omega_g \tag{14}$$

iv. Mathematical Formulation in the (d,q) Reference Frame

Park transformation

To simplify the mathematical model of the Permanent Magnet Synchronous Generator (PMSG), the Park transformation is applied. This mathematical tool converts the three-phase reference frame (a, b, c) into a two-axis orthogonal reference frame (d, q). More specifically, it transforms the three stator windings, separated by 2π/3, into two orthogonal windings in the rotating reference frame, as illustrated in Figure 6 (Darabian et al., 2023).

Figure 6. Transformation from the natural reference frame (a,b,c) to Park (d,q) frame

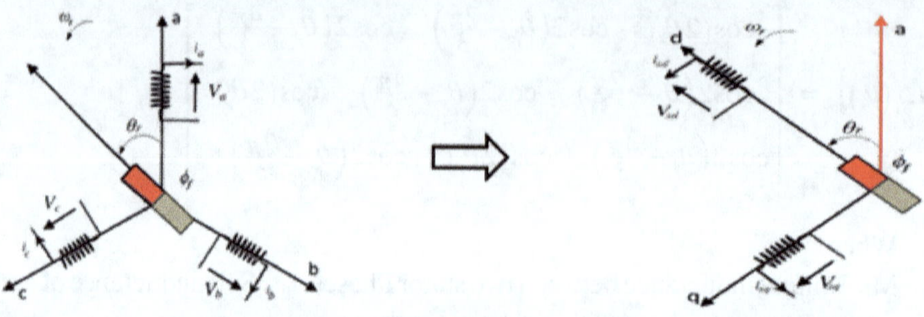

The transformation from the three-phase (a,b,c) system to the two-phase (d, q) system, and its inverse, is achieved through the direct and inverse Park transformation matrices P (θ) and P-1, respectively, as expressed by:

$$\begin{bmatrix} X_d \\ X_q \end{bmatrix} = \mathbf{P}(\theta).\begin{bmatrix} X_a \\ X_b \\ X_c \end{bmatrix} \begin{bmatrix} X_a \\ X_b \\ X_c \end{bmatrix} = \mathbf{P(\theta)}^{-1}.\begin{bmatrix} X_d \\ X_q \end{bmatrix} \quad (15)$$

$$P(\theta) = \sqrt{\frac{2}{3}}\begin{bmatrix} \cos\theta & \cos(\theta - \frac{2\pi}{3}) & \cos(\theta - \frac{4\pi}{3}) \\ \sin\theta & \sin(\theta - \frac{2\pi}{3}) & \sin(\theta - \frac{4\pi}{3}) \\ \frac{1}{\sqrt{2}} & \frac{1}{\sqrt{2}} & \frac{1}{\sqrt{2}} \end{bmatrix} P(\theta)^{-1} = \sqrt{\frac{2}{3}}\begin{bmatrix} \cos\theta & \sin\theta & \frac{1}{\sqrt{2}} \\ \cos(\theta - \frac{2\pi}{3}) & \sin(\theta - \frac{2\pi}{3}) & \frac{1}{\sqrt{2}} \\ \cos(\theta - \frac{4\pi}{3}) & \sin(\theta - \frac{4\pi}{3}) & \frac{1}{\sqrt{2}} \end{bmatrix}$$

Park Model of the PMSG

By applying the Park transformation to Equation (9), the electrical model of the PMSG becomes:

$$\begin{bmatrix} V_d \\ V_q \end{bmatrix} = P[\theta].R_s \begin{bmatrix} i_a \\ i_b \\ i_c \end{bmatrix} + P[\theta].\frac{d}{dt}\begin{bmatrix} \varphi_a \\ \varphi_b \\ \varphi_c \end{bmatrix} \quad (16)$$

Expanding this system (16), the voltage equations in the (dq) reference frame are written as:

$$\begin{cases} V_d = R_s i_d + \frac{d\varphi_d}{dt} - \omega\varphi_q \\ V_q = R_s i_q + \frac{d\varphi_q}{dt} + \omega\varphi_d \end{cases} \quad (17)$$

The flux linkages ϕd and ϕq are expressed as functions of stator currents and inductances:

$$\begin{cases} \varphi_d = L_d i_d + \varphi_f \\ \varphi_q = L_q i_q \end{cases} \quad (18)$$

Substituting (18) into (17), we obtain:

$$\begin{cases} V_d = R_s i_d + L_d \frac{di_d}{dt} - \omega L_q i_q \\ V_q = R_s i_q + L_q \frac{di_q}{dt} + \omega(L_d i_d + \varphi_f) \end{cases} \quad (19)$$

Where:

Ld, Lq: Direct and quadrature inductances (assumed constant and independent of θ)

The electromagnetic torque of the PMSG in the (d, q) frame is expressed as:

$$T_{em} = p.\left[(L_d - L_q) i_d i_q + \varphi_f i_q\right] \quad (20)$$

In our case, the generator is a non-salient pole machine (Ld=Lq), and the electromagnetic torque expression simplifies to: $T_{em} = p.\varphi_f i_q$

e. Modeling of Power Electronic Converters

This section focuses on the modeling of power electronic converters within the wind energy conversion chain, which ensure the conversion and adaptation of the electrical energy produced by the wind generator into a form compatible with the electrical grid. The conversion system consists of several components: a rectifier, a DC–DC boost converter, a DC-link, an inverter, and a step-up transformer. Each component is responsible for energy management and adaptation to guarantee optimal grid integration (De Lellis et al., 2018).

The modeling of these elements is essential for optimizing the overall performance of the conversion chain and ensuring maximum efficiency in the transformation of wind energy into electrical power.

a. Model of the Three-Phase IGBT Rectifier

The rectifier is a key component of the wind energy conversion chain, enabling the conversion of the alternating current supplied by the generator into direct current (Dekali et al., 2021).

This converter transforms the three-phase AC voltage into a DC voltage. It is composed of six Insulated Gate Bipolar Transistors (IGBTs) arranged in a bridge configuration; each connected in parallel with a freewheeling diode. This topology ensures precise control of the output current and voltage.

The conversion process depends on the gating signals that determine the switching instants of the static switches (IGBT + diode). The choice of the control strategy is critical, as it governs the waveform and amplitude of the DC voltage delivered by the rectifier (Dekali et al., 2021).

This model is essential for simulating the dynamic behavior of the rectifier under different operating conditions of the wind energy conversion chain.

Figure 7. Architecture of the three-phase IGBT rectifier connected to the wind generator

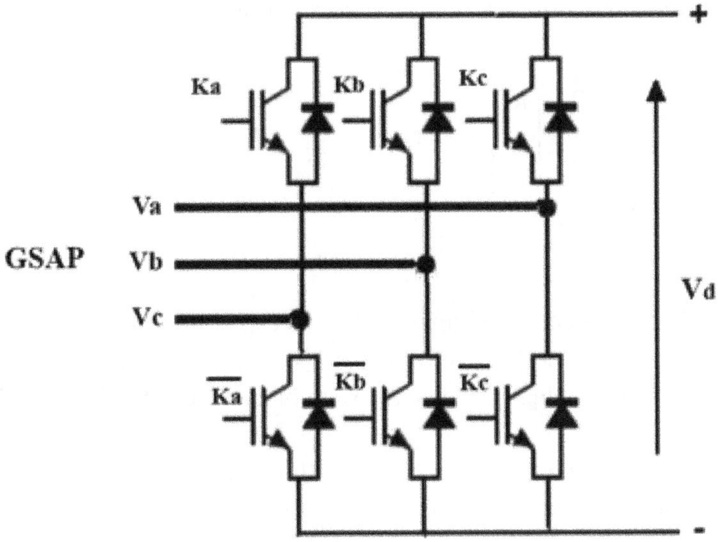

The six-IGBT bridge operates based on a pairwise conduction principle, where two switches are activated simultaneously during each 60° electrical interval.

- One upper switch (T1, T3, T5) and one lower switch (T4, T6, T2), belonging to different phases, are turned on at the same time to allow current flow while avoiding short-circuiting the DC bus.
- This controlled commutation, combined with pulse-width modulation (PWM), enables the regulation of the DC output voltage according to the requirements of the conversion chain.

In modern wind energy systems, the IGBT rectifier is driven by a real-time controller, whose switching signals are determined based on several key parameters:

- The wind turbine rotational speed, to ensure Maximum Power Point Tracking (MPPT) optimization.
- The grid conditions, particularly for power injection and fault management.
- The DC-link voltage, to guarantee system stability and proper feeding of the grid-side inverter.

The input voltages of the rectifier (supplied by the wind generator) are expressed as a function of the DC voltage Vd provided by the rectifier and the switching states of the transistors, using the following matrix relation:

$$\begin{bmatrix} V_a \\ V_b \\ V_c \end{bmatrix} = \begin{bmatrix} \frac{2}{3} & -\frac{1}{3} & -\frac{1}{3} \\ -\frac{1}{3} & \frac{2}{3} & -\frac{1}{3} \\ -\frac{1}{3} & -\frac{1}{3} & \frac{2}{3} \end{bmatrix} \cdot \begin{bmatrix} K_a \\ K_b \\ K_c \end{bmatrix} V_d \tag{21}$$

Where:

Ka, Kb, Kc are the control signals of the corresponding switches.

The control signals for the inverter legs i∈{a,b,c} are determined by the binary variables Ka, Kb, Kc , which indicate the state of each switch (closed if Ki=1open if Ki=0).

The pulse width of the control signals directly influences the switching variables Ka, Kb, Kc, and consequently, the quality of the rectified voltage.

The DC output current flowing to the DC bus is expressed as:

$$i_d = K_a . i_a + K_b . i_b + K_c . i_c \tag{22}$$

Where: (ia, ib, ic) represent the phase currents.

ii. Boost PFC Converter Model

Power quality disturbances, such as voltage sags and instantaneous power interruptions, are among the most common and detrimental phenomena in wind energy systems. To mitigate these issues, a Boost PFC converter (step-up DC-DC converter with Power Factor Correction) is employed in cascade with the PWM rectifier (Gao & Gao, 2016; Girsang et al., 2014).

The Boost converter is regulated using a voltage hysteresis control strategy. The output voltage of the converter is compared to a reference value. When the output voltage exceeds the upper hysteresis limit, the converter is turned off conversely, when it drops below the lower limit, the converter is reactivated. By modulating the on-time of the switch, this control scheme ensures two main objectives:

- The input current of the rectifier remains sinusoidal and in phase with the input voltage
- Power losses are reduced, and the overall system power factor is improved.

The converter, represented by the switch K, operates periodically with a duty cycle α at a switching frequency Fc. This duty cycle is generated by the hysteresis controller to regulate the output voltage VDC.

Figure 8 illustrates the structure of the Boost converter.

Figure 8. Structure of the Boost PFC Converter

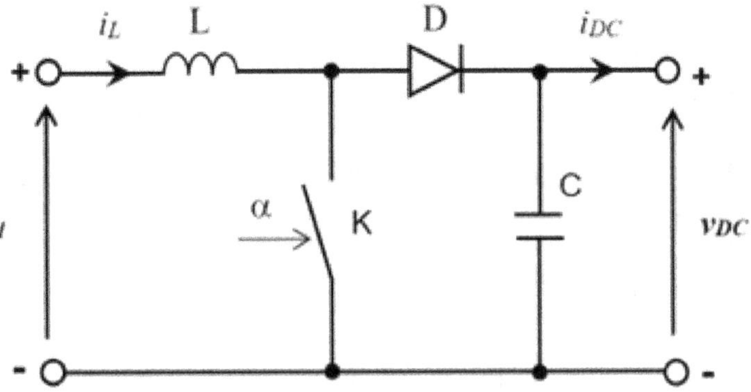

Depending on the state of the switch K, two operating modes can be defined:

✓ **Mode 1** (0 < t < αTc): **K closed, diode blocked**

In this mode, the inductor L is charged, while the capacitor C supplies the load. The voltage-current relationship can be expressed as:

$$L.\frac{di_L(t)}{dt} = v_d(t)\ C.\frac{dv_{DC}(t)}{dt} = -i_{DC}(t) \quad (23)$$

✓ **Mode 2** (αTc < t < Tc): **K open, diode conducting**

In this mode, the inductor L discharges, and the capacitor C is charged. The corresponding relationship is:

$$L\frac{di_L(t)}{dt} = v_d(t) - v_{DC}(t) \quad (24)$$

By analyzing these two configurations, the relationships between the input quantities (Vd, Id) and the output quantities (V_{DC}, I_{DC}) of the Boost converter can be expressed by the following system of equations:

$$\begin{cases} L\frac{di_L(t)}{dt} = v_d(t) - v_{DC}(t).(1-\alpha) \\ i_{DC}(t) = i_d(t).(1-\alpha) \end{cases} \quad (25)$$

Under continuous conduction mode and steady-state conditions ($\frac{diL}{dt} = 0$), the system of equations (25) simplifies to:

$$\begin{cases} V_{DC} = \frac{V_d}{1-\alpha} \\ I_{DC} = I_d.(1-\alpha) \end{cases} \quad (26)$$

Where (V_d,i_d) and (V_{DC},i_{DC}) represent the input and output voltage and current of the Boost converter, respectively, and α is the converter duty cycle.

iii. Dc -Link Model

The DC-link plays a crucial role in the wind energy conversion chain by providing an interface between the rectifier and the inverter. It stabilizes the DC voltage and temporarily stores energy through a capacitor, which smooths out voltage fluctuations caused by variations in load and power generation.

The voltage across the DC-link capacitor is governed by the following equation:

$$V_{DC}(t) = \frac{1}{C}\int i_c.dt \quad (27)$$

Where C is the DC-link capacitor value, and ic represents the capacitor current, defined as:

$$i_C = i_{DC} - i_{inv} \quad (28)$$

This simplified model characterizes the transient behavior of the DC-link and its essential role in regulating the DC bus voltage within the wind energy conversion chain (Guo et al., 2023).

iv. Inverter Model

In the wind energy conversion chain, the inverter is responsible for converting the DC voltage from the DC-link into an AC voltage synchronized with the grid. It also regulates the active and reactive power exchanged with the grid (Hassan et al., 2025). The inverter used in this study is a three-phase, three-leg inverter, consisting of IGBTs with antiparallel diodes, controlled by a Pulse Width Modulation (PWM) strategy.

Each leg comprises two switches, controlled by complementary PWM signals (S1, S2), (S3, S4), (S5, S6) to prevent DC-link short-circuiting. For each switch Ki, two operating states are defined:

- When Si=1, switch ki is in the closed state.
- When Si=0, switch ki is in the open state.

Figure 9 shows the structure of the three-phase inverter.

Figure 9. Structure of the PWM Three-Phase Inverter

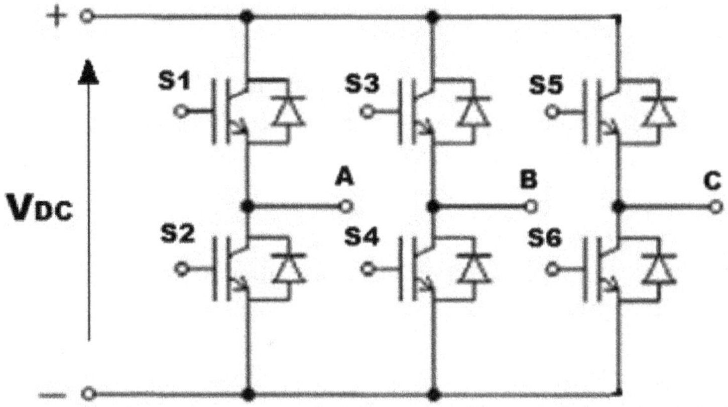

Based on the switching states Si and the DC-link voltage VDC, the output pole voltages of the inverter, measured with respect to the negative DC bus, can be expressed as:

$$\begin{cases} V_{An} = \frac{V_{DC}}{2}(2S_1 - 1) \\ V_{Bn} = \frac{V_{DC}}{2}(2S_3 - 1) \\ V_{Cn} = \frac{V_{DC}}{2}(2S_5 - 1) \end{cases} \quad (29)$$

The line voltages can be derived from the pole voltages as:

$$\begin{cases} U_{AB} = V_{An} - V_{Bn} \\ U_{BC} = V_{Bn} - V_{Cn} \\ U_{CA} = V_{Cn} - V_{An} \end{cases} \quad (30)$$

Finally, the phase voltages are derived from the inverter leg voltages as:

$$\begin{cases} V_A = \frac{1}{3}.[2.V_{An} - V_{Bn} - V_{Cn}] \\ V_B = \frac{1}{3}.[2.V_{Bn} - V_{An} - V_{Cn}] \\ V_C = \frac{1}{3}.[2.V_{Cn} - V_{An} - V_{Bn}] \end{cases} \quad (31)$$

v. Grid Side Filter Model

At the output of the inverter, the current waveform typically exhibits harmonic distortions due to high-frequency switching operations. These distortions can significantly degrade the power quality when injecting energy into the grid. To mitigate this issue and ensure compliance with grid standards, an LCL filter is implemented on the grid side (He et al., 2023).

The LCL filter is specifically designed to attenuate high-order harmonics while preserving the fundamental frequency component. It consists of two series inductors (L1 and L2) and a shunt capacitor (Cf) placed between them. A damping resistor (Rf) is often added in series with the capacitor to improve the filter's stability and prevent resonance phenomena.

The equivalent single-phase representation of the proposed LCL filter configuration is illustrated in Figure 10.

Figure 10. Equivalent single-phase diagram of the LCL filter

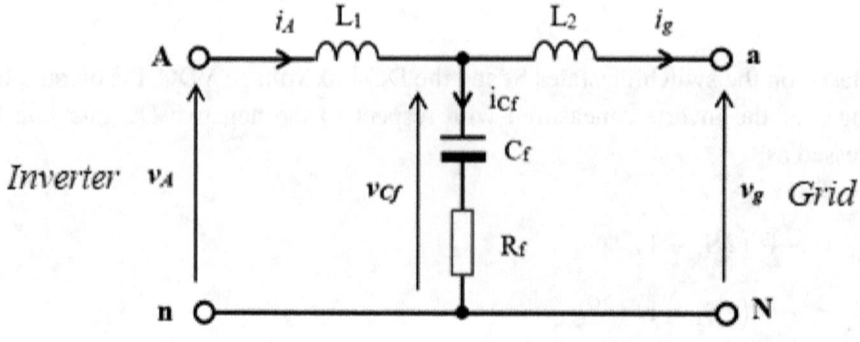

The dynamic equations of the inverter connected to the grid through the LCL filter can be expressed as:

$$\begin{cases} i_A = i_C + i_g \\ L_1 \frac{di_A}{dt} = V_A - V_{Cf} \\ L_2 \frac{di_g}{dt} = V_{Cf} - V_g \\ i_C = C_f \frac{dV_{Cf}}{dt} \end{cases} \quad (32)$$

The transfer function of the LCL filter, relating the grid current ig to the inverter voltage VA, considering the damping resistor Rf, is given by:

$$H_{LCL}(S) = \frac{i_g(S)}{V_A(S)} = \frac{1}{\left(C_f L_1 L_2\right).S^3 + R_f C_f\left(L_1 + L_2\right).S^2 + \left(L_1 + L_2\right).S + R_f C_f} \quad (33)$$

The resonance frequency f0 is calculated as:

$$f_0 = \frac{1}{2\pi} \cdot \sqrt{\frac{L_1 + L_2}{L_1 L_2 C_f}} \quad (34)$$

The LCL filter parameters can be calculated following standard design procedures. The inductances, capacitor, and damping resistance are determined as follows (He et al., 2018):

$$\begin{cases} L_1 = \frac{V_{DC}.V_g}{\sqrt{2}.(60\%).f_{sw}.P_g} \\ L_2 = \frac{2\pi f_g.V_g^2}{(5\%).(2\pi f_{sw})^2.P_0} \cdot \left(1 + \sqrt{\frac{1}{k_\alpha^2}}\right) \\ C_f = (5\%).\frac{P_0}{2\pi f_g.V_g^2} \\ R_f = \frac{2\pi f_g.V_g^2}{3 \times (5\%).2\pi f_0.P_0} \end{cases} \quad (35)$$

Where:
VD: DC bus voltage (V)
Vg: RMS grid voltage (V)
P0: Transferred active power (W)
fg: Fundamental grid frequency (Hz)
fsw: Inverter switching frequency (Hz)

f0: Resonance frequency (Hz)
Kα: Attenuation factor (typically 0.2)
L1, L2: Inverter-side and grid-side inductances (H)
Cf: Filter capacitance (F)
Rf: Damping resistance (Ω)

vi. Grid Model

The electrical grid to which the inverter is connected is assumed to have infinite power compared to the other components. This assumption simplifies the analysis by considering the grid voltage and frequency as constant references, a common approach in grid-connected renewable energy systems (Heier, 2014).

In grid-connected mode, the grid dictates the operating frequency through a Phase-Locked Loop (PLL) for synchronization. The PLL is essential to track the grid phase angle and frequency accurately, ensuring stable and efficient power injection from the inverter (Hossain et al., 2017). The performance and stability of the PLL under grid disturbances such as voltage sags or frequency variations have been extensively studied in recent literature (Ichim-Burlacu et al., 2024).

In the Park reference frame with angular frequency ωg, the grid is modeled by the following system of equations:

$$\begin{cases} V_d = (L_1 + L_2).\frac{di_{dg}}{dt} - \omega_g(L_1 + L_2).i_{qg} + V_{dg} \\ V_q = (L_1 + L_2).\frac{di_{qg}}{dt} + \omega_g(L_1 + L_2).i_{dg} + V_{qg} \end{cases} \quad (36)$$

Where:
Vd: inverter output voltages in the (d,q) frame
Vdg: grid voltages in the (d,q) frame
idg,iqg: grid currents in the (d,q) frame
The active and reactive powers injected into the grid are expressed as:

$$\begin{cases} P_g = \frac{3}{2}(V_{dg}.i_{dg} + V_{qg}.i_{qg}) \\ Q_g = \frac{3}{2}(V_{qg}.i_{dg} - V_{dg}.i_{qg}) \end{cases} \quad (37)$$

This model allows controlling the active and reactive powers delivered to the grid by regulating the (d,q) components of the injected current. Advanced control strategies based on this model enable efficient grid support functionalities, such as reactive power compensation and grid voltage support during disturbances.

2. WIND SYSTEM CONTROL

a. Control Structure

The wind energy conversion system (WECS) investigated in this study employs a Permanent Magnet Synchronous Generator (PMSG) connected to the electrical grid through a dual-stage power conversion system. This configuration includes a machine-side converter (MSC) and a grid-side converter (GSC), coupled via a DC-link. Such an arrangement enables full-speed variable operation and independent control of both the generator and the power injection to the grid. The control structure adopted in this system is illustrated in Figure 11.

Figure 11. Control structure of the wind energy system

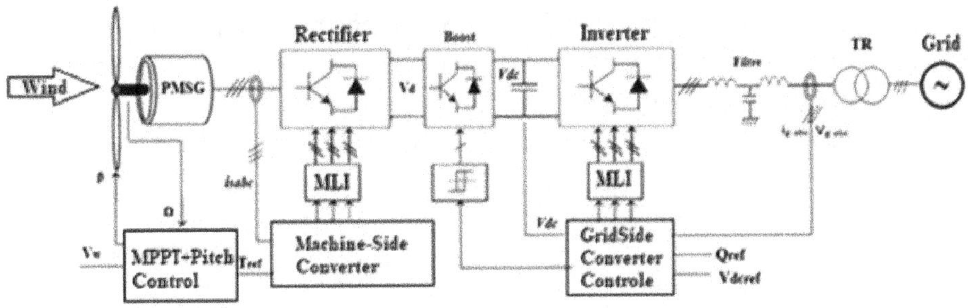

i. Control Strategy of the Wind System

To ensure variable-speed operation of the grid-connected wind system, several control strategies are implemented:

✓ ***Maximum Power Point Tracking (MPPT):***

In below-rated wind speeds, an MPPT algorithm dynamically adjusts the generator speed to track the optimal tip speed ratio, ensuring that the maximum extractable wind power is captured.

✓ ***Pitch angle Regulation:***

When the wind speed exceeds its rated value, the blade pitch angle is adjusted to limit the aerodynamic power captured by the turbine. This prevents overloading the generator and helps maintain power output at the rated value.

✓ ***Control of the Machine-Side Converter (MSC):***

The MSC regulates the electromagnetic torque and rotor speed of the generator. This is typically achieved using field-oriented control (FOC), which decouples the torque and flux components of the stator current.

✓ ***Control of the Grid-Side Converter (GSC):***

The GSC is responsible for maintaining a constant DC-link voltage, ensuring synchronization with the grid via a phase-locked loop (PLL), and regulating the active and reactive power injection. By controlling the d-q components of the grid current, the system can provide dynamic support services such as voltage regulation and low-voltage ride-through capability during grid faults.

As shown in Figure 12, the wind turbine operates across four distinct regions, each with a specific control objective:

Figure 12. Operating regions of a wind turbine system

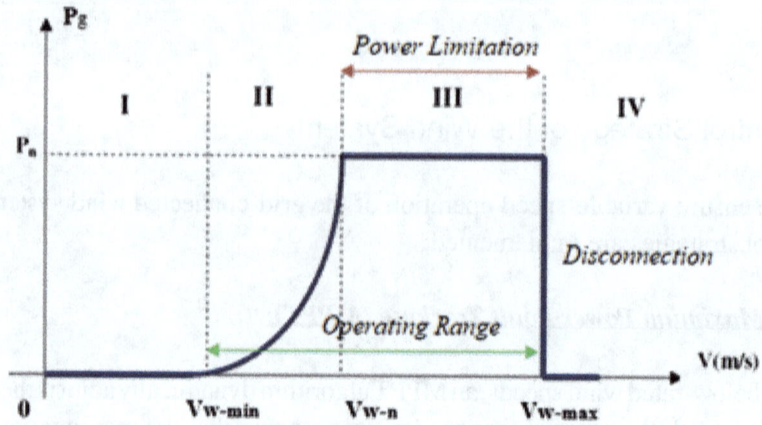

The operating regions are associated with specific control objectives (Jing et al., 2017):

Region I: In this region, the wind speed is below the cut-in speed (Vw-min). The turbine may rotate but is unable to provide sufficient torque; therefore, no energy can be captured, and the output power remains zero.
Region II: In this region, the wind speed reaches the minimum value Vw-min, and the turbine begins generating power. The MPPT algorithm is activated to extract the maximum power at each wind speed until it reaches the rated wind speed Vw-n, corresponding to the rated power Pn.
Region III: Between the rated wind speed and the maximum wind speed Vw-max, the turbine operates at a constant speed through pitch angle control. This adjustment limits the power to its rated value Pn.
Region IV: When the wind speed reaches its maximum value, an emergency mechanism is activated for safety reasons. The wind turbine blades are set to the feathering position ($\beta = 0°$) to prevent any mechanical damage.

b. MPPT Control of the Wind Turbine

MPPT Control (Maximum Power Point Tracking) is essential in a wind energy conversion system to maximize the energy extracted from the wind (Figure 12, Region II). The variable and unpredictable nature of wind causes fluctuations in the generated power. To efficiently operate the wind turbine over a wide range of wind speeds and achieve optimal efficiency, it is necessary to associate the machine-side AC/DC converter with an MPPT controller incorporating current regulation loops. This architecture allows the generator to operate at its maximum power point (Kiessling et al., 2021).

Several MPPT techniques have been proposed in the literature, including (Lammouchi et al., 2024):

- **Perturb and Observe (P&O) method:** based on monitoring power variations.
- **Fuzzy logic control:** which uses a set of logical rules to adjust the operating point.
- **Neural network-based algorithms:** capable of learning and modeling complex behaviors.
- **Sliding mode or torque-based control approaches:** known for their robustness but more complex to implement.

The mechanical power characteristic as a function of turbine speed for different wind speeds (Figure 4) exhibits a bell-shaped curve. The peaks of these curves represent the optimal points, forming the optimal power curve. To ensure the maximum power coefficient Cpmax, it is necessary to maintain an optimal tip speed ratio λ between the blade rotational speed and the wind speed.

This ratio is defined as (Lammouchi et al., 2024):

$$\lambda = \frac{\mathbf{R} \cdot \mathbf{\Omega}_t}{\mathbf{V}_w} \quad (38)$$

The reference speed is calculated from the measured wind speed (Gao & Gao, 2016):

$$\mathbf{\Omega}_{ref} = \frac{\lambda_{opt} \cdot \mathbf{V}_w}{\mathbf{R}} \quad (39)$$

When the turbine operates in the optimal regime, the extracted power is maximum and can be expressed as:

$$P_{max} = \frac{1}{2}.\rho.\pi.R^2.C_{pmax}.V_\omega^3 \quad (40)$$

Using (39), this power can also be expressed as:

$$\mathbf{P}_{max} = \left(\frac{1}{2}.\frac{\rho.\pi.\mathbf{R}^5.\mathbf{C}_{pmax}}{\lambda_{opt}^3}\right).\mathbf{\Omega}_{ref}^3 \quad (41)$$

This reference speed is then used to generate a torque reference using the following relation:

$$\mathbf{T}_{ref} = \mathbf{k}_{opt} \cdot \mathbf{\Omega}_{ref}^2 \quad (42)$$

Where Kopt is a constant related to the aerodynamic characteristics of the turbine.

$$K_{opt} = \frac{1}{2}.\frac{\rho.\pi.R^5.C_{pmax}}{\lambda_{opt}^3}$$

The objective of the control is to adjust the generator speed to maintain this optimal ratio, regardless of wind speed variations. To achieve this goal, the control system must continuously track the optimal power curve. The control strategy depends on the available measurements and whether speed regulation is applied.

The objective of the control is to adjust the generator speed to maintain this optimal ratio, regardless of wind speed variations. To achieve this goal, the control system must continuously track the optimal power curve. The control strategy depends on the available measurements and whether speed regulation is applied.

Two main categories of MPPT approaches are widely discussed in the literature:

✓ **MPPT without knowledge of the Cp (λ, β) characteristic:**

In this class of methods, the aerodynamic model of the turbine is not explicitly known. Control strategies are based on real-time observations of electrical quantities such as power or current. Popular techniques include the Perturb and Observe (P&O) algorithm and artificial intelligence methods such as fuzzy logic controllers or neural networks (Majout et al., 2022; Manwell et al., 2009). These approaches are relatively simple to implement but may suffer from oscillations around the maximum power point and are sensitive to parameter variations.

✓ **MPPT with knowledge of the turbine characteristics:**

When the Cp (λ, β) curve is known, it becomes possible to analytically determine the optimal operating point corresponding to maximum aerodynamic efficiency. In such cases, the control strategy can follow the optimal power curve using estimations or measurements of wind speed and rotor speed. Two distinct structures can be applied:

- **Speed-based control (with mechanical speed feedback):** The generator speed is measured and compared to a reference derived from the optimal TSR. A PI controller then adjusts the electromagnetic torque accordingly.
- **Sensorless control (without mechanical speed feedback):** Instead of measuring the speed directly, the control system estimates it using electrical signals or observer-based algorithms, thereby reducing sensor cost and improving system robustness (Meenakshi et al., 2024; Moutchou & Abbou, 2021).

Each method offers trade-offs between implementation complexity, robustness, and accuracy. In recent research, hybrid approaches that combine estimation techniques with adaptive control have gained attention for their ability to enhance performance under turbulent wind profiles (Nasiri et al., 2020).

i. Control with Speed Regulation

This control approach involves adjusting the electromagnetic torque on the shaft of the PMSG in order to achieve the reference rotational speed.

The speed control scheme is illustrated in Figure 13.

Figure 13. MPPT Control with Speed Regulation

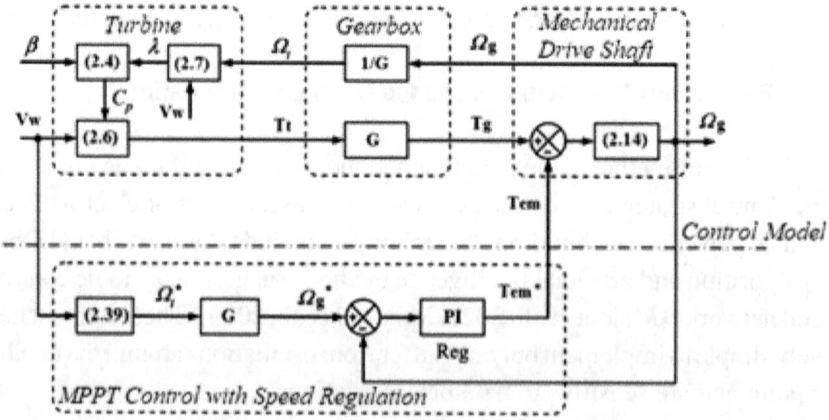

The reference electromagnetic torque Tem*, required to reach a generator speed Ωg equal to its reference Ωg*, is computed by the speed regulator. This regulator, of Proportional-Integral (PI) type, ensures accurate speed control while minimizing the effect of the mechanical torque Tg considered as a disturbance.

The closed-loop transfer function of the speed control is expressed as:

$$\frac{\Omega_g(S)}{\Omega_{g-ref}(S)} = \frac{1 + \frac{k_{p\omega}}{k_{i\omega}}.S}{1 + \left(\frac{f + k_{p\omega}}{k_{i\omega}}\right).S + \frac{J}{k_{i\omega}}S^2} \quad (43)$$

The characteristic equation is identified as:

$$1 + \left(\frac{f + k_{p\omega}}{k_{i\omega}}\right).S + \frac{J}{k_{i\omega}}.S^2 = 1 + \frac{2\xi}{\omega_n}.S + \frac{1}{\omega_n^2}.S^2 \quad (44)$$

The PI gains are calculated using:

$$\begin{cases} k_{p\omega} = 2\xi J.\omega_n - f \\ k_{i\omega} = J.\omega_n^2 \end{cases}$$

Where:

J: System inertia

f: Viscous friction coefficient

ζ: Damping ratio

ωn : Natural frequency

For a given operating condition (constant wind speed), the objective is to achieve maximum mechanical power, corresponding to Cpmax. This is satisfied when the tip speed ratio λ equals its optimal value λopt.

However, using an anemometer to measure wind speed only provides a local estimate, which may not accurately represent the average wind speed at the rotor plane. Any error in this estimation leads to reduced power extraction. For this reason, most modern wind turbines operate without explicit speed regulation.

ii. Control without speed Regulation

This control strategy is based on the assumption that the wind speed varies very little in steady-state operation compared to the electrical time constants of the wind energy system. This implies that the turbine's acceleration torque can be considered negligible. Moreover, by neglecting the viscous friction torque, the following expression can be written:

$$\mathbf{J}.\frac{\mathbf{d\Omega_g}}{\mathbf{dt}} + \mathbf{f}.\mathbf{\Omega_g} = \mathbf{T_g} - \mathbf{T_{em}} \approx 0 \quad (45)$$

Which leads, under steady-state conditions, to:

$$\mathbf{T_g} = \mathbf{T_{em}} \quad (46)$$

An estimation of the turbine rotational speed, denoted as $\Omega t\text{-}est$, is performed based on the measurement of the rotational speed of the PMSG:

$$\mathbf{\Omega_{t-est}} = \frac{\mathbf{\Omega_g}}{\mathbf{G}} \quad (47)$$

Since obtaining an accurate measurement of the wind speed is complex, an estimated wind speed, denoted as Vw-est, is derived from the estimated rotational speed of the turbine:

$$\mathbf{V_{w-est}} = \frac{\mathbf{\Omega_{t-est}}.\mathbf{R}}{\boldsymbol{\lambda}} \quad (48)$$

The wind turbine torque can then be determined from the estimated wind speed. It can be expressed as:

$$\mathbf{T}_{t-est} = \frac{1}{2}.\mathrm{C}_p.\boldsymbol{\rho}.\boldsymbol{\pi}.\mathrm{R}^3.\mathbf{V}^2_{w-est} \quad (49)$$

The reference electromagnetic torque is then calculated from the estimated turbine torque:

$$\mathbf{T}_{em-ref} = \frac{\mathbf{T}_{t-est}}{\mathbf{G}} = \mathbf{k}_t.\boldsymbol{\Omega}^2_g \quad (50)$$

Where kt is a constant defined by:

$$k_t = \frac{K_{opt}}{G^2} \text{ With: } \mathrm{K}_{opt} = \frac{1}{2}.\frac{\rho.\pi.R^5.C_{pmax}}{\lambda^3_{opt}}$$

The principle of the MPPT control strategy without speed regulation is illustrated in Figure 14.

Figure 14. MPPT Control without Speed Regulation

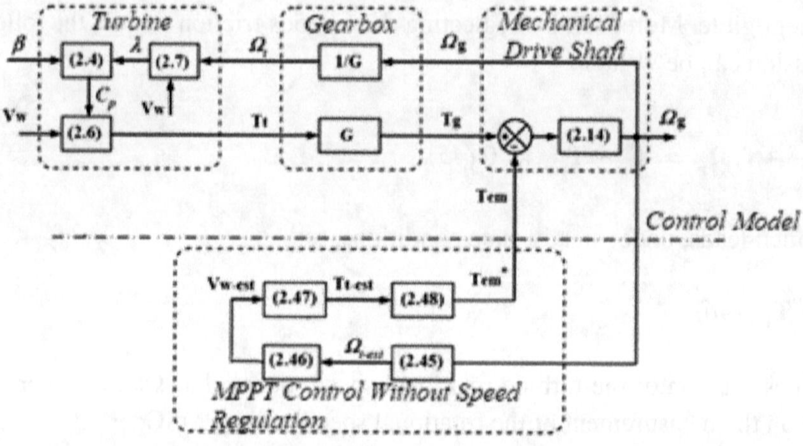

iii. MPPT Control Based on TSR (Tip Speed Raio)

The Maximum Power Point Tracking (MPPT) control based on the Tip Speed Ratio (TSR) aims to optimize the power extracted by a wind turbine by adjusting the rotor's rotational speed according to the wind speed (Ngamroo, 2017). This approach seeks to maintain the TSR at its optimal value, denoted as **λopt** where the aerodynamic efficiency of the turbine reaches its maximum (Patil & Kamble, 2018).

The TSR value is determined using the generator's rotational speed and the wind speed. The error between the actual TSR and the optimal TSR is defined as follows:

$$e_{\lambda}(t) = \lambda_{opt} - \lambda(t) = \lambda_{opt} - \frac{\Omega_g(t) . R}{G . V_w(t)} \quad (51)$$

To maximize the power coefficient Cp, the control objective is to keep the TSR close to **λopt**.

This error is regulated using a Proportional-Integral (PI) controller to determine the optimal torque to be applied.

The algorithm of the TSR-based MPPT control is given by the following relation:

$$T_{em-ref} = K_p . e_{\lambda}(t) + K_i . \int e_{\lambda}(\tau) . d\tau \quad (52)$$

In this case, the applied torque governs the dynamic behavior of the angular velocity based on the system's moment of inertia.

The MPPT control scheme based on TSR regulation is illustrated in Figure15.

Figure 15. MPPT Control Based on TSR

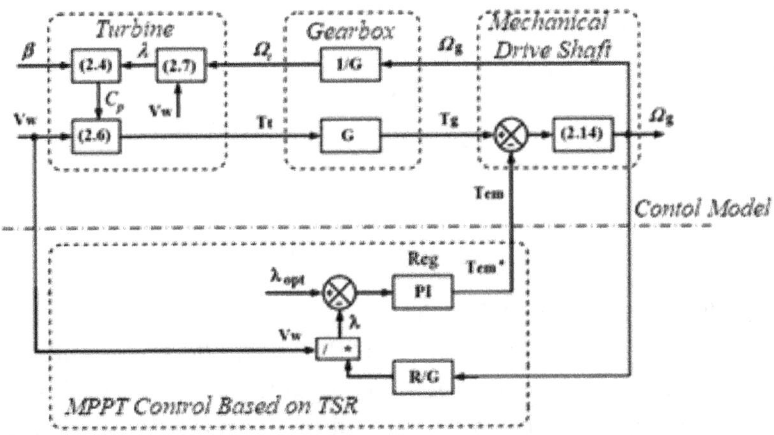

In this case, the low-frequency transfer function of the tip speed ratio λ is expressed as:

$$\frac{\lambda(S)}{\lambda_{opt}(S)} = \frac{1 + \frac{k_{p\lambda}}{k_{i\lambda}} . S}{1 + \left(\frac{f + k_{p\lambda}}{k_{i\lambda}}\right) . S + \frac{J}{k_{i\lambda}} S^2} \quad (53)$$

Where:

$$\begin{cases} k_{p\lambda} = 2\xi J.\omega_n - f \\ k_{i\lambda} = J.\omega_n^2 \end{cases}$$

Figure 16 presents the response of the TSR-based MPPT control under a variable wind speed profile.

Figure 16. Response of the Tip Speed Ratio Control

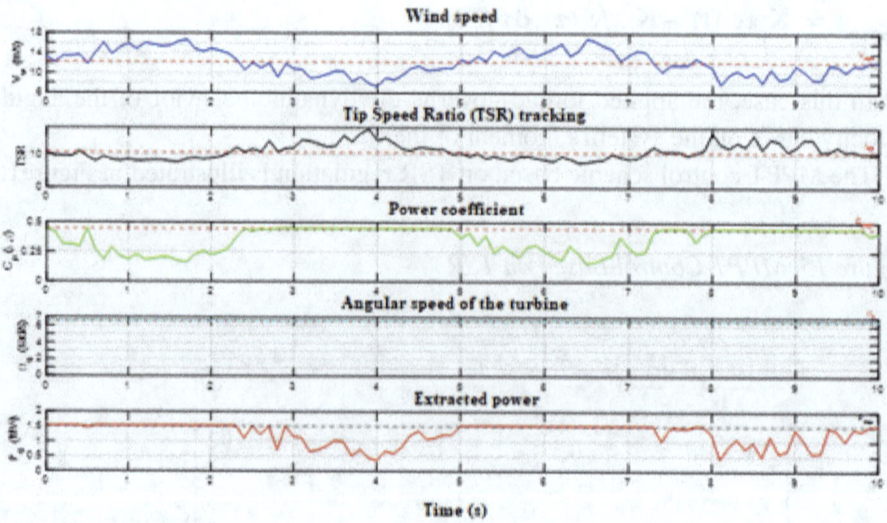

The results in Figure 16 highlight the effectiveness of the MPPT control for wind energy systems. Specifically, the TSR-based MPPT controller successfully optimized the power extraction by adapting the rotor speed to changes in wind speed. It is evident that this approach maintained the TSR at its optimal value **λopt**, ensuring turbine operation at optimal aerodynamic efficiency.

c. Pitch Angle Control

i. Principale

To ensure proper operation of the wind energy system and maintain grid connection during strong wind gusts, a blade pitch control system is activated to regulate blade orientation in order to keep the generated power within the nominal range (Figure 12, Region III). The pitch regulation system in medium and large-scale grid-connected wind turbines is generally based on a hydraulic mechanism controlled by a computer-based control system (Rahimi, 2017). Some manufacturers employ electric or hydraulic servomotors for blade pitch adjustment (Rehouma et al., 2023).

As illustrated in Figure 3, increasing the pitch angle β reduces the amount of aerodynamic energy captured by the turbine. At low wind speeds, the blades face the wind to extract the maximum available power. When the wind speed reaches its nominal value Vw-n the blades start to pitch. At maximum wind speed Vw-max, they reach the feathering position to limit power extraction.

ii. Blade Pitch Angle Control

The blade pitch regulation system generally employs a Proportional–Integral (PI) controller, as shown in figure 17.

The generator speed Ωg is compared to its reference value Ωg-ref. The resulting error signal is then sent to the controller, which generates the desired pitch angle reference βref.

This reference angle is applied to the servomotor, which adjusts the blade position accordingly. As a result, the generator speed is regulated, and the output power is limited.

Figure 17. Pitch Angle Regulation Diagram

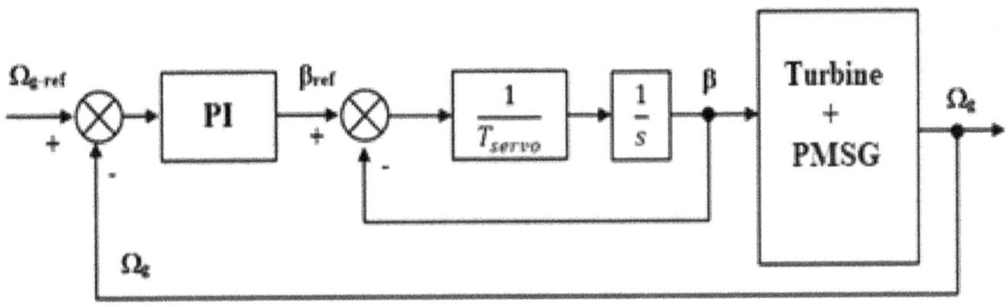

The pitch angle β follows its reference βref through a first-order response characterized by the servo time constant Tservo. This dynamic is taken into account in the control loop to model a realistic system response. This feedback loop enables the regulation of the power coefficient, and hence, the control of the turbine output power.

The pitch control system is represented by the following low-frequency transfer fonction:

$$\frac{\beta(S)}{\beta_{ref}(S)} = \frac{\frac{1 + k_{p\beta}}{k_{i\omega}}.S}{1 + \left(\frac{1 + k_{p\beta}}{k_{i\beta}}\right).S + \frac{T_{servo}}{k_{i\beta}}.S^2} \quad (54)$$

Given a damping ratio ξ and a natural frequency ω_n, the PI controller parameters can be expressed as follows:

$$\begin{cases} k_{p\beta} = 2.\xi.T_{servo}.\omega_n - 1 \\ k_{i\beta} = T_{servo}.\omega_n^2 \end{cases}$$

Figure 18 shows the pitch angle response β controlled by a PI regulator under a variable wind speed profile, along with the corresponding power coefficient and wind turbine output power.

Figure 18. Blade Pitch Angle Regulation Results

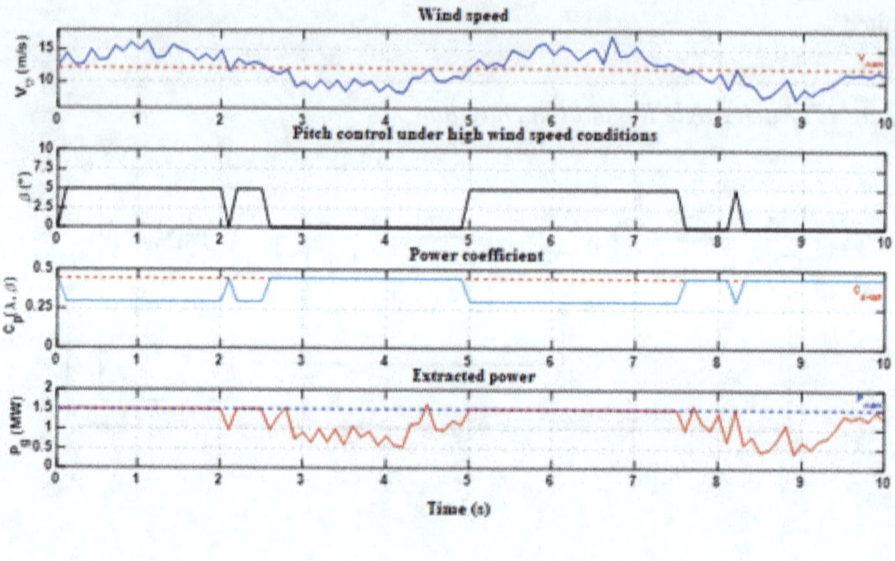

According to the results shown in Figure 18, the controller only activates when the wind speed Vw exceeds the nominal value of 12.1 m/s. In this situation, the control strategy increases the pitch angle β to reduce the power coefficient Cp, thereby adjusting the extracted power to its rated value of 1.5 MW.

For this regulation strategy, the pitch angle is limited between 0° and 5°. When the wind speed is below the nominal value, the pitch angle is set to zero to ensure maximum power production.

d. Machine-Side Converter Control (MSC)

Vector control, also known as Field-Oriented Control (FOC), is an advanced control method widely used to precisely control synchronous machines, particularly the Permanent Magnet Synchronous Generator (PMSG) (Romdhane et al., 2023). This technique enables decoupling between magnetic flux control and electromagnetic torque control, similar to the operation of a separately excited DC machine, thus significantly improving dynamic performance.

i. Principle of Vector Control

The principle of vector control relies on transforming three-phase quantities (voltages, currents, and fluxes) into a rotating reference frame (d, q), aligned with the rotor flux (Romdhane et al., 2023).

In this strategy, the aim is to create a linearized model and transform the PMSG into an equivalent structure resembling that of a separately excited DC machine from a torque perspective. This decoupling is achieved by forcing the direct-axis current id to zero while maintaining constant flux, so that the torque becomes directly proportional to the quadrature-axis current iq.

In this control scheme, the reference stator current along the q-axis is derived from the MPPT control via the reference electromagnetic torque (Sawyer & Tapia, 2005). The reference stator current along the d-axis is set to zero, allowing the generator to operate at zero reactive power (i.e., with unity power factor).

Figure 19 shows the block diagram of the machine-side converter control. This control strategy enables independent regulation of the stator d- and q-axis currents, and hence, the active and reactive powers delivered by the generator.

Figure 19. Block diagram of machine-side vector control (MSC)

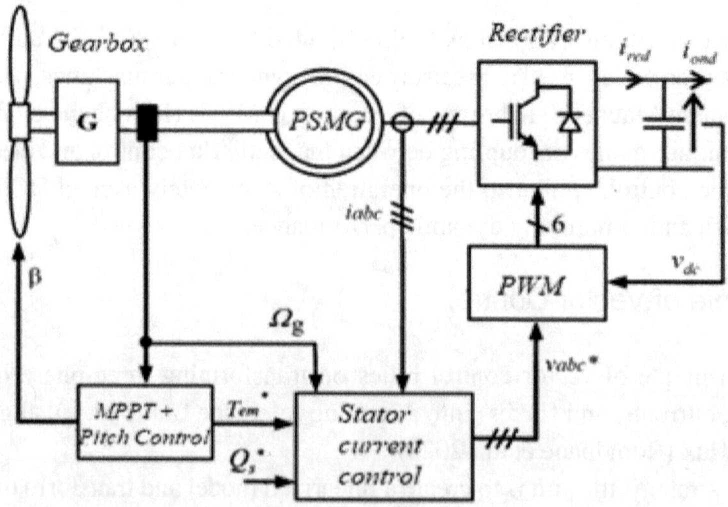

ii. Generator Current Regulation

To meet the control objectives, the following setpoints are typically imposed:

$$\mathbf{i_d^*} = 0 \ \mathbf{i_q^*} = \frac{\mathbf{T^*_{em}}}{\mathbf{p} \cdot \boldsymbol{\varphi}_\mathbf{f}} \quad (55)$$

The machine-side converter (rectifier) is controlled using the PWM (Pulse Width Modulation) technique. This method requires the reference stator voltages (Varef, Vbref and Vcref), which are obtained from the d- and q-axis reference voltages Vd* and Vq* via the inverse park transformation. These reference voltages are computed by the inner current control loops for id and iq.

The transfer function of the PMSG current loop is typically expressed as:

$$\mathbf{G_{idq}(p)} = \frac{\mathbf{I_{d,q}(S)}}{\mathbf{V_{d,q}(S)}} = \frac{1}{\mathbf{R_S} + \mathbf{L_S}.\mathbf{S}} \quad (56)$$

The corresponding control loops are illustrated in Figure 20.

Figure 20. Current regulation loops for the PMSG

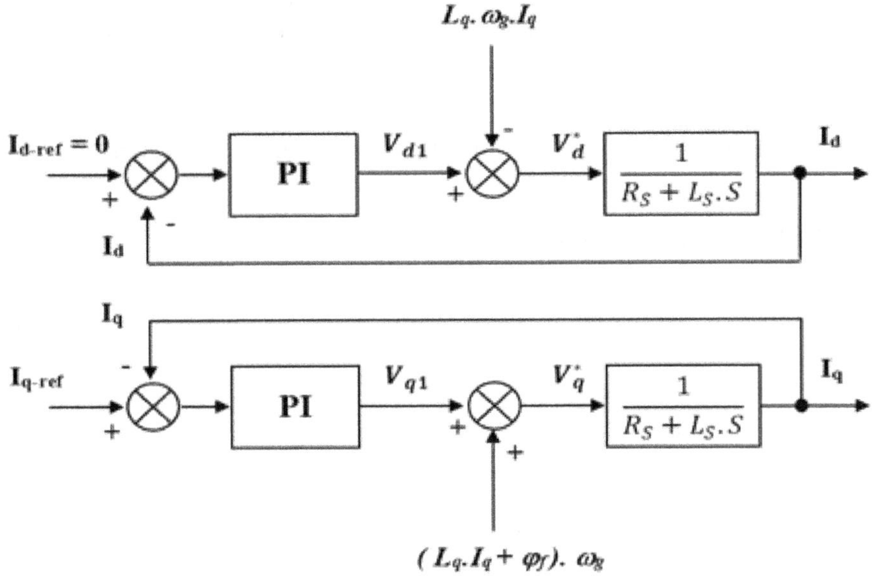

Both id and iq control loops use identical PI controllers, described by the general form:

$$\mathbf{C}_{\mathbf{dq-m}}(\mathbf{S}) = \mathbf{k}_{\mathbf{p-im}} + \frac{\mathbf{k}_{\mathbf{i-im}}}{\mathbf{S}} \quad (57)$$

The closed-loop system is represented by the following low-frequency transfer function:

$$\mathbf{H}_{\mathbf{im}}(\mathbf{S}) = \frac{\mathbf{C}_{\mathbf{im}}(\mathbf{S}).\mathbf{G}_{\mathbf{idq}}(\mathbf{S})}{1+\mathbf{C}_{\mathbf{im}}(\mathbf{S}).\mathbf{G}_{\mathbf{idq}}(\mathbf{S})} = \frac{\mathbf{k}_{\mathbf{i-im}} + \mathbf{k}_{\mathbf{p-im}}.\mathbf{S}}{\mathbf{L}_{\mathbf{S}}.\left(\mathbf{S}^2 + \frac{\mathbf{R}_{\mathbf{S}} + \mathbf{k}_{\mathbf{p-im}}}{\mathbf{L}_{\mathbf{S}}}.\mathbf{S} + \frac{\mathbf{k}_{\mathbf{i-im}}}{\mathbf{L}_{\mathbf{S}}}\right)} \quad (58)$$

For a given damping ratio ξ and natural frequency $\omega_{n\text{-}im}$, the PI controller parameters are defined as:

$$\begin{cases} k_{p-im} = 2.\xi.\omega_{n-im}.L_s - R_S \\ k_{i-im} = L_s.\omega_{n-im}^2 \end{cases}$$

The machine-side current control is validated under a variable wind speed profile, using the selected reference current values. The current responses are designed to be damped transients, based on the criteria: ($\omega_{n\text{-}im}$ = 628 rad/s, and ξ = 0.707).

The resulting PI controller gains for the current regulators are: ($k_{p\text{-}im}$=0.2825 and $k_{i\text{-}im}$=128)

Figure 21 presents the simulation results of the machine-side current regulation.

Figure 21. Simulation results of generator current regulation

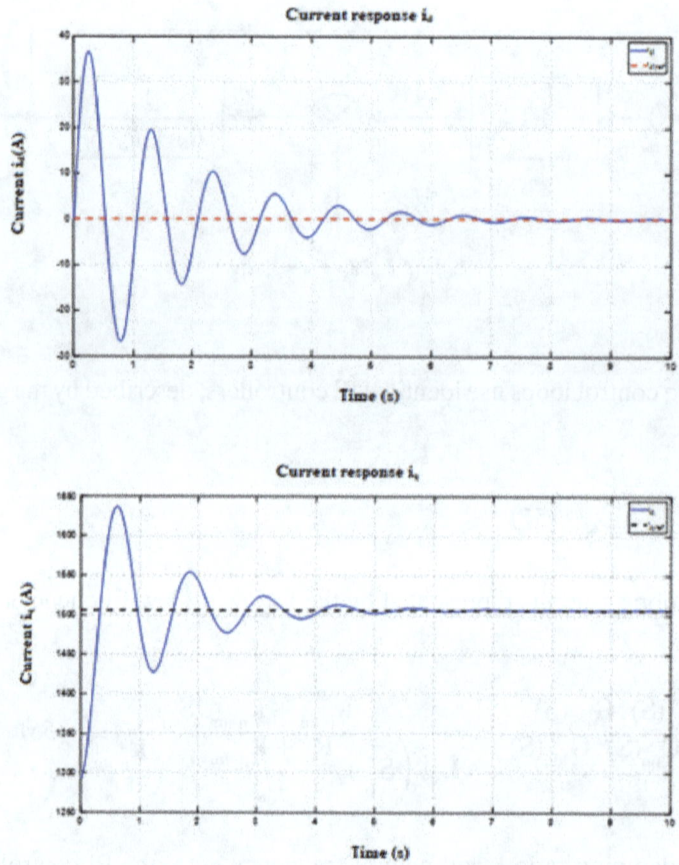

The results demonstrate that the generator currents are properly regulated with acceptable ripple levels.

In particular, the choice of a zero id-ref component enables power transfer from the generator at unity power factor. Moreover, these results confirm effective control of the variable-speed PMSG.

e. Gride-Side Converter Control (GSC)

The wind energy conversion system is connected to the three-phase grid through a PWM inverter. The role of this converter is to transform the DC power available across the DC link into an AC form suitable for grid injection (Sharma et al., 2023). The power flow through the grid-side converter is controlled in such a way as to maintain the DC link voltage at a constant reference value (Shayeghi et al., 2021).

An increase in the output power, rather than the input power at the DC bus (extracted power), causes a decrease in the DC voltage V_{DC}, and vice versa. Therefore, the output power is regulated to maintain constant V_{DC}. Additionally, the reactive power injected into the grid is controlled to be zero, ensuring unity power factor operation.

The control structure of the grid-side converter is illustrated in Figure 22.

Figure 22. Grid-side vector control structure (GSC)

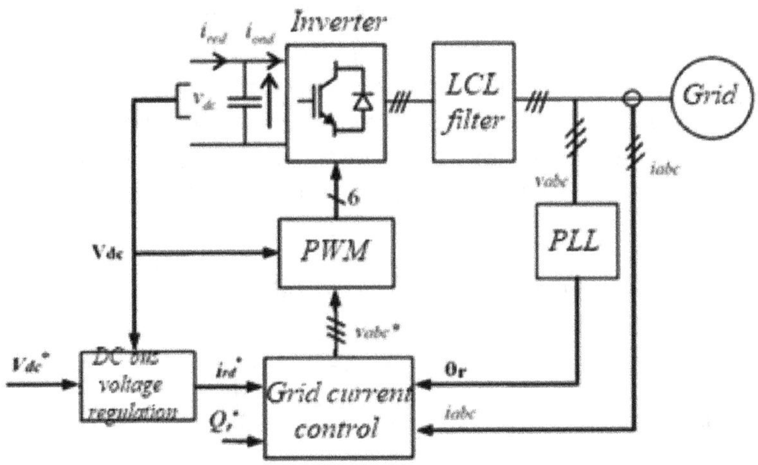

The control of active and reactive power on the grid side is generally achieved by regulating the direct and quadrature components of the currents injected into the grid.

This method uses two cascaded control loops:

- An outer voltage regulation loop for VDC is used to define the reference for the active power to be injected.
- An inner current control loop is used to regulate power. This loop provides the reference values for the direct and quadrature grid currents, which are

then used after regulation to reconstruct the reference voltages for the PWM block.

i. DC Link Voltage Control

The DC bus level is controlled in order to maintain the DC-link voltage VDC constant. As a first step, it is important to set the reference voltage V^*_{DC}. This reference must be higher than the peak value of the AC output line-to-line voltage from the inverter (Sawyer & Tapia, 2005).

For a given modulation index mmm, the expression of V^*_{DC} is given by:

$$\mathbf{V}^*_{\mathbf{DC}} = \frac{2\sqrt{2}.\mathbf{U_S}}{\mathbf{m}.\sqrt{3}} \quad (59)$$

where US is the RMS line voltage at the inverter output.

To maintain VDC constant, a PI controller is used to generate the DC current reference I^*_{DC}. This voltage control loop is slower than the current control loop (typically 10 times slower), since voltage regulation operates at a higher level in the control hierarchy and should not disturb the fast current dynamics.

Assuming negligible converter losses, the control diagram shown in Figure 23 is adopted. The transfer function of the rectifier is assumed to be unity (Wais, 2017). The active power reference P^*_{res} to be injected into the grid is:

Figure 23. Block diagram of DC link voltage regulation

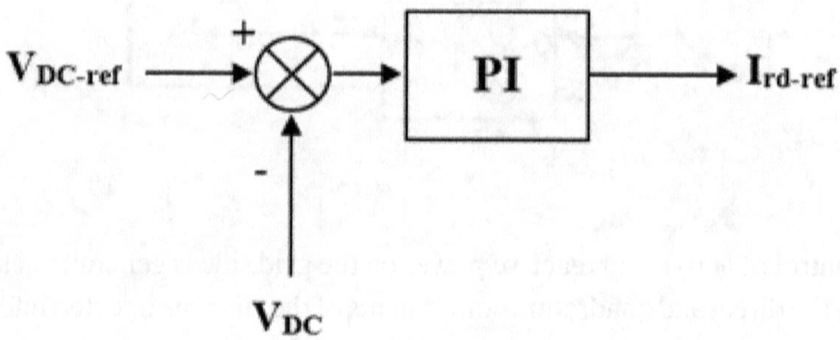

The PI controller is expressed as:

$$C_{dc}(S) = k_{p-dc} + \frac{k_{i-dc}}{S} \quad (60)$$

The low-frequency transfer function of the DC bus is given by:

$$H_{dc}(S) = \frac{C_{dc}(S).G_{dc}(S)}{1 + C_{dc}(S).G_{dc}(S)} = \frac{k_{i-dc} + k_{p-dc}.S}{C.\left(S^2 + \frac{k_{p-dc}}{C}.S + \frac{k_{i-dc}}{C}\right)} \quad (61)$$

For a given damping ratio ξ and natural frequency ωn-dc, the PI controller parameters are computed as:

$$\begin{cases} k_{p-dc} = 2.C.\xi.\omega_n \\ k_{i-dc} = C.\omega_{n-dc}^2 \end{cases}$$

To test the performance of the DC voltage regulation, a variable wind speed profile is applied with a reference voltage VDC* = 1150 V. The PI parameters (kp-dc=0.45 et ki-dc=9.87) are chosen based on a damped transient response with (ωn-dc=31 rad/s et ξ=0.707).

ii. Active and Reactive Power Control

This control strategy is based on regulating the powers injected into the grid by acting on the d-axis ird and q-axis irq current components (Wang & Chang, 2004). The direct-axis current reference ird-ref representing the active power P^*_{res} is provided by the DC voltage regulation loop. The quadrature-axis current reference irq-ref (corresponding to reactive power) is set to zero (Q^*_{res} = 0) to ensure unity power factor.

The current control loops on the grid side use these current references. In our system, we used classical PI regulators along with vector control and decoupling compensation to eliminate cross-coupling effects between current components.

In the Park reference frame, the instantaneous active and reactive powers delivered to the grid are expressed as:

$$\begin{cases} P_{res} = \frac{3}{2}\left(V_{rd}.i_{rd} + V_{rq}.i_{rq}\right) \\ Q_{res} = \frac{3}{2}\left(V_{rq}.i_{rd} - V_{rd}.i_{rq}\right) \end{cases} \quad (62)$$

The vector control strategy, aligned with the grid voltage vector, is formulated as:

$$\begin{cases} V_{rd} = V_r \\ V_{rq} = 0 \end{cases} \quad (63)$$

Taking Equation (2.63) into account, the power expressions become:

$$\begin{cases} P_{res} = \frac{3}{2}.V_r.i_{rd} \\ Q_{res} = -\frac{3}{2}V_r.i_{rq} \end{cases} \tag{64}$$

Thus, the current references ird-ref and irq-ref are imposed by the reference powers P^*_{res} and Q^*_{res}:

$$\begin{cases} i_{rd-ref} = \frac{2}{3.V_r}P^*_{res} \\ i_{rq-ref} = -\frac{2}{3.V_r}Q^*_{res} = 0 \end{cases} \tag{65}$$

iii. Grid Current Control

Using the grid voltage vector, the current control model can be written as:

$$\begin{cases} V_{rd} = R_r.i_{rd} + L_r.\frac{di_{rd}}{dt} - e_{rq} + V_r \\ V_{rq} = R_r.i_{rq} + L_r.\frac{di_{rq}}{dt} + e_{rd} \end{cases} \tag{66}$$

Where:

$\begin{cases} e_{rq} = L_r.\omega_r.i_{rq} \\ e_{rd} = L_r.\omega_r.i_{rd} \end{cases}$ are the coupling components.

Rr and Lr: are the resistance and inductance of the transmission line, respectively, including the filter parameters.

$_r$: is the grid angular frequency.

Similarly to the machine current control, the controlled system is represented by the following low-frequency transfer function:

$$H_{ir}(S) = \frac{C_{ir}(S).G_{idq}(S)}{1 + C_{ir}(S).G_{idq}(S)} = \frac{k_{i-ir} + k_{p-ir}.S}{L_r.\left(S^2 + \frac{R_r + k_{p-ir}}{L_r}.S + \frac{k_{i-ir}}{L_r}\right)} \tag{67}$$

Given a damping ratio ξ and natural frequency ω_{n-ir}, the PI controller parameters are calculated as:

$$\begin{cases} k_{p-ir} = 2.\xi.\omega_{n-ir}.L_r - R_r \\ k_{i-ir} = L_r.\omega^2_{n-ir} \end{cases}$$

The performance of the grid-side current control is tested using the same wind speed profile and the previously established current references.

Transient responses for the current are designed to be damped, based on the following criteria: ($\omega_{n\text{-}ir}$ = 157 rad/s et ξ = 0.707)

Accordingly, the parameters of the grid current PI controllers are (kp-ir=0.023 et ki-ir=1.2455).

Figure 24 illustrates the regulation responses of the DC bus voltage and the grid-side currents.

Figure 24. DC Bus Voltage and Grid-Side Current Regulation Results

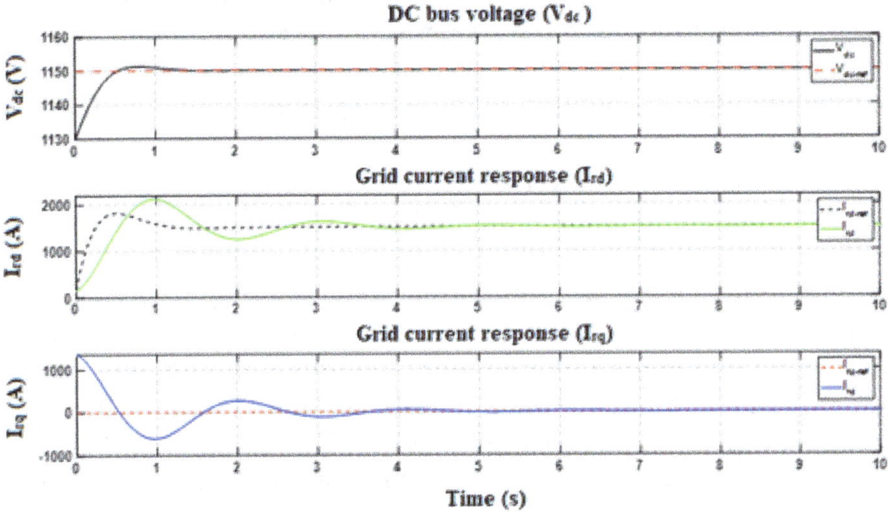

Based on the obtained results, it can be observed that the DC bus voltage VDC gradually converges to its reference value VDC-ref with a damped transient response. This steady convergence indicates that the control system is stable and effective, successfully minimizing oscillations around the setpoint, which is essential to ensure the reliability and quality of the delivered energy.

Furthermore, the analysis of the same figure shows that the grid-side currents are well regulated, exhibiting low ripple. This characteristic is crucial because currents with low ripple help reduce losses and reduce electromagnetic disturbances on electrical equipment, while improving the quality of power injected into the grid.

The choice of imposing a zero-quadrature current reference component (Irq-ref =0) ensures active power transfer with a unity power factor. In other words, this means that reactive power is minimized, optimizing the system's efficiency and avoiding penalties related to poor power factor management in electrical networks.

Finally, these results confirm that the implemented control strategy enables effective energy transfer from the variable-speed wind turbine equipped with a permanent magnet synchronous generator (PMSG) connected to the grid. This demonstrates the system's ability to meet the dynamic requirements of the grid while maximizing energy extraction, thereby ensuring a smooth and efficient integration of wind power generation into the electrical network.

CONCLUSION

This chapter presented a detailed modeling and control strategy for a wind energy conversion system based on a Permanent Magnet Synchronous Generator (PMSG). The study incorporated aerodynamic, mechanical, electrical, and power electronic aspects, providing an accurate and comprehensive representation of the overall system behavior.

The modeling framework included the wind profile, turbine dynamics, PMSG, static converters, and filtering stages, enabling effective dynamic analysis and simulation of the system.

A control strategy was then developed to optimize energy extraction. On the machine side, a Maximum Power Point Tracking (MPPT) method based on the Tip Speed Ratio (TSR) was adopted, combined with a vector control approach to precisely regulate torque and stator currents. On the grid side, a vector control strategy was also implemented to regulate the DC-link voltage, control the active and reactive power exchanged with the grid, and ensure synchronization through a Phase-Locked Loop (PLL).

The effectiveness of the modeling and control strategies was validated through MATLAB/Simulink simulations under variable wind speed conditions. The results demonstrated the system's ability to track the maximum power point, maintain DC-link voltage stability, and meet power quality requirements for grid integration.

This work provides a strong foundation for the development of efficient wind energy systems and opens avenues for future research into advanced control strategies compatible with smart grid requirements.

REFERENCES

Abogrean, E., Elhatmi, E. A., & Saad, J. A. (2022). Simulation of MPPT with a PMSG-based wind energy conversion system considering variable wind speed. *Scientific Journal of University of Benghazi*, *35*(2). Advance online publication. DOI: 10.37376/sjuob.v35i2.3807

Ailliot, P., & Monbet, V. (2012). Markov-switching autoregressive models for wind time series. *Environmental Modelling & Software*, *30*, 92–101. DOI: 10.1016/j.envsoft.2011.10.011

Allal, A., Lammouchi, Z., Khechekhouche, A., Naoui, M., Alqunun, K., & AlGhadhban, A. (2024). Advanced residual optimal mapping approach for precise detection of stator faults in induction motors. *IEEE Access : Practical Innovations, Open Solutions*, *12*, 143515–143530. DOI: 10.1109/ACCESS.2024.3442671

Alqunun, K., Bechiri, M. B., Naoui, M., Khechekhouche, A., Marouani, I., Guesmi, T., Alshammari, B. M., AlGhadhban, A., & Allal, A. (2025). An efficient bearing fault detection strategy based on a hybrid machine learning technique. *Scientific Reports*, *15*, 18739.

Aoumri, M., Harrouz, A., Yaichi, I., Baala, S., & Wira, P. (2024). Robustness investigation of FOC of PMSG-based variable-speed wind turbine connected on grid. In *Proceedings of the 3rd International Conference on Advanced Electrical Engineering (ICAEE)*, Sidi-bel-Abbès, Algeria. https://doi.org/DOI: 10.1109/ICAEE61760.2024.10783398

Baba, M. A., Naoui, M., & Cherkaoui, M. (2023). Fault-tolerant control strategy for Hall sensors in BLDC motor drive for electric vehicle applications. *Sustainability*, *15*(13), 10430. DOI: 10.3390/su151310430

Benbouhenni, H., & Bizon, N. (2021). Third-order sliding mode applied to the direct field-oriented control of the asynchronous generator for variable-speed contra-rotating wind turbine generation systems. *Energies*, *14*(18), 5877. DOI: 10.3390/en14185877

Chen, R., Qin, D., & Liu, C. (2023). Dynamic modelling and dynamic characteristics of wind turbine transmission gearbox-generator system electromechanical-rigid-flexible coupling. *Alexandria Engineering Journal*, *65*, 307–325. DOI: 10.1016/j.aej.2022.10.036

Chen, Z., Guerrero, J. M., & Blaabjerg, F. (2009). A review of the state of the art of power electronics for wind turbines. *IEEE Transactions on Power Electronics*. Advance online publication. DOI: 10.1109/TPEL.2009.2017082

Chowdary, V. G., Sankar, V. U., & Busawon, K. (2018). A review on various MPPT techniques for wind energy conversion system. In *Proceedings of the International Conference on Computing, Power, Energy, Information and Communication (IC-CPEIC)*, Chennai, India. https://doi.org/DOI: 10.1109/ICCPEIC.2018.8525219

Collier, D. A. F., & Heldwein, M. L. (2011). Modeling and design of a micro wind energy system with a variable-speed wind turbine connected to a permanent magnet synchronous generator and a PWM rectifier. In *Proceedings of the XI Brazilian Power Electronics Conference (COBEP)*, Natal, Brazil. https://doi.org/DOI: 10.1109/COBEP.2011.6085301

Dani, A., Benlamlih, M., Mekrini, Z., & Boulaala, M. (2024). Wind energy conversion technologies and control strategies: A review. *International Journal of Renewable Energy Research*, *14*(1), 140–154. DOI: 10.20508/ijrer.v14i1.14296.g8869

Dannehl, J., Liserre, M., & Fuchs, F. W. (2011). Filter-based active damping of voltage source converters with LCL filter. *IEEE Transactions on Industrial Electronics*, *58*(8), 3623–3633. DOI: 10.1109/TIE.2010.2081952

Darabian, R., Yaghini, A., & Tarafdar Hagh, M. (2023). Design of a new robust controller for voltage regulation of DC link in wind farm side converter integrated with a hybrid AC and DC large scale power system. *IET Generation, Transmission & Distribution*, *17*(5), 1561–1572. DOI: 10.1049/gtd2.12780

De Lellis, M., Reginatto, R., Saraiva, R., & Trofino, A. (2018). The Betz limit applied to airborne wind energy. *Renewable Energy*, *127*, 32–40. DOI: 10.1016/j.renene.2018.04.034

Dekali, Z., Baghli, L., & Boumediene, A. (2021). Improved hardware implementation of a TSR-based MPPT algorithm for a low-cost connected wind turbine emulator under unbalanced wind speeds. *Energy*, *232*, 121039. DOI: 10.1016/j.energy.2021.121039

Gao, R., & Gao, Z. (2016). Pitch control for wind turbine systems using optimization, estimation and compensation. *Renewable Energy*, *91*, 501–515. DOI: 10.1016/j.renene.2016.01.057

Girsang, I. P., Dhupia, J. S., Muljadi, E., Singh, M., & Pao, L. Y. (2014). Gearbox and drivetrain models to study dynamic effects of modern wind turbines. *IEEE Transactions on Industry Applications*, *50*(6), 3777–3786. DOI: 10.1109/TIA.2014.2321029

Guo, X., Zhu, D., Hu, J., & Zou, X.. (2023). Inertial PLL of grid-connected converter for fast frequency support. *CSEE Journal of Power and Energy Systems*, *9*(4), 1594–1599. DOI: 10.17775/CSEEJPES.2021.08650

Hassan, A., Ahmad, G., Shafiullah, M., Islam, A., & Alam, M. S. (2025). Review of the intelligent frameworks for pitch angle control in wind turbines. *IEEE Access : Practical Innovations, Open Solutions, 13*, 29864–29885. DOI: 10.1109/ACCESS.2025.3540367

He, J., Huang, L., Wu, D., Zhu, C., & Xin, H. (2018). Frequency support from PMSG-based wind turbines with reduced DC-link voltage fluctuations. *CES Transactions on Electrical Machines and Systems, 2*(3), 296–302.

He, W., Yan, J., Ortega, R., & Zonetti, D. (2023). A new adaptive phase-locked loop for synchronization of a grid-connected voltage source converter: Simulation and experimental results. *arXiv preprint arXiv:2309.08276*

Heier, S. (2014). *Grid integration of wind energy: Onshore and offshore conversion systems*. John Wiley & Sons., DOI: 10.1002/9781118703274

Hossain, E., Hossain, J., Sakib, N., & Bayindir, R. (2017). Modelling and simulation of permanent magnet synchronous generator wind turbine: A step to microgrid technology. *International Journal of Renewable Energy Research, 7*(1).

Ichim-Burlacu, C., Rat, C.-L., & Prostean, O. (2024). Buck-boost converter modeling and control based on its role within a microgrid. In *Proceedings of the 6th Global Power, Energy and Communication Conference (GPECOM)*, Budapest, Hungary. https://doi.org/DOI: 10.1109/GPECOM61896.2024.10582595

Jing, Y., Sun, H., Zhang, L., & Zhang, T. (2017). Variable speed control of wind turbines based on the quasi-continuous high-order sliding mode method. *Energies, 10*(10), 1626. DOI: 10.3390/en10101626

Kiessling, J., Ström, E., & Tempone, R. (2021). Wind field reconstruction with adaptive random Fourier features. *Proceedings - Royal Society. Mathematical, Physical and Engineering Sciences, 477*(2252), 20210236. DOI: 10.1098/rspa.2021.0236

Lammouchi, Z., Naoui, M., Bekakra, Y., Khechekhouche, A., & Sbita, L. (2024). Model-free predictive control based on the execution time of active vectors using ultra-local model. *Arabian Journal for Science and Engineering, 50*, 5885–5898. DOI: 10.1007/s13369-024-085317

Majout, B., El Alami, H., Salime, H., Laabidine, N. Z., El Mourabit, Y., Motahhir, S., Bouderbala, M., Karim, M., & Bossoufi, B. (2022). A review on popular control applications in wind energy conversion system based on permanent magnet generator PMSG. *Energies, 15*(17), 6238. DOI: 10.3390/en15176238

Manwell, J. F., McGowan, J. G., & Rogers, A. L. (2009). *Wind energy explained: Theory, design and application* (2nd ed.). John Wiley & Sons.

Meenakshi, V., Kaveri, V. V., Antony, A. S. M., Radhika, S., Kavitha, M., & Pushpavalli, M. (2024). IoT enhanced optimization of wind energy systems using LCL filter. In *Proceedings of the International Conference on Intelligent Computing and Control Systems (ICICCS)*. IEEE. https://doi.org/DOI: 10.1109/ACCAI61061.2024.10601921

Moutchou, R., & Abbou, A. (2021). Control of grid side converter in wind power-based PMSG with PLL method. *International Journal of Power Electronics and Drive Systems*, *12*(4), 2191–2200. DOI: 10.11591/ijpeds.v12.i4.pp2191-2200

Nasiri, M., Mobayen, S., Faridpak, B., Fekih, A., & Chang, A. (2020). Small-signal modeling of PMSG-based wind turbine for low voltage ride-through and artificial intelligent studies. *Energies*, *13*(24), 6685. DOI: 10.3390/en13246685

Ngamroo, I. (2017). Review of DFIG wind turbine impact on power system dynamic performances. *IEEJ Transactions on Electrical and Electronic Engineering*, *12*(4), 401–411. DOI: 10.1002/tee.22379

Patil, S. A., & Kamble, S. S. (2018). A review on various MPPT techniques for wind energy conversion system. In *Proceedings of the International Conference on Computing, Power, Energy, Information and Communication (ICCPEIC)*. IEEE. https://doi.org/DOI: 10.1109/ICCPEIC.2018.8525219

Rahimi, M. (2017). Modeling, control and stability analysis of grid-connected PMSG-based wind turbine assisted with diode rectifier and boost converter. *International Journal of Electrical Power & Energy Systems*, *93*, 84–96. DOI: 10.1016/j.ijepes.2017.05.019

Rehouma, Y., Hamida, M. A., Naoui, M., Sbita, L., Taibi, D., & Amara, Y. (2023). Single-stage standalone photovoltaic water pumping system using predictive torque control (PTC) of induction machine. *Electric Power Components and Systems*. Advance online publication. DOI: 10.1080/15325008.2023.2278624

Romdhane, M., Naoui, M., & Mansouri, A. (2023). PMSM inter-turn short circuit fault detection using the fuzzy-extended Kalman filter in electric vehicles. *Electronics (Basel)*, *12*(18), 3758. DOI: 10.3390/electronics12183758

Sawyer, S., & Tapia, A. (2005). The sociotechnical nature of mobile computing work: Evidence from a study of policing in the United States. *International Journal of Technology and Human Interaction*, *1*(3), 1–14.

Sharma, R., Sahay, K., & Singh, S. (2023). Control aspects for PMSG-based grid-connected wind energy conversion system – A review. *SSRG International Journal of Electrical and Electronics Engineering*, *10*(12), 72–86. DOI: 10.14445/23488379/IJEEE-V10I12P108

Shayeghi, H., Pourjafar, S., & Sedaghati, F. (2021). A buck-boost converter: Design, analysis and implementation suggested for renewable energy systems. *Iranian Journal of Electrical and Electronic Engineering, 17*(2), 1862. DOI: 10.22068/IJEEE.17.2.1862

Wais, P. (2017). A review of Weibull functions in wind sector. *Renewable & Sustainable Energy Reviews, 70*, 1099–1107. DOI: 10.1016/j.rser.2016.12.014

Wang, Q., & Chang, L. (2004). An intelligent maximum power extraction algorithm for inverter-based variable speed wind turbine systems. *IEEE Transactions on Power Electronics, 19*(5), 1242–1249. DOI: 10.1109/TPEL.2004.833459 [illegible] H., [illegible] S., & [illegible] (2017). A [illegible] converter: Design analysis and implementation [illegible]. *Journal of [illegible]*, [illegible]. DOI: 10.1016/[illegible]

Wais, P. (2017). A review of Weibull functions in wind sector. *Renewable & Sustainable Energy Reviews*, 70, 1099–1107. DOI: 10.1016/j.rser.2016.12.014

Wang, Q., & Chang, L. (2004). An intelligent maximum power extraction algorithm for inverter-based variable speed wind turbine systems. *IEEE Transactions on Power Electronics*, 19(5), 1242–1249. DOI: 10.1109/TPEL.2004.833459

Chapter 12

Cybersecurity Risks in Human-in-the-Loop (HITL) System: Emerging Technologies and Integration

Nidhya Rathinavel
https://orcid.org/0000-0002-9962-5478
PSG Institute of Technology and Applied Research, India

C. Monishkka
https://orcid.org/0009-0002-9374-0747
PSG Institute of Technology and Applied Research, India

K. G. Sukirda
https://orcid.org/0009-0002-9374-0747
PSG Institute of Technology and Applied Research, India

ABSTRACT

This chapter examines cybersecurity risks in Human-in-the-Loop (HITL) systems within emerging technologies and civil infrastructure, focusing on energy systems. As digital transformation accelerates through AI, IoT, and cyber-physical systems, the role of human operators remains critical yet increasingly vulnerable. The chapter explores how human-machine interfaces expand the attack surface, highlights real-world incidents, and analyzes emerging threats targeting both technical and cognitive aspects. By integrating insights from human factors, cybersecurity frameworks, and

DOI: 10.4018/979-8-3373-5621-1.ch012

infrastructure design, this chapter provides strategies for resilience, governance, and risk mitigation ensuring that HITL systems remain secure, adaptive, and robust in an evolving threat landscape.

INTRODUCTION

The digitization of the civil infrastructure is bringing into being an age where automation and Intelligent systems become vital to efficiency, safety, and sustainability. In this new context, Human-in-the-Loop (HITL) systems have become an important paradigm that combines human judgment with machine intelligence in sophisticated decision-making scenarios. In contrast to fully autonomous systems, HITL systems maintain a core role for the human operator, who remains actively involved in oversight, control, and intervention functions. This is especially crucial in high-consequence civil infrastructure applications such as power grids, water treatment, transportation systems and urban planning, where failures have the potential to cause widespread societal, economic and environmental impacts (Figure 1)(Khan et al., 2024; Natarajan et al., 2025) .

Figure 1. Humans-in-the-Loop Systems in Civil Infrastructure

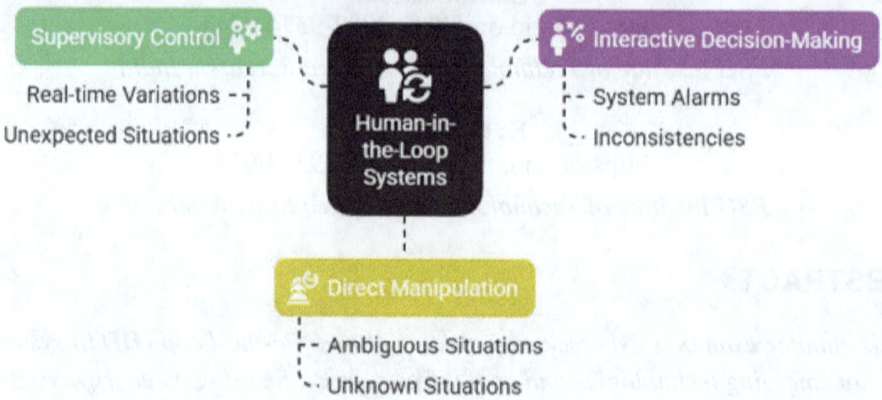

Human-in-the-Loop systems are computational or cyber-physical systems with human decision-making as an inherent aspect. Unlike fully automated systems, which apply pre-authorized responses or machine learning outputs, HITL systems depend on dynamic interaction between computational components and human agents. Such interaction can be adaptive and iterative in form, allowing the system to respond to real-time variations and unexpected situations through a combination of algorithmic efficiency and human judgment(Gaham et al., 2015).

Supervisory Control: Humans guide and adjust system operations without being responsible for all low-level work. Interactive Decision-Making: Humans provide critical input after system alarms, inconsistencies. Direct Manipulation: Operators step in when automatic systems are confronted with ambiguous or unknown situations. In civil infrastructure, the modes guarantee that human judgment is at the core of achieving operational resilience and safety.

INCREASING ROLE OF HITL IN CIVIL INFRASTRUCTURE

The use of artificial intelligence (AI) and the Internet of Things (IoT) combined in cyber-physical interactions is becoming increasingly common and efficient in civil infrastructure projects, but requires human involvement in dealing with a crisis and requiring contextual judgment. Human-in-the-loop architectures support flexibility and resilience when responding to a crisis, keeping the infrastructure functioning under both normal and exceptional conditions, or the right balance of automation and human cognitive effort.(Al-Turjman, 2019).

2.1 Hitl System In Civil Infrastructure

A common Human-in-the-Loop (HITL) system in civil infrastructure operates through a multi-layered architecture designed to enhance decision-making by combining human insight with automated processes. The sensing layer forms the foundation, comprising IoT devices, weather sensors, and real-time data acquisition systems that collect environmental and operational information. This raw data is then processed in the processing layer, where machine learning algorithms, data analytics platforms and decision-support tools analyze the inputs and generate actionable insights. These insights are presented to operators through the interface layer, which includes visual dashboards, alerts and interactive tools that facilitate informed human decisions. Following this, the control layer executes either automated or manually approved instructions based on the evaluated data. Finally, the feedback mechanism plays a crucial role by capturing the outcomes of actions and incorporating both system performance and operator feedback into future iterations. This continuous learning loop between human and machine fosters adaptive and resilient infrastructure management, enhancing both system performance and safety over time.

2.2 Challenges In Hitl Implementation

The application of HITL systems in civil infrastructure presents a number of technical, organizational, as well as human-oriented challenges:

- Cognitive Load and Usability: HITL systems needs confidence in displaying the information they present in a manner that reduces cognitive load and facilitates rapid and accurate decision making. Inadequate interface design can complicate the operators, leading to poor performance.
- Trust and Transparency: The operators need to have faith in the recommendations produced by the system. Lack of transparency in AI models, the "black box" problem, can decrease confidence and result in over-reliance or dismissal of system recommendations.

2.3 Hitl In Energy Infrastructure

One of the most pervasive uses of HITL systems is energy infrastructure, specifically power grid operation and monitoring. With the advent of distributed energy resources (DERs), renewable sources of generation, and advanced metering infrastructure (AMI), grid operators must now balance supply and demand through a mixture of automated control and human action(Vlachos & Biskas, 2010).

Examples of HITL applications in energy systems are:

- Tracking AI-predicted faults and cross-verifying with contextual information.
- Acting on security alerts the human validation before initiating containment procedures.

Within this environment, HITL systems avoid blackouts, improve energy distribution and grid resilience against cyber-physical attacks.

2.4 Hitl In Transportation And Water Systems

In transport systems, HITL systems are applied in traffic signal management, public transport scheduling and autonomous vehicle management. For instance, a controller in a traffic control center might be required to override AI-controlled traffic lights during a large-scale event or emergency to allow for first responders.

In the same manner, water treatment plant operators track SCADA systems to control water quality and to identify contamination. HITL systems enables abnormality detection by AI but depend on human expertise to validate results and apply corrective actions. The Flint, Michigan, water crisis emphasized the importance of human accountability and oversight, even where there are highly automated systems(Żywiec et al., 2024).

2.5 Human-Ai Collaboration And The Evolving Role Of Operators

With increasingly sophisticated AI and ML technologies being integrated into HITL systems, the function of human operators is being transformed from low-level task accomplishment to high-level strategic decision-making and oversight. Human operators are turning into supervisors, interpreters of data and ethical guardians of automated procedures(Chen, 2023)(Segalas, n.d.).

Moving advanced HITL systems from the lab into practice will also require the operators adopt new skills, including an understanding of the workings of AI models, the interpretation of complex data visualizations and how to manage probabilistic reasoning. Managing the ability to cope with uncertainty and incomplete information during decision making is equally important. Consequently, the effective utilization of HITL systems requires more than just technology; it requires that human cognitive skills, organizational form and work processes match the capacity of intelligent and adaptive systems (Chaki, 2023).

2.6 Toward A Human-Centered Infrastructure Future

Finally, HITL systems embody a human-centered philosophy of infrastructure design a philosophy that appreciates the special talents of humans and utilizes the strengths of machines. This philosophy understands that in high-risk, high-consequence environments, autonomous systems are not yet feasible or desirable.

Design principles for Human in the Loop (HITL) systems emphasize transparency, resiliency, universality and scalability in order to seamlessly involve human judgement in automated decision making. Efficient transparent systems will communicate the rationale for their decisions. Resilient designs adapt behaviours when under stress, while universal and scalable design will ensure usability and effectiveness across domains of behaviour. The additional strength and insight of supervised human judgement leveraging machine intelligence drastically increases the security, recognizes adaptability, and increases fault-tolerance of civil infrastructure systems. Human judgement will remain a necessary aspect of the management of intelligent human involvement in social-technical systems when new technology and cyber-threats rapidly evolve uncertainty and vulnerability (Alqunun et al., 2025)(Lammouchi et al., 2025) (Youssef et al., 2023)

HITL CYBERSECURITY RISK LANDSCAPE

As systems with "human-in-the-loop" (HITL) components become increasingly important in civilian infrastructure, they also become desirable targets for cyber attackers. The inclusion of a human operator in a cyber-physical system creates new risk vectors that span both the human decision-making processes and the actions of machines. This sub-section of the chapter will look at the diversity of cybersecurity risks unique to systems with HITL components, specifically focusing on energy, transportation, and water resource infrastructure(Gaham et al., 2015).

3.1 Nature Of Cybersecurity Threats In Hitl Systems

While there are similarities between HITL systems and traditional IT systems, HITL systems are designed to coordinate actions in real-time between the actions of a human operator and machine(s), often in fast-paced or high-stake situations. The real-time coordination of human operator and machine action increases incidence of possible harm on both sides and widens the possible attack surface (e.g., with HITL systems, consider)

User interfaces, AI-based decision tools, and avenues for communication within HITL systems comprise significant vulnerabilities to hacking, data poisoning, or sabotage. When human factors, including phishing, fatigue, or misinterpretation, outcomes can be confounded, coordination can be interrupted, and security and operational effectiveness can be compromised.

The interconnection of these variables makes HITL systems especially susceptible to compound, multi-layered attacks that exploit both technical vulnerabilities, and human vulnerabilities.(Allal et al., 2024)(Romdhane et al., 2023)(Ahmed Baba et al., 2023)

3.2 Cyber Threats In Hitl Environments

Cybersecurity threats to HITL systems can be categorized as:

3.2.1. Human-Targeted Attacks

Cyber Threats Specifically Aimed at Humans in HITL Systems: Social engineering encompasses phishing and other psychological manipulation methods to create a false sense of trust in the operator, resulting in the operator revealing credentials or undercutting safety-theoretic controls. Interface manipulation involves the intentional manipulation of dashboards, alerts or visual cues to mislead operators, acting to misguide the operator's decision making. Information overload occurs when cyber attackers intentionally backfill systems with excessive or conflicting information so as to degrade the operator's appropriate assessment of the situation and degrade their response effectiveness. Collectively, and at the roots, these threats suffer from cognitive and perceptual vulnerabilities, thereby degrading the reliability of the human and the resiliency of the system.

3.2.2 System-Targeted Attacks

- Malware Intrusion: When adversaries use ransomware, worms, or viruses to hack control systems to cause unintended interruptions of operations.
- Data Manipulation: Changing sensor inputs or other decision-support data so that the operator has no choice but to draw the wrong conclusion.
- Denial of Service (DoS): systems requiring manual actions to achieve objectives during times of high-stress operations.

3.2.3 Hybrid Threats

Advanced Persistent Threats (APTs): Concealed campaigns with persistence, compromising human credentials and systems' integrity.Man-in-the-Loop Attacks: Adversaries intercept and modify communications between human and automation components in real time.

Table 1. Cybersecurity Threats, Targeted HITL Components, and Mitigation Strategies

Threat Type	Targeted HITL Component	Potential Consequence	Mitigation Strategy
Social Engineering	Human operator	Credential theft, privilege escalation	Awareness training, phishing simulations
Malware/Ransomware	Control systems, workstations	System downtime, data loss	Endpoint protection, regular backups
Man-in-the-Middle (MITM) Attack	Communication interface (e.g., SCADA)	Data interception, false data injection	Network segmentation, encrypted communication
Interface Spoofing	Visual dashboards, control GUIs	Misinterpretation by operator	Interface validation, redundancy checks
Insider Threats	Internal users	Deliberate sabotage or unauthorized changes	Role-based access, activity logging
Zero-Day Exploits	Unpatched software	Full system compromise	Threat intelligence integration, timely patching.

Bottom of Form

3.3 Case Studies: Actual Events

Several prominent examples illustrate the risks of cyber-attacks on HITL systems:

- Colonial Pipeline Ransomware Attack (2021): Although this event was not specifically an HITL breakdown, it highlighted how human operational influence decisions made under duress (for example, shut down the flow of fuel) were done in response to a cyber-event in IT systems associated with control of the critical infrastructure service.
- Ukrainian Power Grid Attack (2015): The attackers initiated a scheme to steal legitimate operator credentials that were subsequently used to modify SCADA interface operation in order to create outages. This event featured the use of coordinated attack methods that exploited weaknesses in the human and technical systems.
- Water Treatment Facility Breach in Florida (2021): Hackers illegally accessed a water treatment control system and attempted to adjust chemical levels to extremely high levels. The event was ultimately stopped by a human employee who noticed the irregularity and took action to reverse.

These examples demonstrate how HITL systems can be manipulated by either machines or human operators in the loop process.

3.4 Cognitive And Organizational Vulnerabilities

The human portion of HITL systems is both an asset and a liability. Humans are able to adaptively and ethically respond to new situations; they are also predictable in the cognitive failures they make:

Overreliance on automation in HITL systems can cause operators to depend excessively on machine decisions, overlook anomalies, and suffer alert fatigue from repeated low-priority warnings. Misaligned mental models, inadequate training, weak security culture, and poor coordination among operational and cyber teams further reduce situational awareness and system performance.

3.5 New Threat Vectors

As technologies continue to evolve, new and sophisticated threat vectors are emerging that pose significant risks to human-in-the-loop (HITL) systems. One such threat is adversarial AI, where attackers intentionally introduce deceptive inputs designed to confuse or mislead machine learning algorithms that underpin anomaly detection or decision-support functions. Similarly, synthetic media and deepfakes present alarming possibilities, as they can be used to impersonate system administrators or generate fake alerts, especially during high-stress scenarios, leading to confusion and potentially harmful actions. Supply chain attacks are also on the rise, targeting third-party components or software integrated into HITL systems, thereby exploiting trusted dependencies to gain unauthorized access or control. Furthermore, insider threats remain a persistent challenge, with employees whether disgruntled or possessing untrustworthy backgrounds potentially abusing their access privileges to sabotage operations, leak sensitive information or manipulate system behavior from within. Each of these vectors contribute towards the difficulty of ultimately protecting HITL systems, especially when adversaries are using mixed social and technical vectors(Lin & Biggio, 2021).

3.6 Risk Magnifiers In Civil Infrastructure

Many aspects of civil infrastructure create a more extensive threat landscape for cybersecurity: Human-in-the-loop (HITL) systems in civil infrastructure often face persistent cybersecurity challenges rooted in their operational context. Legacy systems, which are commonly found in older infrastructure, typically lack modern security protections and may not support timely updates or patches, leaving them vulnerable to exploitation. Additionally, the high availability requirements of civil systems such as power grids or water supply networks mean they must operate continuously, leaving little room for downtime needed for security maintenance or upgrades. These systems are often geographically dispersed, with components located in remote or physically unprotected areas, increasing the risk of localized attacks on hardware, software, or communication links. Moreover, the involvement of multiple stakeholders, spanning public agencies, private contractors, and third-party vendors, adds a layer of complexity to governance, accountability and incident response, making coordination and unified cybersecurity strategies more difficult to implement effectively. These factors contribute to a cyber risk landscape in which traditional perimeter defences are often inadequate, while adaptive and multi-layered approaches become necessary.

3.7 Threat Actors And Motivations

Malicious actors who are making use of HITL systems include nation-states with the objective of geopolitical disruption, cybercriminals with the objective of profit, hacktivists acting on ideological means, and insiders abusing skills and privilege. Low-skill attackers can also pose a threat. The threat of these actors is compounded by the human element, the complexity of integrating data, and the fact that a breach in the digital world can have real-world, physical consequences.

3.8 Challenges In The Detection And Response Process

There a number of challenges associated with detecting and responding to human-in-the-loop systems:

Real-Time Constraints:

Time-lags between detection and response can trigger real world consequences. Human decisions may have only seconds, or less, to detect and react to anomalies. Complicated attacks can be designed to exploit these time constraints. Anomaly vs Intrusion Confusion. Not all anomalies are attacks. Both false positives (e.g., actual system faults falsely identified as attacks) and false negatives (e.g., attack being failed to be recognized) can result in poor preparedness.

Chain of Command Disruption:

There may degrade in the chain-of-command, because of attack-induced panic, communication failure or being locked out of command, during an attack. Contingency procedures for operating in a degraded mode should be included in the human-machine decision loop.

3.9 Threats To Cybersecurity From Human-In-The-Loop (Hitl)

As human-in-the-loop (HITL) technology continues to evolve, so too do the threats that target these systems, revealing emerging and increasingly sophisticated attack trends. One notable development is the rise of AI-driven attacks, where adversaries employ machine learning models to bypass traditional detection methods or craft hyper-personalized phishing messages tailored to exploit individual vulnerabilities. Similarly, the use of deepfake technology in social engineering is gaining traction, with synthetic audio and video being used to convincingly impersonate supervisors or emergency personnel, thereby manipulating operators into taking harmful actions. Another concerning trend involves hybrid multi-vector attacks, which combine physical sabotage with coordinated cyber intrusions to amplify the damage inflicted. In addition, the emergence of autonomous threat agent's bots capable of making real-time decisions to evade defenses, mislead operators or interfere with human-machine collaboration represents a new frontier of risk. These complex and adaptive threats underscore the urgent need for proactive cybersecurity strategies that can anticipate and counteract this evolving threat landscape.(Ghazi-Tehrani & Pontell, 2022)(Zandi et al., 2024)

Table 2. Categorization of Cybersecurity Risks Across HITL Civil Infrastructure Domains

Infrastructure Domain	Typical HITL Application	Cybersecurity Risks	Real-World Relevance
Smart Grids	Load balancing, outage management	Phishing attacks, SCADA hijacking, AI alteration	Colonial Pipeline, attacks on Ukraine power grid
Intelligent Transportation Systems	Traffic control, surveillance, routing	GPS spoofing, tampered sensors Saving time for other drivers or hitting pedestrians	Traffic signal hijacking, stealing money from toll booths
Smart Cities	Urban operations, emergency response	Privacy breach, captured surveillance (lost privacy)	replay of command Privacy fugitive incident, emergency dispatching
Healthcare Infrastructure	Remote monitoring, robotic surgery	DDoS attacks, ransomeware	misdiagnoses due to data changes Widespread ransomware incidents across hospitals
Water and Waste Management	Chemical dosing, pump control	Remote manipulation, spoofed interface access, unauthorized access	Oldsmar incident (2021)

Cyber security threats in Human-in-the-Loop (HITL) systems fall under both the digital and the human aspects. HITL systems face cyber security threats that need both a technically robust security framework and spatial awareness of the psychology of the human operator. Identifying vulnerabilities such as phishing, cognitive overload and adversarial AI, enables systems and operators to be robust in a multi-layered defence continuum. Future work will include exploring further advanced technology sectors and industry use cases, countermeasures, and, frameworks of mitigation, within HITL systems to bolster resilience against the evolving cyber threat landscape.

AI/ML IN HITL SYSTEMS

AI/ML is being increasingly used to help human operators with anomaly detection, predictive maintenance, demand forecasting, and decision making in real-time. Generally, these technologies are used as decision support systems (DSS) that provide recommendations or flag outliers driven by data-based models.

- Adversarial attacks: AI/ML models can also be fooled by specially constructed inputs designed to produce erroneous outputs; this is sometimes referred to as adversarial machine learning. In HITL systems, erroneous alerts can be generated, cause false classifications of faults or confuse the operator's actions.

- Model poisoning: Attackers can have an impact on model behavior by corrupting training data and this becomes an on-going vulnerability in systems that re-train continuously on run-time data.
- Transparency and trust: When operators lack confidence in a model's actions because of lack of explainability (the "black box" problem), operators cannot trust it and rely on proper human intervention is compromised, which is particularly deadly under stressed conditions

4.1 Digital Twins

Digital twins are software representations of physical systems that can be utilized for simulation, monitoring and optimization. They are commonly used in smart grids, water systems and transportation networks to provide real-time system analysis and planning support.

- Data Integrity Attacks: A digital twin's accuracy relies on trustworthy, real-time sensor data streams. Spoofed or corrupted sensor data can lead to bad models.
- Disclosure of Operational Information: A vulnerable digital twin provides adversaries with potentially sensitive operational information and could allow adversaries to learn of weaknesses in simulation before launch(Serpanos, 2022).

4.2 Distributed Architectures And Edge Computing

Edge computing uses data processing capacity closer to the data source such as sensors and actuators (that is) computing nodes are out in the field rather than a centralized one. This architecture has lower latency, utilizes our capacity better and allows for rapid local decision making as needed as part of a critical infrastructure.

Security Considerations:

- More Attack Surfaces: The additional distributed computing nodes means that with each node represents another option for an attack
- Local Compromises leading to Global Effects: Attacks on edge devices can send upstream command requests(Raghav & Kait, 2024).

4.3 Internet Of Things (Iot) & Sensor Networks

IoT technologies can add ubiquitous monitoring and control capabilities to any infrastructure system via networked devices. Connected devices can range from smart meters [in power grids] to SCADA sensors [in industrial systems].

Security Implications:

Most IoT devices do not demand a high level of authentication, encryption, or require patching regularly, and are therefore the first choice for attackers. Hacked Internet of Things devices can be placed in botnets (e.g. as evidenced by attacks using the Mirai malware), that can later be used to conduct denial-of-service (DoS) activity. An attacker may be able to inject fake sensor readings to change the perception of the operator or even incorrectly change manual interventions based on their perception changes.

4.4 Automation And Robotics

Automated control systems such as robots in water treatment or reconfiguring the grid, do not only allow for a level of control that minimizes human effort but may also allow for precision beyond human capacity. Alternatively, even with an HITL configuration, humans are able to monitor and intervene in edge situations or emergencies.

Security Implications:

- Human Override Complexity: Human override means to respond to an emergency with something other than the automated systems that are usually used. Such human overrides should be safe and recognizable but are sometimes poorly designed or tested.
- Simultaneous Multi-Channel Attack: The ability to connect with both machine automation and the human interface means that an incident can be responded to with confusion or contradiction.

4.5 Cloud-Based Scada And Virtualization

Modern SCADA increasingly relies on cloud-based systems for analytics and centralized management but virtualization also allows for near-infinite scaling and remote access.

Security Issues:

- Credential theft: Access in the cloud means that when credentials are compromised or stolen, administrative credentials are extremely valuable because they can facilitate total and complete access to critical infrastructure.
- Cross-tenant vulnerabilities: In cloud systems that are multi-tenancy, vulnerabilities in one tenant's security isolation can lead to an exposure of data or privilege escalation to one or more other tenants(Sajid et al., 2016)(Wali & Alshehry, 2024).

4.6 5g And Next-Generation Connectivity

The consequences of implementing 5G is that it facilitates ultra-low-latency, high-bandwidth communication with operators, decision engines and sensors in real-time human-in-the- loop (HITL) environments. Some security implications include:

- Network slicing threat: While introducing flexibility, poorly configured network slices can provide lateral mobility with aspects of the network.
- Vendor Lock-in and proprietary protocols: Use of proprietary solutions obscures security practices and delays patches. (Figure 2)

Figure 2. Technology's role in civil infrastructure, from human-led to fully automated

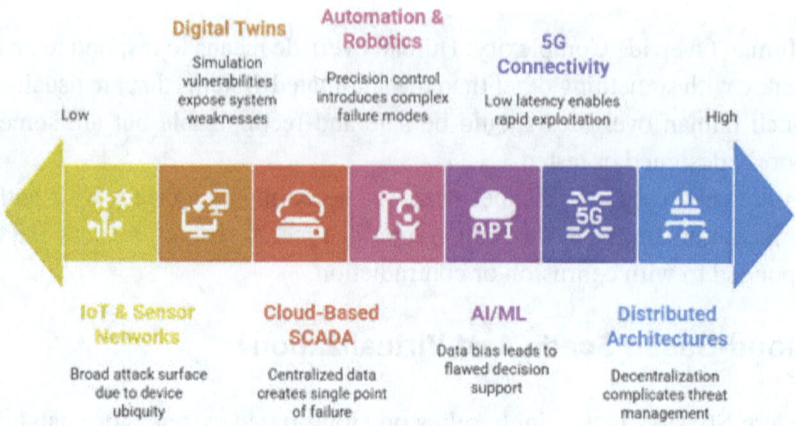

4.7 Human-Centered Impacts Of New Technologies

New technologies are mostly not separate from human operators. Their introduction Transforms workflows, expectations and decision-making models.

Security Risk:

- Training Gaps: New technologies may require new skills but training programs may fall short of achieving this because the rollout of new technology had taken priority, increasing operational risk.
- De-Skilling Risks: Operators do not upskill if they are over-reliant on automation. In time, unless operators use those skills, those operational skills can erode and operators are less able to respond to anomalies. The integration of a heterogeneous suite of new technologies within legacy systems will challenge interoperability.

Security Implications:

Differences in communication protocols create potential vulnerabilities to system security. New emerging technologies may not integrate completely with centralized authentication or monitoring systems creating blind spots. Inconsistent security standards across venders complicate and enforce unified policies. (Strobhar, 1995)

SECTORAL CASE STUDIES IN CIVIL INFRASTRUCTURE

This will provide a series of sectorally-centric case studies, showing the real-world implications of cybersecurity threats in human-in-the-loop (HITL) systems. These case studies will indicate how HITL dynamics play out in several civil infrastructure sectors, indicate typical vulnerabilities and review actual events and countermeasures. These case studies will support theoretical understanding in operational contexts while reinforcing the nuanced relationships human operators, technological systems and cybersecurity threats.

Smart Grids and Demand-Response Systems Smart grids are examples of HITL systems that rely on digital communication, automation, and energy management. Human operators provide management of grid stability, perform system health checks and respond to anomalies, often with the help of AI-based analytics and digital twins(Eskandar et al., 2020).

Cybersecurity Concerns:

- False Data Injection Attacks (FDIA): Adversaries may be able to inject conflicting readings into smart meters or sensors and improperly distort demand estimates which could enormously affect load balancing and improved measurements in demand forecasting.
- Cyber-Physical Risks of Decentralized Control: Decentralized systems like solar panels and batteries require human monitoring however these resources add several control nodes that could be attacked(Dawson et al., 2021).

Case Example: A European utility faced a multi-stage attack in 2021 from compromised IoT (internet of things) devices with respect to load readings being reported as faulty. Where the automated load shed mechanism started to load shed incorrectly due to the 'clean' load measurements, a human operator identified it as an anomaly after seeing the operation involved several smart meters as compromised and then avoided any unnecessary blackouts.

Lesson learned: Realizable HITL (Human in the Loop) frameworks will require well-designed (unambiguous), realistic analytics and training to ensure that operator's actions would not have cascaded together (cyber-physical interventions will have to be avoided).

5.1 Transportation Systems And Traffic Management Centers

Today's transportation system relies to a great degree on HITL systems, especially in traffic control centers, autonomous vehicle systems and intelligent highway systems. Human operators monitor the traffic stream, respond to warnings and take action in system failures.

Case example: A city in the U.S. experienced a ransomware attack to the traffic control system, which halted signal timing coordination for a short period. Fortunately, skilled operators used their manual override capabilities to preserve access for emergency vehicles and minimize congestion.

Cybersecurity threats involving human-in-the-loop (HITL) systems - especially in civil infrastructure - arise due to a myriad of reasons, such as many SCADA protocols that have not been updated to include encryption, sensor manipulation and messaging issues that compromise traffic reliability and operator stress when a crisis occurs that initiates social engineering susceptibility. The fix is to have a mix of IT-based cybersecurity protocols, along with human-factor protocols to ascertain reliability and safety.

Lesson learned: Transportation systems need good manual override capabilities and rigorous authentication methods to maintain HITL integrity.

5.2 Water Treatment Plants And Automated Control Systems

Water infrastructure primarily uses SCADA systems, and it is the human operators that ensure safe and effective chemical dosing and pressure regulation while managing fault events. HITL is a significant feature because provision of clean water is necessary.

Cybersecurity Problems:

HITL systems experience security risks related to remote access via opened networks, insider threats, and delayed incident detection. This highlights the need for better security protocols, the need for continuous observation of the systems, and the need to train operators.

Case Example: In 2021, a Florida water treatment facility was attacked by an individual who attempted to remotely increase the sodium hydroxide in the water supply. A vigilant operator noticed the significant change and quickly rolled the adjustment back before it could become dangerous.

Lesson learned: Understanding your surroundings and noticing a noteworthy change quickly is crucial. HITL resiliency is dependent on training operators to recognize patterns quickly(Cervini et al., 2022).

5.3 Integrated Urban Infrastructure (Smart Cities):

Smart cities connect multiple infrastructure systems (for example transportation, energy, water, public safety) with a shared digital platform. When HITL interaction is possible, the operator role is now set across multiple infrastructure domains and that complexity is growing.

- Inter-Dependencies: A compromise in a single system (i.e., video camera surveillance) can propagate and affect multiple systems (i.e., transport/ utility systems).
- Gaps in Inter-Agency Coordination: Poor information dissemination among departments will limit a collective response to incidents(Almeida, 2023) (Mijwil et al., 2022).

Case example: A table-top exercise with a large smart city identified the realization of a single coordinated cyberattack against the city energy management system. The cyberattack led to the energy system oscillating caused both traffic lights and the emergency communications systems to all fail simultaneously. Furthermore, the operators could not link two systems together because they did not realize the two systems were eventually connected.

Lesson Learned: Interdependent urban infrastructure has collective risks which means that holistic risk assessments across disparate operational functionality, shared situational awareness capabilities, information-sharing policies for inter-department coordination amongst agencies are all required for sustaining basic function of HITL systems.

5.4 Comparative Considerations And Lessons Learned

There are patterns that recur across all areas:

- Human personnel remain the ultimate backstop particularly with uncertain or changing events.
- Human performance in high stress environments where physical and cyber systems interact has several critical variables; the most important to a successful outcome are usability and training(Boyce et al., 2011).

These case studies demonstrate the need to create systems that augment human decision-making rather than replace it with automation. Cybersecurity policies need to incorporate the psychological, procedural and technological contexts of HITL environments. The next section examines how the concepts of systemic resilience and human factors can be built upon to improve the HITL security posture for key civil infrastructure.

HUMAN FACTORS AND SYSTEM RESILIENCE

Human-in-the-loop (HITL) system resilience in civilian infrastructures is intimately connected to the performance of human operators. Additional human factors impacted by increasingly advanced cyber attacks may be the difference between success and failure of an infrastructure system, particularly with regard to human situational awareness, high stress decision-making, and the ergonomics of the HITL systems they are using. This section addresses the complicated human factors influencing operational resilience of HITL systems and identifies means to increase operational resilience through human-centered design, training and organizational culture.

6.1 Interface Design And Cognitive Ergonomics

A well-designed interface is necessary for maintaining operator performance. Poorly designed user interfaces even may obscure critical alerts while increasing cognitive load, or permitting errors in high-stress situations.

Cognitive Load Management: Operators generally need to monitor multiple continuous data streams and make real-time decisions. Cognitive overload, or the situation where the volume of information exceeds what an operator can comprehend, can lead to too little-too late or inappropriate decisions. Interfaces should prioritize clarity, show the hierarchy of data and avoid visual clutter(Stone et al., 2005)(Sridevi, 2014).

Human-Centered Design (HCD) principles play a vital role in ensuring that Human-in-the-Loop (HITL) systems are both usable and effective. These principles emphasize the need for intuitive organization and the use of visual symbolism to help operators quickly understand complex information. Consistent use of symbols and color schemes for recurring elements is crucial, as inconsistencies can delay interpretation and decision-making. Additionally, anomaly detection tools should be integrated to process and simplify vast and overwhelming data streams into actionable insights for the operator. Feedback mechanisms are equally important in HCD. Interactive systems must include clear feedback loops that confirm the operator's input has been received and acted upon. Real-time status messages and alerts should effectively differentiate between routine warnings and critical system failures to help guide appropriate and timely responses from the human operator.

6.2 Operator Training And Awareness Programs

Training is critical in developing a workforce with the ability to identify and respond to cyber- physical anomalies. The resilience of HITL is founded on technical skillsets as well as acute awareness of human-centric attack vectors.

Cyber Hygiene and Social Engineering Defense: Operators need training in the identification of phishing, pretexting and other social engineering attacks. Realistic examples, such as emails appearing to be from IT administrators or alerts to urge

Scenario-based training: Simulation-based training approaches (e.g., red teaming, tabletop training, virtual reality simulations) can educate operators as they respond to technical anomalies and ambiguous threats. Training in simulation with several distinct scenarios throughout the year is an excellent method for increasing both procedural memory and confidence under stress.

Continuous learning models: New threats are continuously developing and evolving, which means operator training should be ongoing and flexible enough to incorporate recent developments. Continuously adding lessons learned from recent attacks, briefings about cybersecurity and hands-on technical training is an easy-to-administer way to keep operators and personnel up to datently install software updates, need to be integrated into training and simulations(Almutairi & Alghamdi, 2022).

6.3 Organizational Culture And Communication Channels

Individual skillsets are not the only elements of human resilience, organizational culture and communications channels matter.

Culture of Awareness: A resilient organization has a culture of awareness regarding close calls or near misses, a shared responsibility for cybersecurity and openness around error instead of retribution. A culture like this allows these organizations to identify emerging threats before damage occurs.

Cross-Team Coordination: Infrastructure networks usually cross department boundaries(i.e.,) IT, operations, engineering. Effective rates of cybersecurity response require inter-departmental coordination and communication. Pathways for escalation and response should be established.

Leadership Position: Cybersecurity must be given consideration at the leadership level so that adequate resources, monitoring and support for operator activities are available. Leaders should promote cybersecurity as part of operational safety and risk management(Vasudevan et al., 2022).

6.4 Situation Awareness And Decision Support Tools

Situational awareness is the ability to perceive, comprehend and project the present state and future state of a system. HITL systems can enable situational awareness even further using effective decision support tools.

Visualization Dashboards: Sensor data, threat intelligence and performance metrics, can come together under dashboards to provide operators with a holistic, real-time view of the state of the system. The tools should be user-friendly and configurable.

AI-Augmented Monitoring: AI algorithms can assist in the discovery of anomalies, predict failures in equipment or identify incidents of unauthorized security breaches. However, operators should be able to follow what and why the AI produced a certain output that affects their trust and action.

Confidence Calibration: Operators need to be educated not to completely rely on automated systems nor completely segregate automated systems. Human intuition complements machine analysis and produces greater resilience when used in concert(Yang et al., 2018)(Stanton et al., 2001).

6.5 Red Teaming And Human-Centered Risk Management

Red teaming, the use of simulated attacks to assess a system's defenses, can potentially reveal human and organizational weaknesses that a physical and logical audit may miss.

Human-Centered Red Teaming: In a human-centered red teaming approach, the process not only evaluates the technical vulnerability to an attack but can also assess, how quickly does the human operator recognize a fake interface? Does the phishing campaign eventuate in credential compromise? How does alert fatigue contribute to recognizing the threat?

Behavioural Risk Modeling: Organizations could also exploit their behavioural analytics to model the interrelationships of human factors such as fatigue, stress or overconfidence, to errors or delays in identifying threats. Models could provide direction as to appropriate mitigations that could include redesigning a shift, stress management or additional training(Raney et al., 2024).

6.6 Optimizing Human-In-The-Loop For Improving Systemic Resilience

Systemic resilience in civil infrastructure requires an integrated framework connecting human capabilities and technical solutions. Human-In-The-Loop systems must still prevent redundancy, flexibility and fault tolerance.

Safe Failures Built In: Systems should return to safe failure states once an algorithm cannot support performance and whenever safe to do so. Human operators should have override access, but limited to safe failure mode.

Resilient System Architecture: Distributing control functions throughout the system can eliminate single point failures, although decentralized systems require additional levels of human coordination, and procedural discipline.

Continuous Improvement: Not just technical systems, but human responses should not be lessened by lessons learned from disasters. Feedback loops that allow for improvement to training content, interface and organizational processes represent an iterative improvement process to systemic resilience(Reed et al., 2011).

MITIGATION STRATEGIES AND POLICY IMPLICATIONS

As public infrastructure among-which HITL integrated Systems are becoming encourage, it is important to develop the security and resiliency of Human-in-the-Loop Systems in an ethical and multi-leveled manner. Effective risk mitigation strategies must address technical risk, human behavior, organizational design and consider the emerging regulatory landscape. This section will discuss strategic options to reduce cyber risks in HITL systems and the necessity to address cybersecurity-by-design, best practices, regulatory requirements and the current regulatory landscape.

7.1 Cybersecurity By Design In Hitl Systems

The foundation for all mitigation strategies will begin when designing a HITL system that includes cybersecurity principles in the design, development and deployment of HITL systems.

Secure interface design is a critical component in safeguarding Human-in-the-Loop (HITL) systems. User interfaces must be developed with security principles embedded from the outset. This includes ensuring that all user actions produce real-time, visually explicit feedback, allowing operators to confirm that their commands have been received and executed appropriately. Implementing role-based access controls within graphical user interface (GUI) components helps ensure that users only interact with system functions relevant to their responsibilities. Moreover, alerts should be tiered according to urgency and severity, enabling operators to prioritize their responses effectively. Complementing secure design is the Principle of Least Privilege (PoLP), which dictates that users should only be granted access rights strictly necessary for their operational duties. Granting excessive privileges increases the likelihood of misuse, especially if an operator falls victim to phishing or if an insider intentionally compromises access. Together, secure design and strict access control form a robust defense strategy for HITL systems.

Modular and Isolated Architectures:

If system components are decoupled and a microservices architecture or containerization is employed, the spread of an attack can be diminished. From a safety standpoint, separating the user-facing components of the system from the critical control logic provides multiple layers of protection(Udayakumar, 2023).

7.2 Secure Defaults and Fail-safes

Systems must be designed to fail secure. An example may be for a system that is exhibiting unusual behavior by an operator, or an inability to communicate with the system, the software could "fail" to a secure operational mode until the presence of human acknowledgement is restored.

7.3 Using Frameworks and Best Practices

There are several cybersecurity frameworks that can help secure HITL settings. Frameworks help you manage risk, build protections, and respond to incidents.

The NIST Cybersecurity Framework (CSF) provides a structured approach to managing cybersecurity risks across critical infrastructure sectors and is particularly relevant to Human-in-the-Loop (HITL) systems. It organizes its strategy into five core functions: Identify, Protect, Detect, Respond, and Recover. Within HITL systems, these functions take on specific characteristics that reflect the integration of human operators with machine systems. For instance, the Identify function can include periodic cognitive stress tests to assess operator readiness and highlight potential vulnerabilities in human performance under stress. The Protect function involves developing and delivering training modules that are tailored to the specific needs and roles of operators, helping them build awareness and resilience to cyber threats. The Detect function extends to operator-assisted threat detection, where human insight complements automated detection systems to identify anomalies or suspicious activity more accurately. By adapting these core CSF functions to the HITL context, organizations can create a more resilient and responsive cybersecurity posture.

7.4 MITRE ATT&CK for ICS:

The MITRE ATT&CK knowledge base, and more recently, its extension on industrial control systems (ICS) offers real adversary tactics and techniques to take place against control environments. HITL has use cases that can benefit from mapping operator action and interface vulnerabilities against adversary tactics for comparable attack vectors.

7.5 IEC 62443 Standards

This suite of standards offers usable guidance for the cybersecurity of industrial automation and control systems. It stresses zones and conduits to facilitate separating and protecting various components of the infrastructure. This is particularly important because of the need to separate HITL components from generalized threats of the network(Al-Sada et al., 2024).

7.6 Workforce Enablement and Operational Controls

The effectiveness of Human-in-the-Loop (HITL) systems is tied to the proficiency and preparedness of operators. Regular and role-based training programs which reinforce safe use of the interface, situational awareness of social engineering, and emergency processes should be established. Certification programs consistent with standards such as CISSP, GICSP, or NERC CIP may enhance operator proficiency. Additionally, dual authorizations for critical system functions, along with behavioural monitoring through Security Information and Event Management (SIEM) tools, can enhance accountability, identify anomalies, and effectively reduce insider and operational cyber security risk. (Figure 3)

Figure 3. Cybersecurity stratergies for HITL Systems

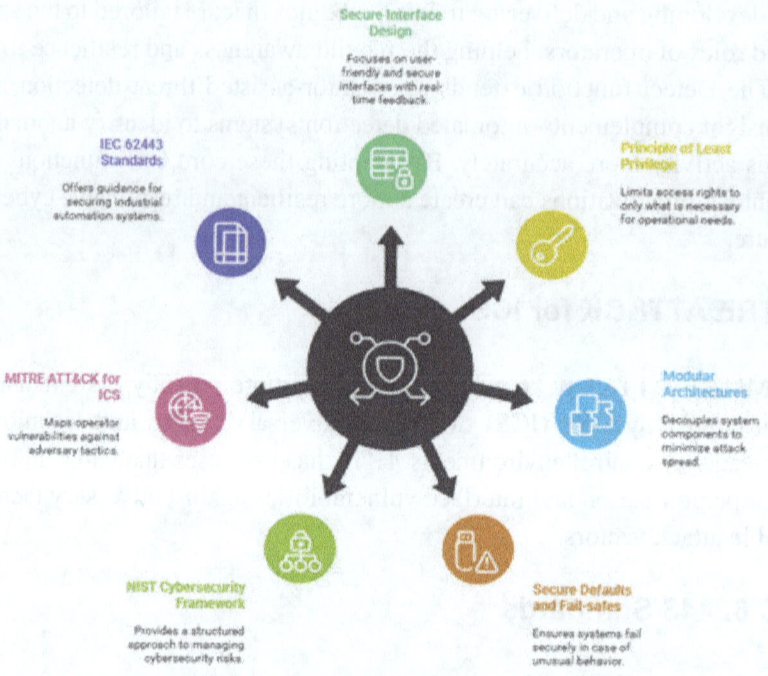

7.7 Institutionaling Governance And Risk Management

Cyber risk in HITL systems needs to be addressed not merely as a technical problem but as an organizational and strategic issue.

Enterprise Risk Integration:

Cyber risk needs to be integrated into enterprise-wide risk management (ERM) processes. HITL-specific risks, e.g., operator fatigue or excessive dependence on automation, need to be addressed in conjunction with financial and operating threats.

Board-Level Oversight:

Senior management and boards of directors need to be sensitized to HITL cybersecurity issues. Periodic reporting of HITL system status, incident statistics and success of training ensures accountability and emphasis on cyber resilience.

7.8 Policy And Regulatory Considerations

As the sophistication and frequency of cyber attacks increase, regulatory bodies are more interested in requiring strict security measures on critical infrastructure. HITL systems fall squarely within such regulatory frameworks.

Sector-Specific Compliance Standards: The NERC CIP and the EU NIS2 Directive reflect two different but complementary views of cybersecurity governance in critical infrastructure. The NERC CIP framework, developed by the North American Electric Reliability Corporation, is primarily sector-specific and compliance-focussed, mainly for electric utilities, and mandates the implementation of strict cybersecurity controls for SCADA and energy management systems reliant on Human-in-the-Loop (HITL) operations. Its prescriptive requirements centre on accountability and system reliability and is technical compliance based to protect the bulk electric system. By contrast, the EU NIS2 Directive adopts a broader, risk-based, and cooperative model that to not only incorporates a wide range of obligated sectors, including energy, water, and transport, but it also advances a collective based obligation to provide uniform security measures for its member states and promotes cross-border collaboration, information sharing. Hence, the NERC CIP strengthens cybersecurity making the participant comply with regulatory requirements in a specific sector, whereas, the NIS2 Directive advances in a coordinated governance framework aimed at strengthening collective resilience across multiple infrastructure's and national borders (Chowdhury & Gkioulos, 2021)(Malatji et al., 2022).

7.9 Data Privacy And Ethics

Both cyber behavioural monitoring software and decision-support analysis of HITL systems all have privacy and ethical issues. When storing and recording operator interaction data, lawful compliance with regimes such as the General Data

Protection Regulation (GDPR) and the California Consumer Privacy Act (CCPA) should be assured.

7.10 Funding And Incentives:

Government and regulatory authorities can contribute proactively by:

- Providing grants for infrastructure modernization initiatives with in-built cybersecurity specifications.
- Tax credits for complying with state-of-the-art HITL security practices.
- Promoting public-private interaction in sharing threat information.

7.11 International Sharing And Cooperation

As cyber threats are global, especially those against civilian infrastructure, international cooperation is vital in the creation of solid HITL systems.

Table 3. Governance and Risk Management Strategies in HITL Cybersecurity

Category	Key Elements	Description / Strategic Focus
Enterprise Risk Integration	Integration with ERM	link HITL cybersecurity risks (e.g., human fatigue, over-reliance on automation) to financial and operational enterprise-wide risk frameworks.
Incident Response Planning	Human-Centric IR Plans for HITL	Develop decision-tree–based incident response and post-analysis framework.
Policy and Regulatory Compliance	Sector-Specific working groups	Embed HITL operator compliance with global cybersecurity regulations.
Data Privacy and Ethics	Regulatory Adherence	Ensure HITL systems comply with global data protection regulations.
Funding and Incentives	Governmental assistance	Leverage incentives for cybersecurity in infrastructure modernization.
International Collaboration	Threat Intelligence Sharing	Promote global and public-private partnerships against cyber threats.

Bottom of Form

Cross-Border Incident Coordination:

Cyberattacks prefer to cross borders. Having global standards for exchanging attack vectors, defense strategies, and incident response procedures enhances overall resilience.

Capacity Building in Emerging Economies: Developing nations are rapidly deploying digital infrastructure with HITL components but lack the capability to track associated cyber threats. Targeted training, technology transfer, and policy support can raise the level of global infrastructure security(Creese et al., 2019).

INNOVATIONS IN TRAINING AND CAPACITY BUILDING

As human-in-the-loop (HITL) systems become more complex, training must also progress through adaptive, gamified, and immersive learning via virtual/augmented reality (VR/AR) to simulate cyberattacks and develop a better situational awareness of threats. Just-in-time learning modules, certifications and continuous education of cybersecurity and human factors will be an important part of developing an expert human capital pool capable of mitigating operational risks.(Martinie et al., 2022) (Verma et al., 2003).

CONCLUSION

The transition to intelligent, interconnected energy systems has augmented the benefits and risks posed by human-in-the-loop (HITL) systems. The chapter endeavored to merge methods in engineering, cybersecurity, and human factors to forge a framework for the secure and human-centric design of systems including embedding cybersecurity considerations in design interfaces, building systems and procedures on adaptive automation principles, and fostering organizational cultures to build reputational knowledge and embed pre-emptive risk management. Next steps should include facilitating greater resilience in Human-AI collaboration, developing systems that incorporate behavioural analytics that may optimize usable security to support anti-threat predictive threat analysis and a system of policies that intersection both cyber and resiliency objectives. At the end of the day, securing HITL energy systems will be based upon a socio-technical approach that converts the humanity of HITL system defect into a strength within the energy system.

FUNDING INFORMATION

This research received no specific grant from any funding agency in the public, commercial, or not-for-profit sectors.

REFERENCES

Ahmed Baba, M., Naoui, M., & Cherkaoui, M. (2023). Fault-tolerant control strategy for Hall sensors in BLDC motor drive for electric vehicle applications. *Sustainability, 15*(13), 10430.

Al-Sada, B., Sadighian, A., & Oligeri, G. (2024). MITRE ATT&CK: State of the art and way forward. *ACM Computing Surveys, 57*(1), 1–37.

Al-Turjman, F. (2019). *Artificial intelligence in IoT*. Springer.

Allal, A., Lammouchi, Z., Khechekhouche, A., Mohamed, N., Alqunun, K., Alghadhban, A., Marouani, I., & Guesmi, T. (2024). Advanced Residual Optimal Mapping Approach for Precise Detection of Stator Faults in Induction Motors. *IEEE Access : Practical Innovations, Open Solutions, 12*, 143515–143530.

Almeida, F. (2023). Prospects of cybersecurity in smart cities. *Future Internet, 15*(9), 285.

Almutairi, B. S., & Alghamdi, A. (2022). The role of social engineering in cybersecurity and its impact. *Journal of Information Security, 13*(4), 363–379.

Alqunun, K., Bechiri, M. B., Naoui, M., Khechekhouche, A., Marouani, I., Guesmi, T., Alshammari, B. M., AlGhadhban, A., & Allal, A. (2025). An efficient bearing fault detection strategy based on a hybrid machine learning technique. *Scientific Reports, 15*(1), 18739. PMID: 40437009

Boyce, M. W., Duma, K. M., Hettinger, L. J., Malone, T. B., Wilson, D. P., & Lockett-Reynolds, J. (2011). Human performance in cybersecurity: A research agenda. *Proceedings of the Human Factors and Ergonomics Society Annual Meeting, 55*(1), 1115–1119.

Cervini, J., Rubin, A., & Watkins, L. (2022). Don't drink the cyber: Extrapolating the possibilities of Oldsmar's water treatment cyberattack. *International Conference on Cyber Warfare and Security, 17*(1), 19–25.

Chaki, J. (2023). Applications of different methods to handle uncertainty in artificial intelligence. In *Handling uncertainty in artificial intelligence* (pp. 93–101). Springer.

Chen, X. (2023). Ethical Governance of AI: An Integrated Approach via Human-in-the-Loop Machine Learning. *Computer Sciences & Mathematics Forum, 8*(1), 29.

Chowdhury, N., & Gkioulos, V. (2021). Cyber security training for critical infrastructure protection: A literature review. *Computer Science Review, 40*, 100361.

Creese, S., Shillair, R., Bada, M., & Dutton, W. H. (2019). Building the Cybersecurity Capacity of Nations. *Society and the Internet: How Networks of Information and Communication Are Changing Our Lives*, 165.

Dawson, M., Bacius, R., Gouveia, L. B., & Vassilakos, A. (2021). Understanding the challenge of cybersecurity in critical infrastructure sectors. *Land Forces Academy Review, 26*(1), 69–75.

Eskandar, S., Wang, J., & Razavi, S. (2020). Human-in-the-loop cyber-physical systems for construction safety. In *Cyber-physical systems in the built environment* (pp. 161–173). Springer.

Gaham, M., Bouzouia, B., & Achour, N. (2015). Human-in-the-loop cyber-physical production systems control (HiLCP2sC): a multi-objective interactive framework proposal. In *Service orientation in holonic and multi-agent manufacturing* (pp. 315–325). Springer.

Ghazi-Tehrani, A. K., & Pontell, H. N. (2022). Phishing evolves: Analyzing the enduring cybercrime. In *The new technology of financial crime* (pp. 35–61). Routledge.

Khan, M. W., Saad, S., Ammad, S., Rasheed, K., & Jamal, Q. (2024). Smart infrastructure and AI. In *AI in Material Science* (pp. 193–215). CRC Press.

Lammouchi, Z., Naoui, M., Bekakra, Y., Khechekhouche, A., & Sbita, L. (2025). Model-free predictive control based on the execution time of active vectors using ultra-local model. *Arabian Journal for Science and Engineering, 50*(8), 5885–5898.

Lin, H.-Y., & Biggio, B. (2021). Adversarial machine learning: Attacks from laboratories to the real world. *Computer, 54*(5), 56–60.

Malatji, M., Marnewick, A. L., & Von Solms, S. (2022). Cybersecurity capabilities for critical infrastructure resilience. *Information and Computer Security, 30*(2), 255–279.

Martinie, C., Navarre, D., Palanque, P., Barboni, E., & Steere, S. (2022). Engineering operations-based training. *Proceedings of the ACM on Human-Computer Interaction, 6*(EICS), 1–25.

Mijwil, M., Doshi, R., Hiran, K. K., Al-Mistarehi, A.-H., & Gök, M. (2022). Cybersecurity challenges in smart cities: An overview and future prospects. *Mesopotamian Journal of Cybersecurity*, *2022*, 1–4.

Natarajan, S., Mathur, S., Sidheekh, S., Stammer, W., & Kersting, K. (2025). Human-in-the-loop or AI-in-the-loop? Automate or Collaborate? *Proceedings of the AAAI Conference on Artificial Intelligence*, *39*(27), 28594–28600.

Raghav, Y. Y., & Kait, R. (2024). Edge computing empowering distributed computing at the edge. In *Emerging Trends in Cloud Computing Analytics, Scalability, and Service Models* (pp. 67–83). IGI Global Scientific Publishing.

Raney, A., Bendelac, S., Manville, K., Tan, M., & Yamaguchi, K. (2024). An AI red team playbook. *Assurance and Security for AI-Enabled Systems*, *13054*, 71–97.

Reed, D. A., Zabinsky, Z. B., & Boyle, L. N. (2011). A framework for optimizing civil infrastructure resiliency. *Structures Congress 2011*, 2104–2112.

Romdhane, M., Naoui, M., & Mansouri, A. (2023). PMSM Inter-Turn Short Circuit Fault Detection Using the Fuzzy-Extended Kalman Filter in Electric Vehicles. *Electronics (Basel)*, *12*(18), 3758.

Sajid, A., Abbas, H., & Saleem, K. (2016). Cloud-assisted IoT-based SCADA systems security: A review of the state of the art and future challenges. *IEEE Access : Practical Innovations, Open Solutions*, *4*, 1375–1384.

Segalas, J. (n.d.). Ethical Governance of AI: An Integrated Approach via Human-in-the-Loop. *Machine Learning*.

Serpanos, D. (2022). False Data Injection Attacks on Sensor Systems. *Proceedings of the 41st IEEE/ACM International Conference on Computer-Aided Design*, 1–5.

Sridevi, S. (2014). User interface design. *International Journal of Computer Science and Information Technology Research*, *2*(2), 415–426.

Stanton, N. A., Chambers, P. R. G., & Piggott, J. (2001). Situational awareness and safety. *Safety Science*, *39*(3), 189–204.

Stone, D., Jarrett, C., Woodroffe, M., & Minocha, S. (2005). *User interface design and evaluation*. Elsevier.

Strobhar, D. A. (1995). Evolution of operator decision making. *ISA Transactions*, *34*(4), 405–409.

Udayakumar, P. (2023). Design and deploy security for infrastructure, data, and applications. In *Design and Deploy a Secure Azure Environment: Mapping the NIST Cybersecurity Framework to Azure Services* (pp. 75–148). Springer.

Vasudevan, S., Piazza, A., & Carr, M. (2022). Qualitative factors in organizational cyber resilience. *2022 International Conference on Cyber Resilience (ICCR)*, 1–5.

Verma, D., Farr, J., & Johannesen, L. H. (2003). System training metrics and measures: A key operational effectiveness imperative. *Systems Engineering*, *6*(4), 238–248.

Vlachos, A. G., & Biskas, P. N. (2010). Balancing supply and demand under mixed pricing rules in multi-area electricity markets. *IEEE Transactions on Power Systems*, *26*(3), 1444–1453.

Wali, A., & Alshehry, F. (2024). A survey of security challenges in cloud-based SCADA systems. *Computers*, *13*(4), 97.

Yang, A.-M., Li, S.-S., Ren, C.-H., Liu, H.-X., Han, Y., & Liu, L. (2018). Situational awareness system in the smart campus. *IEEE Access : Practical Innovations, Open Solutions*, *6*, 63976–63986.

Youssef, R., Hamida, M. A., Mohamed, N., Lassaad, S., Djamel, T., & Yasmine, A. (2023). Single-stage standalone photovoltaic water pumping system using predictive torque control (PTC) of induction machine. *Electric Power Components and Systems*, •••, 1–17.

Zandi, G., Yaacob, N. A., Tajuddin, M., & Nik Abdul Rahman, N. K. (2024). Artificial intelligence and the evolving cybercrime paradigm: Current threats to businesses. *Journal of Information Technology Management*, *16*(4), 162–170.

Żywiec, J., Tchórzewska-Cieślak, B., & Sokolan, K. (2024). Assessment of Human Errors in the Operation of the Water Treatment Plant. *Water (Basel)*, *16*(17), 2399.

Compilation of References

Abdalla, S. A., Abdullah, S. S., & Kassem, A. M. (2025). An adaptive frame and intelligent control approach for an autonomous hybrid renewable energy technology consisting of PV, wind, and fuel cell innovation. *Alexandria Engineering Journal, 114*, 279–291. DOI: 10.1016/j.aej.2024.10.057

Abdel Hakeem, S. A., & Kim, H. (2025). Advancing intrusion detection in V2X networks: A comprehensive survey on machine learning, federated learning, and edge AI for V2X security. *IEEE Transactions on Intelligent Transportation Systems.*

Abdelsalam, A. K., Massoud, A. M., Ahmed, S., & Enjeti, P. N. (2011). High-performance adaptive perturb and observe MPPT technique for photovoltaic-based microgrids. *IEEE Transactions on Power Electronics, 26*(4), 1010–1021.

Abeleira, M. G., Pérez, C. S., Sánchez, F. J. R., Sánchez, P. M., Santiso Gómez, E., & Calvo, J. A. J. (2024). Aperiodic two-layer energy management system for community microgrids based on blockchain strategy. *Energy Reports, 10*, 1234–1245. DOI: 10.1016/j.egyr.2024.01.034

Abogrean, E., Elhatmi, E. A., & Saad, J. A. (2022). Simulation of MPPT with a PMSG-based wind energy conversion system considering variable wind speed. *Scientific Journal of University of Benghazi, 35*(2). Advance online publication. DOI: 10.37376/sjuob.v35i2.3807

Abolfazl Hadavi, M. T. (2025). Critical inertia thresholds for frequency stability in renewable Energy-Integrated power systems. *International Journal of Electrical Power & Energy Systems, 169*, 110733.

Abouyehia, M., Egea-Àlvarez, A., & Ahmed, K. H. (2025). Evaluating inertia estimation methods in low-inertia power systems: A comprehensive review with analytic hierarchy process-based ranking. *Renewable & Sustainable Energy Reviews, 217*, 115794.

Abut, T. (2016). Modeling and optimal control of a DC motor. *International Journal of Engineering Trends and Technology*, *32*(3), 146–150.

Addoweesh, K. E., & Noman, A. M. (2016). Simulation and dSPACE hardware implementation of the MPPT techniques using buck boost converter. *International Journal of Engineering Research and Applications*, *6*(3), 47–53.

Affanni, A.. (n.d.). EV battery state of charge: Neural network based estimation. *Proceedings of the International Electric Vehicle Conference*.

Agarwal, S., & Jain, A. (2024). Blockchain Applications in Smart Grid: Challenges and Opportunities. In *2023 IEEE International Conference on Blockchain and Distributed Systems Security (ICBDS)* (Proceedings). DOI: DOI: 10.1109/ICBDS58040.2023.10346363

Ahmad, S., Hasan, S. M. N., Hossain, M. S., Uddin, R., Ahmed, T., Mustayen, A. G. M. B., Hazari, M. R., Hassan, M., Parvez, M. S., & Saha, A. (2024). A review of hybrid renewable and sustainable power supply system: Unit sizing, optimization, control, and management. *Energies*, *17*(23), 6027. Advance online publication. DOI: 10.3390/en17236027

Ahmad, Z., Singh, S. N., & Kumar, A. (2018). Design and implementation of real-time monitoring based on Android and HMI of PV tower system for laboratory scale. *Indonesian Journal of Electrical Engineering and Computer Science*, *11*(1), 187–194.

Ahmed Baba, M., Naoui, M., & Cherkaoui, M. (2023). Fault-tolerant control strategy for Hall sensors in BLDC motor drive for electric vehicle applications. *Sustainability*, *15*(13), 10430.

Ahmed Baba, M., Naoui, M., & Cherkaoui, M. (2023). Fault-Tolerant Control Strategy for Hall Sensors in BLDC Motor Drive for Electric Vehicle Applications. *Sustainability*, *15*(13), 10430. DOI: 10.3390/su151310430

Ahmed, M. M. R., Mirsaeidi, S., Koondhar, M. A., Karami, N., Tag-Eldin, E. M., & Ghamry, N. A. (2024). Mitigating uncertainty problems of renewable energy resources through efficient integration of hybrid solar PV/wind systems into power networks. *IEEE Access: Practical Innovations, Open Solutions*, *12*, 30311–30328. DOI: 10.1109/ACCESS.2024.3370163

Ahmed, M., Abdelrahem, M., Harbi, I., & Kennel, R. (2020). An adaptive model-based MPPT technique with drift-avoidance for grid-connected PV systems. *Energies*, *13*(24), 6656. Advance online publication. DOI: 10.3390/en13246656

Ahmed, N. A., & Miyatake, M. (2008). A novel maximum power point tracking for photovoltaic applications under partially shaded insolation conditions. *Electric Power Systems Research, 78*(5), 777–784.

Ailliot, P., & Monbet, V. (2012). Markov-switching autoregressive models for wind time series. *Environmental Modelling & Software, 30*, 92–101. DOI: 10.1016/j.envsoft.2011.10.011

Ajiboye, O. K., Ochiegbu, C. V., Ofosu, E. A., & Gyamfi, S. (2023). A review of hybrid renewable energies optimisation: Design, methodologies, and criteria. *International Journal of Sustainable Energy, 42*(1), 648–684. DOI: 10.1080/14786451.2023.2227294

Aleksandra, Z., Monika, P., & Ireneus, Z. (2021). Application of LCA to Determine Environmental Impact of Concentrated Photovoltaic Solar Panels—State-of-the-Art. *Energies, 14*(31), 1–20.

Aleksandra, Z., Monika, P., & Ireneus, Z. (2023). Cooling of Concentrated Photovoltaic Cells—A Review and the Perspective of Pulsating Flow Cooling. *Energies, 16*(2), 1–23.

Alhamami, A. H., Abba, S. I., Musa, B., Dodo, Y. A., Salami, B. A., Dodo, U. A., & Alyami, S. H. (2024). Off-grid multi-region energy system design based on energy load demand estimation using hybrid nature-inspired optimization algorithms. *Energy Conversion and Management, 315*, 118766.

Ali, M. H. (2021). Exploring cybersecurity issues in 5G enabled electric vehicle. *IET Generation, Transmission & Distribution, 15*(24), 3435–3449.

Ali, M. H., Zakaria, M., & El-Tawab, S. (2025). A comprehensive study of recent maximum power point tracking techniques for photovoltaic systems. *Scientific Reports, 15*(1), 14269. PMID: 40274881

Ali, M. U., Zafar, A., Nengroo, S. H., Hussain, S., Alvi, M. J., & Kim, H.-J. (2019). Towards a smarter battery management system for electric vehicle applications: A critical review of lithium-ion battery state of charge estimation. *Energies, 12*(3), 446. Advance online publication. DOI: 10.3390/en12030446

Allal, A., Lammouchi, Z., Khechekhouche, A., Mohamed, N., Alqunun, K., Alghadhban, A., & Guesmi, T. (2024). Advanced Residual Optimal Mapping Approach for Precise Detection of Stator Faults in Induction Motors. *IEEE Access: Practical Innovations, Open Solutions, 12*, 143515–143530.

Allal, A., Lammouchi, Z., Khechekhouche, A., Mohamed, N., Alqunun, K., Alghadhban, A., Marouani, I., & Guesmi, T. (2024). Advanced Residual Optimal Mapping Approach for Precise Detection of Stator Faults in Induction Motors. *IEEE Access : Practical Innovations, Open Solutions*, *12*, 143515–143530.

Allal, A., Mansouri, N., & Naoui, M. (2024). Advanced residual optimal mapping approach for precise detection of stator faults in induction motors. *IEEE Access: Practical Innovations, Open Solutions*, *12*, 143515–143530. DOI: 10.1109/ACCESS.2024.3442671

Almeida, F. (2023). Prospects of cybersecurity in smart cities. *Future Internet*, *15*(9), 285.

Almutairi, B. S., & Alghamdi, A. (2022). The role of social engineering in cybersecurity and its impact. *Journal of Information Security*, *13*(4), 363–379.

Aloo, L. A., Kihato, P. K., Kamau, S. I., & Orenge, R. S. (2023). Modeling and control of a photovoltaic-wind hybrid microgrid system using GA-ANFIS. *Heliyon*, *9*(4), e14678. Advance online publication. DOI: 10.1016/j.heliyon.2023.e14678 PMID: 37025884

Alqunun, K., Bechiri, M. B., Naoui, M., Khechekhouche, A., Marouani, I., Guesmi, T., & Allal, A. (2025). An efficient bearing fault detection strategy based on a hybrid machine learning technique. *Scientific Reports*, *15*(1), 18739. PMID: 40437009

Al-Sada, B., Sadighian, A., & Oligeri, G. (2024). MITRE ATT&CK: State of the art and way forward. *ACM Computing Surveys*, *57*(1), 1–37.

Alsaigh, R., Mehmood, R., & Katib, I. (2022). AI explainability and governance in smart energy systems: A review. *arXiv*. https://doi.org//arXiv.2211.00069DOI: 10.48550

Alsharafi, A. S., Besheer, A. H., & Emara, H. M. (2018). Primary frequency response enhancement for future low inertia power systems using hybrid control technique. *Energies*, *11*(4), 699.

Al-Turjman, F. (2019). *Artificial intelligence in IoT*. Springer.

Aluko, A. O., Dorrell, D. G., Pillay Carpanen, R., & Ojo, E. E. (2020). Heuristic optimization of virtual inertia control in grid-connected wind energy conversion systems for frequency support in a restructured environment. *Energies*, *13*(3), 564.

Al-Wesabi, I., Fang, Z., Cai, J., Hussein Farh, H. M., Aboudrar, I., & Dagal, I. (2024). Fast DC-link voltage control based on power flow management using linear ADRC combined with hybrid salp particle swarm algorithm for PV/wind energy conversion system. *International Journal of Hydrogen Energy*, *61*, 688–709. DOI: 10.1016/j.ijhydene.2024.02.325

Amditis, A., Kubmann, H., Polychronopoulos, A., Engstrom, J., & Andreone, L. (2006). System architecture for integrated adaptive HMI solutions. *2006 IEEE Intelligent Vehicles Symposium*, 388–393.

Anika, T., Syed, N. S., Mohammad, J. H., Kaiser, A. R., & Abu, K. (2025). Advances in multijunction solar cells: An overview. *Solar Energy Advances*, *5*, 1–14.

Anis, W. R., Mertens, R. P., & Van Overstraeten, R. J. (1985). Coupling of a volumetric pump to a photovoltaic array. Solar cells, 14(1), 27-42.

Aoumri, M., Harrouz, A., Yaichi, I., Baala, S., & Wira, P. (2024). Robustness investigation of FOC of PMSG-based variable-speed wind turbine connected on grid. In *Proceedings of the 3rd International Conference on Advanced Electrical Engineering (ICAEE)*, Sidi-bel-Abbès, Algeria. https://doi.org/DOI: 10.1109/ICAEE61760.2024.10783398

Appasani, B., Mishra, S. K., Jha, A. V., Mishra, S. K., Enescu, F. M., Sorlei, I. S., Bîrleanu, F. G., Takorabet, N., Thounthong, P., & Bizon, N. (2022). Blockchain-Enabled Smart Grid Applications: Architecture, Challenges, and Solutions. *Sustainability*, *14*(14), 8801. DOI: 10.3390/su14148801

Ashfaq, T., Khalid, M. I., Ali, G., Affendi, M. E., Iqbal, J., Hussain, S., Ullah, S. S., Yahaya, A. S., Khalid, R., & Mateen, A. (2022). An Efficient and Secure Energy Trading Approach with Machine Learning Technique and Consortium Blockchain. *Sensors (Basel)*, *22*(19), 7263. DOI: 10.3390/s22197263 PMID: 36236363

Ashish Shrestha, F. G.-L. (2021). Parametric Sensitivity Analysis of Rotor Angle Stability Indicators. *Energies*, *14*(16), 5023.

Azom, M. A., Hossain, M. S., & Khan, M. Y. A. (2025). Review of Electrical and Thermal Modeling Techniques for Three-Phase PMSM Drives. *Control Systems and Optimization Letters*, *3*(1), 75–83.

Bach, T. A., Kristiansen, J. K., Babic, A., & Jacovi, A. (2024). Unpacking human-AI interaction in safety-critical industries: A systematic literature review. *IEEE Access: Practical Innovations, Open Solutions*.

Badrudeen, T. U., Nwulu, N. I., & Gbadamosi, S. L. (2024). Low-inertia control of a large-scale renewable energy penetration in power grids: A systematic review with taxonomy and bibliometric analysis. *Energy Strategy Reviews*, *52*, 101337.

Bai, X.. (2022). Electric vehicle charging station planning with dynamic prediction of elastic charging demand: A hybrid particle swarm optimization algorithm. *Complex & Intelligent Systems*, *8*(2), 1035–1046. DOI: 10.1007/s40747-021-00575-8

Baldini, A., Ciabattoni, L., Felicetti, R., Ferracuti, F., Freddi, A., & Monteriù, A. (2018). Dynamic surface fault tolerant control for underwater remotely operated vehicles. *ISA Transactions*, *78*, 10–20. PMID: 29503009

Bálint Hartmann, I. V. (2019). Effects of decreasing synchronous inertia on power system dynamics—Overview of recent experiences and marketisation of services. *International Transactions on Electrical Energy Systems*, *29*(04).

Bansal, P. (2015). Charging of electric vehicles: Technology and policy implications. *Journal of Science Policy & Governance*, *6*(1), 1–38.

Barać, B., Krpan, M., Capuder, T., & Kuzle, I. (2021). Modeling and initialization of a virtual synchronous machine for power system fundamental frequency simulations. *IEEE Access: Practical Innovations, Open Solutions*, *9*, 160116–160134.

Barker, L. D., Jakuba, M. V., Bowen, A. D., German, C. R., Maksym, T., Mayer, L., & Whitcomb, L. L. (2020). Scientific challenges and present capabilities in underwater robotic vehicle design and navigation for oceanographic exploration under-ice. *Remote Sensing*, *12*(16), 2588.

Baškarad, T., Holjevac, N., & Kuzle, I. (2024). A new perspective on frequency control in conventional and future interconnected power systems. *International Journal of Electrical Power & Energy Systems*, *156*, 109731.

Baškarad, T., Kuzle, I., Đaković, J., & Ilak, P. (2018, November). Analysis of the wind generation impact on inertial and primary frequency response of the croatian electric power system. In *Mediterranean Conference on Power Generation, Transmission, Distribution and Energy Conversion (MEDPOWER 2018)* (p. 93). Stevenage UK: IET.

Bassey, K. E. (2023). Hybrid renewable energy systems modeling. *Engineering Science and Technology Journal*, *4*(6), 571–588. DOI: 10.51594/estj.v4i6.1255

Bechekir[1], M. B. S., Brahami[1], M., Bendaho, H., & Brahimi, A. (2025). Comparative Analysis of MPPT Algorithms: P&O and Inc for Optimizing PV Systems. Smart Computing and Control Renewable Energy Systems: Advanced Computational Techniques for Wireless Network Optimization, 1238, 60.

Bechiri, M. B.. (2024). Effective diagnosis approach for broken rotor bar fault using Bayesian-based optimization of machine learning hyperparameters. *Journal of Electrical Engineering*, *12*(October), 139923–139936.

Benbouhenni, H., & Bizon, N. (2021). Third-order sliding mode applied to the direct field-oriented control of the asynchronous generator for variable-speed contra-rotating wind turbine generation systems. *Energies*, *14*(18), 5877. DOI: 10.3390/en14185877

BENMEDDAH. A., & LAIDI, K. (2025). Optimisation de l'algorithme MPPT pour un système photovoltaïque en intégrant de nouvelles techniques d'intelligence artificielle (Doctoral dissertation, Faculté des sciences et de la technologie).

Bhattacharyya, R., & Bhattacharyya, A. (2020). Smart Grid Demand-Side Management by Machine Learning. *Proceedings of the Global AI Congress 2019*, 633–644.

Bidram, A., Davoudi, A., & Balog, R. S. (2012). Control and circuit techniques to mitigate partial shading effects in photovoltaic arrays. *IEEE Journal of Photovoltaics*, *2*(4), 532–546.

Blaabjerg, F., Yang, Y., Yang, D., & Xu, B. (2017). Distributed power-generation systems and protection. *Proceedings of the IEEE*, *105*(7), 1311–1331. DOI: 10.1109/JPROC.2017.2696878

Blair, C. (1975). *Silent victory: the US submarine war against Japan*. No Title.

Bouraiou, A., Neçaibia, A., Motahhir, S., Bouakkaz, M. S., Attoui, I., Boutasseta, N., & Ziane, A. (2021, January). Supervision and monitoring of photovoltaic systems using siemens PLC and HMI. In *International Conference on Digital Technologies and Applications* (pp. 1147-1157). Cham: Springer International Publishing.

Boyce, M. W., Duma, K. M., Hettinger, L. J., Malone, T. B., Wilson, D. P., & Lockett-Reynolds, J. (2011). Human performance in cybersecurity: A research agenda. *Proceedings of the Human Factors and Ergonomics Society Annual Meeting*, *55*(1), 1115–1119.

Brezovec, M., Kuzle, I., & Krpan, M. (2015). Detailed mathematical and simulation model of a synchronous generator. *Journal of Energy: Energija*, *64*(1-4), 102–129.

Browne, D. (2016). *Adaptive user interfaces*. Elsevier.

Brown, S., Sinha, S., & Schaefer, A. J. (2024). Markov decision process design: A framework for integrating strategic and operational decisions. *Operations Research Letters*, *54*, 107090. PMID: 38560724

Brunton, S. L., Rowley, C. W., Kulkarni, S. R., & Clarkson, C. (2010). Maximum power point tracking for photovoltaic optimization using ripple-based extremum seeking control. *IEEE Transactions on Power Electronics*, *25*(10), 2531–2540.

Cannan, J., & Hu, H. (2011). Human-machine interaction (HMI): A survey. *University of Essex*, *27*, 46–64.

Cao, P., ... (2018). An IPT system with constant current and constant voltage output features for EV charging. *Proceedings: IECON 2018 - 44th Annual Conference of the IEEE Industrial Electronics Society, 1*, 4775–4780.

Cao, J. Y., Liang, J., & Cao, B. G. (2005, August). *Optimization of fractional order PID controllers based on genetic algorithms. In 2005 international conference on machine learning and cybernetics* (Vol. 9). IEEE.

Cao, X., Zheng, Y., Xiao, H., & Xiao, M. (2025). Maximum power point tracking control of offshore wind-photovoltaic hybrid power generation system with crane-assisted. *CMES-Computer Modeling in Engineering & Sciences*, *143*(1), 289–334. DOI: 10.32604/cmes.2025.063954

Carlson, W. B. (1988). Thomas Edison as a manager of R&D: The case of the alkaline storage battery, 1898-1915. *IEEE Technology and Society Magazine*, *7*(4), 4–12.

Cervini, J., Rubin, A., & Watkins, L. (2022). Don't drink the cyber: Extrapolating the possibilities of Oldsmar's water treatment cyberattack. *International Conference on Cyber Warfare and Security, 17*(1), 19–25.

Chaichan, W., Waewsak, J., Naklua, Y., Ali, F., Mueanmas, C., Nikhom, R., Kongruang, C., & Gagnon, Y. (2025). Sustainability analysis of hybrid renewable-based power generation with battery energy storage system for remote islands: Application to Koh Hang, Thailand. *Cleaner Energy Systems*, *12*, 100203. Advance online publication. DOI: 10.1016/j.cles.2025.100203

Chaki, J. (2023). Applications of different methods to handle uncertainty in artificial intelligence. In *Handling uncertainty in artificial intelligence* (pp. 93–101). Springer.

Chappel, A. R., & Mitchell, C. M. (1996). Use of Visualization and Contextualization in Training Operators of Complex Systems. *Proceedings of the Human Factors and Ergonomics Society Annual Meeting*, *40*(4), 264–268.

Chedid, R., & Saliba, Y. (1996). Optimization and control of autonomous renewable energy systems. *International Journal of Energy Research*, *20*(7), 609–624.

Cheng, M., Sun, L., Buja, G., & Song, L. (2015). Advanced electrical machines and machine-based systems for electric and hybrid vehicles. *Energies*, *8*(9), 9541–9564.

Chen, R., Qin, D., & Liu, C. (2023). Dynamic modelling and dynamic characteristics of wind turbine transmission gearbox-generator system electromechanical-rigid-flexible coupling. *Alexandria Engineering Journal*, *65*, 307–325. DOI: 10.1016/j.aej.2022.10.036

Chen, X. (2023). Ethical Governance of AI: An Integrated Approach via Human-in-the-Loop Machine Learning. *Computer Sciences & Mathematics Forum*, *8*(1), 29.

Chen, X., Qin, P., Chen, Y., Zhao, J., Li, W., Mao, Y., & Zhao, T. (2022). Inter-turn short circuit fault diagnosis of PMSM. *Electronics (Basel)*, *11*(10), 1576.

Chen, Z., Guerrero, J. M., & Blaabjerg, F. (2009). A review of the state of the art of power electronics for wind turbines. *IEEE Transactions on Power Electronics*. Advance online publication. DOI: 10.1109/TPEL.2009.2017082

Chethan, G. N. (2019, August). Performance analysis of PMSM drive with Space-vector PWM and sinusoidal PWM fed VSI. *In 2019 International Conference on Power Electronics Applications and Technology in Present Energy Scenario (PETPES)* (pp. 1-6). IEEE.

Chin, C. S., Neelakantan, P., Yoong, H. P., & Teo, K. T. K. (2015). Cell modelling and model parameters estimation techniques for photovoltaic simulator application: A review. *Applied Energy*, *154*, 500–519.

Choobineh, F., Arif, S., & Hamed, M. (2023). Blockchain technology in energy systems: A state-of-the-art review. *IET Renewable Power Generation*. Advance online publication. DOI: 10.1049/blc2.12020

Chowdary, V. G., Sankar, V. U., & Busawon, K. (2018). A review on various MPPT techniques for wind energy conversion system. In *Proceedings of the International Conference on Computing, Power, Energy, Information and Communication (ICCPEIC)*, Chennai, India. https://doi.org/DOI: 10.1109/ICCPEIC.2018.8525219

Chowdhury, N., & Gkioulos, V. (2021). Cyber security training for critical infrastructure protection: A literature review. *Computer Science Review*, *40*, 100361.

Chrifi-Alaoui, L., Drid, S., Ouriagli, M., & Mehdi, D. (2023). Overview of photovoltaic and wind electrical power hybrid systems. *Energies*, *16*(12), 4778. Advance online publication. DOI: 10.3390/en16124778

Chromik, M., & Butz, A. (2021). Human-XAI interaction: a review and design principles for explanation user interfaces. *IFIP Conference on Human-Computer Interaction*, 619–640.

Collier, D. A. F., & Heldwein, M. L. (2011). Modeling and design of a micro wind energy system with a variable-speed wind turbine connected to a permanent magnet synchronous generator and a PWM rectifier. In *Proceedings of the XI Brazilian Power Electronics Conference (COBEP)*, Natal, Brazil. https://doi.org/DOI: 10.1109/COBEP.2011.6085301

Coppolino, L., Nardone, R., Petruolo, A., & Romano, L. (2023). Building cyber-resilient smart grids with digital twins and data spaces. *Applied Sciences (Basel, Switzerland), 13*(24), 13060. Advance online publication. DOI: 10.3390/app132413060

Covic, G. A., Boys, J. T., Budhia, M., & Huang, C. Y. (2011). Electric vehicles—Personal transportation for the future. *World Electric Vehicle Journal, 4*(1), 693–704.

Creese, S., Shillair, R., Bada, M., & Dutton, W. H. (2019). Building the Cybersecurity Capacity of Nations. *Society and the Internet: How Networks of Information and Communication Are Changing Our Lives*, 165.

D'Arco, S., Suul, J. A., & Fosso, O. B. (2015). A Virtual Synchronous Machine implementation for distributed control of power converters in SmartGrids. *Electric Power Systems Research, 122*, 180–197.

Dani, A., Benlamlih, M., Mekrini, Z., & Boulaala, M. (2024). Wind energy conversion technologies and control strategies: A review. *International Journal of Renewable Energy Research, 14*(1), 140–154. DOI: 10.20508/ijrer.v14i1.14296.g8869

Dannehl, J., Liserre, M., & Fuchs, F. W. (2011). Filter-based active damping of voltage source converters with LCL filter. *IEEE Transactions on Industrial Electronics, 58*(8), 3623–3633. DOI: 10.1109/TIE.2010.2081952

Darabian, R., Yaghini, A., & Tarafdar Hagh, M. (2023). Design of a new robust controller for voltage regulation of DC link in wind farm side converter integrated with a hybrid AC and DC large scale power system. *IET Generation, Transmission & Distribution, 17*(5), 1561–1572. DOI: 10.1049/gtd2.12780

Daw, M. O., Shalem, S. M., Habeeb, L. J., & Faraj, A. O. (2025). A Comparative Study of Perturb & Observe and ANN-Based MPPT Algorithms Under Various Environmental Conditions. *Scientific Journal of University of Benghazi, 38*(1), 62–85.

Dawson, M., Bacius, R., Gouveia, L. B., & Vassilakos, A. (2021). Understanding the challenge of cybersecurity in critical infrastructure sectors. *Land Forces Academy Review, 26*(1), 69–75.

De Lellis, M., Reginatto, R., Saraiva, R., & Trofino, A. (2018). The Betz limit applied to airborne wind energy. *Renewable Energy, 127*, 32–40. DOI: 10.1016/j.renene.2018.04.034

Dekali, Z., Baghli, L., & Boumediene, A. (2021). Improved hardware implementation of a TSR-based MPPT algorithm for a low-cost connected wind turbine emulator under unbalanced wind speeds. *Energy*, *232*, 121039. DOI: 10.1016/j.energy.2021.121039

Deng, H., Wu, J., Luo, J., Yan, R., Zhang, C., Li, P., & Zheng, R. (2021). Research on a Power Grid Cascading Failure Prevention and Control Method considering WSN. *Wireless Communications and Mobile Computing*, *2021*(1), 9439977.

Deng, X., Mo, R., Wang, P., Chen, J., Nan, D., & Liu, M. (2023). Review of RoCoF estimation techniques for low-inertia power systems. *Energies*, *16*(9), 3708.

Deshpande, S., & Bhasme, N. R. (2017). A review of topologies of inverter for grid connected PV systems. 2017 Innovations in Power and Advanced Computing Technologies (i-PACT), 1-6.

Desideri, R., Francesc, M. P., Alicia, C., Jérôme, B., Montse, V., & Joan, R. U. (2025). Enhancing efficiency of dense array CPV receivers with controlled DC-DC converters and adaptive microfluidic cooling under non-uniform solar irradiance. *Solar Energy Materials and Solar Cells*, *279*, 1–12.

Divyasharon, R., Banu, R. N., & Devaraj, D. (2019, April). Artificial neural network based MPPT with CUK converter topology for PV systems under varying climatic conditions. In *2019 IEEE International Conference on Intelligent Techniques in Control, Optimization and Signal Processing (INCOS)* (pp. 1-6). IEEE.

Djelamda, I., & Bouchareb, I. (2022). Field-oriented control based on adaptive neuro-fuzzy inference system for PMSM dedicated to electric vehicle. *Bulletin of Electrical Engineering and Informatics*, *11*(4), 1892–1901.

Dong, H., Tran, T. T. M., Bazilinskyy, P., Hoggenmüller, M., Dey, D., Cazacu, S., Franssen, M., & Gao, R. (2023). Holistic HMI Design for Automated Vehicles: Bridging In-Vehicle and External Communication. *Adjunct Proceedings of the 15th International Conference on Automotive User Interfaces and Interactive Vehicular Applications*, 237–240.

dSpace GmbH. (2021). Real-Time Interface Implementation Guide [Technical manual].

EFEMWENKIEKIE, U. K., BABALOLA, O. P., DIRISU, O., & Oyedepo, O. S. (2020). IMPACT OF ELECTRIC VEHICLE IN 21 ST CENTURY DEVELOPING COUNTRIES. [IJMPERD]. *International Journal of Mechanical and Production Engineering Research and Development*, *10*(3), 11797–11806.

Elenga Baningobera, B., Oleinikova, I., Uhlen, K., & Pokhrel, B. R. (2024). Challenges and solutions in low-inertia power systems with high wind penetration. *IET Generation, Transmission & Distribution, 18*(24), 4221–4244.

Elgendy, M. A., Zahawi, B., & Atkinson, D. J. (2013). Assessment of the incremental conductance maximum power point tracking algorithm. *IEEE Transactions on Sustainable Energy, 4*(1), 108–117.

Energies. (2022). Towards Blockchain-Based Energy Trading: A Smart Contract Implementation of Energy Double Auction and Spinning Reserve Trading. *Energies, 15*(11), 4084. https://doi.org/DOI: 10.3390/en15114084

Energy Informatics. (2024). Optimizing power system trading processes using smart contract algorithms. *Energy Informatics*. Advance online publication. DOI: 10.1186/s42162-024-00457-6

Enslin, J. H. R., Wolf, M. S., Snyman, D. B., & Swiegers, W. (1997). Integrated photovoltaic maximum power point tracking converter. *IEEE Transactions on Industrial Electronics, 44*(6), 769–773.

Eom, H., & Lee, S. H. (2022). Mode confusion of human–machine interfaces for automated vehicles. *Journal of Computational Design and Engineering, 9*(5), 1995–2009.

Eskandar, S., Wang, J., & Razavi, S. (2020). Human-in-the-loop cyber-physical systems for construction safety. In *Cyber-physical systems in the built environment* (pp. 161–173). Springer.

Esram, T., & Chapman, P. L. (2007). Comparison of photovoltaic array maximum power point tracking techniques. *IEEE Transactions on Energy Conversion, 22*(2), 439–449. DOI: 10.1109/TEC.2006.874230

European Commission. (2021). *Regulatory paradigm and challenge for blockchain integration of energy systems*. MDPI., DOI: 10.3390/su15032571

Fabbri, G.. (2014). Impact of V2G/G2V technologies on distributed generation systems. *IEEE International Symposium on Industrial Electronics*, 1677–1682.

Faisal, M., Hannan, M. A., Ker, P. J., Hussain, A., Mansor, M. B., & Blaabjerg, F. (2020). Review of energy storage system technologies in microgrid applications: Issues and challenges. *IEEE Access: Practical Innovations, Open Solutions, 8*, 218685–218693. DOI: 10.1109/ACCESS.2020.3036556

Faiz, J., & Nejadi-Koti, H. (2019). Eccentricity fault diagnosis indices for permanent magnet machines: State-of-the-art. *IET Electric Power Applications*, *13*(9), 1241–1254.

Fang, J., Sun, Y., Wang, Y., Wei, B., & Hang, J. (2019). Improved ZSVC-based fault detection technique for incipient stage inter-turn fault in PMSM. *IET Electric Power Applications*, *13*(12), 2015–2026.

Faria, J., Pombo, J., Calado, M. R., & Mariano, S. (2019). Power management control strategy based on artificial neural networks for standalone PV applications with a hybrid energy storage system. *Energies*, *12*(5), 902. Advance online publication. DOI: 10.3390/en12050902

Femia, N., Petrone, G., Spagnuolo, G., & Vitelli, M. (2005). Optimization of perturb and observe maximum power point tracking method. *IEEE Transactions on Power Electronics*, *20*(4), 963–973. DOI: 10.1109/TPEL.2005.850975

Femia, N., Petrone, G., Spagnuolo, G., & Vitelli, M. (2013). Perturb and observe MPPT algorithm with a current controller based on the sliding mode. *International Journal of Electrical Power & Energy Systems*, *44*(1), 346–356.

Fie, L. (2022). Blockchain in energy systems: Values, opportunities, and limitations. *Frontiers in Energy Research*. Advance online publication. DOI: 10.1007/s11708-022-0818-8

Fossen, T. I. (2011). *Handbook of marine craft hydrodynamics and motion control*. John Wiley & Sons.

Foster, R., Ghassemi, M., & Cota, A. (2009). *Solar energy: renewable energy and the environment*. CRC Press.

Fotopoulou, M., & Rakopoulos, D. (2021). Day ahead optimal dispatch schedule in a smart grid containing distributed energy resources and electric vehicles. *Sensors (Basel)*, *21*(21), 7295. Advance online publication. DOI: 10.3390/s21217295 PMID: 34770613

Frontiers in Blockchain. (2025). *Revolutionizing the energy sector: exploring diversified blockchain applications in production, distribution, trading, and management of energy*. Frontiers in Blockchain., DOI: 10.3389/fbloc.2025.1544770

Fu, S., Tan, Y., & Xu, Z. (2023). Blockchain-Based Renewable Energy Certificate Trade for Low-Carbon Community of Active Energy Agents. *Sustainability*, *15*(23), 16300. DOI: 10.3390/su152316300

Gaham, M., Bouzouia, B., & Achour, N. (2015). Human-in-the-loop cyber-physical production systems control (HiLCP2sC): a multi-objective interactive framework proposal. In *Service orientation in holonic and multi-agent manufacturing* (pp. 315–325). Springer.

Gao, R., & Gao, Z. (2016). Pitch control for wind turbine systems using optimization, estimation and compensation. *Renewable Energy*, *91*, 501–515. DOI: 10.1016/j.renene.2016.01.057

Garud, K. S., Jayaraj, S., & Lee, M. Y. (2021). A review on modeling of solar photovoltaic systems using artificial neural networks, fuzzy logic, genetic algorithm and hybrid models. *International Journal of Energy Research*, *45*(1), 6–35.

Ghazi-Tehrani, A. K., & Pontell, H. N. (2022). Phishing evolves: Analyzing the enduring cybercrime. In *The new technology of financial crime* (pp. 35–61). Routledge.

Girija, G. S. P., & Sahoo, A. K. (2025). An adaptive energy monitoring system for a hybrid power plant using renewable energy. *Transactions on Energy Systems and Engineering Applications*, *6*(1), 1–15.

Girsang, I. P., Dhupia, J. S., Muljadi, E., Singh, M., & Pao, L. Y. (2014). Gearbox and drivetrain models to study dynamic effects of modern wind turbines. *IEEE Transactions on Industry Applications*, *50*(6), 3777–3786. DOI: 10.1109/TIA.2014.2321029

González-Longatt, F. M. (2005). Model of photovoltaic module in Matlab. In II CIBELEC (pp. 1–6). Venezuela.

Gonzalez-Sanchez, J., López-Pimentel, J.-C., & Morales-Rosales, L. A. (2025). A Digital Identity Blockchain Ecosystem: Linking Government-Certified and Uncertified Tokenized Objects. *Applied Sciences (Basel, Switzerland)*, *15*(15), 8577. DOI: 10.3390/app15158577

Goyal, S., Sharma, R., & Kumar, A. (2023). Blockchain-based microgrid solutions for rural electrification in India: Opportunities and challenges. *Renewable & Sustainable Energy Reviews*, *167*, 112–123. DOI: 10.1016/j.rser.2023.112345

Green, M. A., Dunlop, E. D., Hohl-Ebinger, J., Yoshita, M., Kopidakis, N., & Hao, X. (2022). Solar cell efficiency tables (Version 60). *Progress in Photovoltaics: Research and Applications*, *30*(7), 687–701. DOI: 10.1002/pip.3512

Gruhn, P. (2011). Human machine interface (HMI) design: the good, the bad, and the ugly (and what makes them so). *66th Annual Instrumentation Symposium for the Process Industries*, 25.

Grunditz, E. A., & Thiringer, T. (2016). Performance analysis of current BEVs based on a comprehensive review of specifications. *IEEE Transactions on Transportation Electrification*, *2*(3), 270–289.

Guner, S., & Ozdemir, A. (2017). Stochastic energy storage capacity model of EV parking lots. *IET Generation, Transmission & Distribution*, *11*(7), 1754–1761.

Guo, X., Zhu, D., Hu, J., & Zou, X.. (2023). Inertial PLL of grid-connected converter for fast frequency support. *CSEE Journal of Power and Energy Systems*, *9*(4), 1594–1599. DOI: 10.17775/CSEEJPES.2021.08650

Gupta, B. B., Tewari, A., & Chaudhary, P. (2023). Blockchain-based identity management systems: A review. *Journal of Network and Computer Applications*, *202*, 103404. DOI: 10.1016/j.jnca.2023.103404

Gupta, J., Jain, S., Chakraborty, S., Panchenko, V., Smirnov, A., & Yudaev, I. (2023). Advancing Sustainable Energy Transition: Blockchain and Peer-to-Peer Energy Trading in India's Green Revolution. *Sustainability*, *15*(18), 13633. DOI: 10.3390/su151813633

Guttromson, R. T., Schur, A., Greitzer, F. L., & Paget, M. L. (2007). *Human factors for situation assessment in grid operations*. Pacific Northwest National Lab.(PNNL), Richland, WA (United States).

Gu, Y., & Green, T. C. (2022). Power system stability with a high penetration of inverter-based resources. *Proceedings of the IEEE*, *111*(7), 832–853.

Habib, S., Liu, H., Tamoor, M., Zaka, M. A., Jia, Y., Hussien, A. G., & Kamel, S. (2023). Technical modelling of solar photovoltaic water pumping system and evaluation of system performance and their socio-economic impact. *Heliyon*, *9*(5). PMID: 37229163

Hamed, S. B., Abid, A., Hamed, M. B., Sbita, L., Bajaj, M., Ghoneim, S. S., & Kamel, S. (2023). A robust MPPT approach based on first-order sliding mode for triple-junction photovoltaic power system supplying electric vehicle. *Energy Reports*, *9*, 4275–4297.

Han, Y., Lin, M., Li, N., Qi, Q., Li, J., & Liu, Q. (2024). DCWPSO: Particle swarm optimization with dynamic inertia weight updating and enhanced learning strategies. *PeerJ. Computer Science*, *10*, e2253. PMID: 39314689

Hassan, A., Ahmad, G., Shafiullah, M., Islam, A., & Alam, M. S. (2025). Review of the intelligent frameworks for pitch angle control in wind turbines. *IEEE Access : Practical Innovations, Open Solutions*, *13*, 29864–29885. DOI: 10.1109/ACCESS.2025.3540367

Hayder, W., Abid, A., Ben Hamed, M., & Sbita, L. (2020). Improved PSO algorithms in PV system optimisation. *European Journal of Electrical Engineering and Computer Science*, *4*(1), 1–6.

Hayder, W., Ogliari, E., Dolara, A., Abid, A., Ben Hamed, M., & Sbita, L. (2020). Improved PSO: A comparative study in MPPT algorithm for PV system control under partial shading conditions. *Energies*, *13*(8), 2035. Advance online publication. DOI: 10.3390/en13082035

Hayder, W., Séra, D., Ogliari, E., Lashab, A., Spataru, S., Guerrero, J. M., & Dolara, A. (2022). On improved PSO and neural network P&O methods for PV system under shading and various atmospheric conditions. *Energies*, *15*(20), 7668. Advance online publication. DOI: 10.3390/en15207668

He, W., Yan, J., Ortega, R., & Zonetti, D. (2023). A new adaptive phase-locked loop for synchronization of a grid-connected voltage source converter: Simulation and experimental results. *arXiv preprint arXiv:2309.08276*

Heier, S. (2014). *Grid integration of wind energy: Onshore and offshore conversion systems*. John Wiley & Sons., DOI: 10.1002/9781118703274

He, J., Huang, L., Wu, D., Zhu, C., & Xin, H. (2018). Frequency support from PMSG-based wind turbines with reduced DC-link voltage fluctuations. *CES Transactions on Electrical Machines and Systems*, *2*(3), 296–302.

Hilarydoss, S. (2023). Suitability, sizing, economics, environmental impacts and limitations of solar photovoltaic water pumping system for groundwater irrigation—A brief review. *Environmental Science and Pollution Research International*, *30*(28), 71491–71510. PMID: 33575942

Hmidet, A., Rebei, N., & Hasnaoui, O. (2015, March). Experimental studies and performance evaluation of MPPT control strategies for solar-powered water pumps. In 2015 Tenth international conference on ecological vehicles and renewable energies (EVER) (pp. 1-12). IEEE.

Hmidet, A., Subramaniam, U., Elavarasan, R. M., Raju, K., Diaz, M., Das, N., & Boubaker, O. (2021). Design of Efficient Off-Grid Solar Photovoltaic Water Pumping System Based on Improved Fractional Open Circuit Voltage MPPT Technique. *International Journal of Photoenergy*, *2021*(1), 4925433.

Hohm, D. P., & Ropp, M. E. (2003). Comparative study of maximum power point tracking algorithms. *Progress in Photovoltaics: Research and Applications*, *11*(1), 47–62.

Honari, K., Zhou, X., Rouhani, S., Dick, S., Liang, H., & Miller, J. (2022). A Scalable Blockchain-based Smart Contract Model for Decentralized Voltage Stability Using Sharding Technique. arXiv. https://arxiv.org/abs/2206.13776

Hongchun, SHU, H. D. (2025). Estimation of inertia demand in power systems considering wind power virtual inertia. *International Journal of Electrical Power & Energy Systems*, *167*, 110597.

Hossain, E., Hossain, J., Sakib, N., & Bayindir, R. (2017). Modelling and simulation of permanent magnet synchronous generator wind turbine: A step to microgrid technology. *International Journal of Renewable Energy Research*, *7*(1).

Hossain, M. I., Steigner, T., Hussain, M. I., & Akther, A. (2024). Enhancing Data Integrity and Traceability in Industry Cyber Physical Systems (ICPS) through Blockchain Technology: A Comprehensive Approach. *arXiv*. https://arxiv.org/abs/2405.04837

Hossain, M. M., Ortmeyer, T., & Hall, E. (2023). Human-centered design, development, and evaluation of an interface for a microgrid controller. *Advances in Science, Technology and Engineering Systems Journal*, *8*(3), 80–88. DOI: 10.25046/aj080310

Hua, C., & Lin, J. (2001). Fully digital control of distributed photovoltaic power systems. In *Proceedings of the IEEE International Symposium on Industrial Electronics* (pp. 1–6).

Hua, C., & Lin, J. (2003). An on-line MPPT algorithm for rapidly changing illuminations of solar arrays. *Renewable Energy*, *28*(7), 1129–1142.

Huang, J., & Koroteev, D. D. (2021). Artificial intelligence for planning of energy and waste management. *Sustainable Energy Technologies and Assessments*, *47*, 101426.

Huang, W.. (2022). A hybrid model-based diagnosis approach for open-switch faults in PMSM drives. *IEEE Transactions on Power Electronics*, *37*(4), 3728–3732.

Hudaya, R. (2015). Pengembangan Perangkat Lunak HMI/Scada Mandiri pada Lingkungan Networked Control Systems. *Prosiding Industrial Research Workshop and National Seminar*, *6*, 1–8.

Hunter, A. (2023). Interactions with AI systems: trust and transparency. *2023 IEEE Engineering Informatics*, 1–6.

Hussain, A., Eswarakrishnan, V., Aslam, A., & Tripura, S. (2025). Charging stations demand forecasting using LSTM based hybrid transformer model (pp. 1–27). *Scientific Reports*, *15*(1), 36639. Advance online publication. DOI: 10.1038/s41598-025-20421-y PMID: 41120429

Hussain, M. T., Sarwar, A., Tariq, M., & Urooj, S., BaQais, A., & Hossain, M. A. (2023). An evaluation of ANN algorithm performance for MPPT energy harvesting in solar PV systems. *Sustainability*, *15*(14), 11144.

Hwang, J. J.. (2013). Lifecycle performance assessment of fuel cell/battery electric vehicles. *International Journal of Hydrogen Energy*, *38*(8), 3433–3446. DOI: 10.1016/j.ijhydene.2012.12.148

Ibrahim, M. M. (2024). Energy management strategies of hybrid renewable energy systems: A review. *Wind Engineering*, *48*(1), 133–161. DOI: 10.1177/0309524X231200010

Ichim-Burlacu, C., Rat, C.-L., & Prostean, O. (2024). Buck-boost converter modeling and control based on its role within a microgrid. In *Proceedings of the 6th Global Power, Energy and Communication Conference (GPECOM)*, Budapest, Hungary. https://doi.org/DOI: 10.1109/GPECOM61896.2024.10582595

IEEE Standards Association. (2018). IEEE Std 1547-2018: IEEE standard for interconnection and interoperability of distributed energy resources.

Ilioudi, A. (2025). Deep-learning-based computer vision for underwater environments.

Irisawa, K., Saito, T., Hosokawa, K., & Takyu, S. (2000). Maximum power point tracking control of photovoltaic generation system under non-uniform insolation by means of monitoring cells. In *Proceedings of the IEEE Power Electronics Specialists Conference* (pp. 948–953).

Jafari, M., Gharehpetian, G. B., & Anvari-Moghaddam, A. (2024). On the role of virtual inertia units in modern power systems: A review of control strategies, applications and recent developments. *International Journal of Electrical Power & Energy Systems*, *159*, 110067.

Jain, S., Karampuri, R., & Chinthamalla, R. (2018). Solar PV Powered Standalone Water Pumping Systems: Modeling, Design, and Analysis. In Handbook of Research on Power and Energy System Optimization (pp. 139-190). IGI Global Scientific Publishing.

Jain, S., & Agarwal, V. (2007). Comparison of the performance of maximum power point tracking schemes applied to single-stage grid-connected photovoltaic systems. *IET Electric Power Applications*, *1*(5), 753–762.

Jander, H., & Alfredson, J. (2013). Human-machine interaction evaluation framework. *International Conference on Human-Computer Interaction*, 221–230.

Javaid, S., Javaid, N., & Alhussein, M. (2023). Towards efficient human–machine interaction for home energy management with seasonal scheduling using deep fuzzy neural optimizer. *Cognition Technology and Work*, *25*, 291–304. DOI: 10.1007/s10111-023-00733-5

Jiang, H., Tjandra, R., Soh, C. B., Cao, S., Soh, D. C. L., Tan, K. T., Tseng, K. J., & Krishnan, S. B. (2023). Digital twin of microgrid for predictive power control to buildings. *Sustainability*, *16*(2), 482. Advance online publication. DOI: 10.3390/su16020482

Jiang, J. A., Huang, T. L., Hsiao, Y. T., & Chen, C. H. (2005). Maximum power tracking for photovoltaic power systems. *Tamkang Journal of Science and Engineering*, *8*(2), 147–153.

Jiang, Z., Jia, Q.-S., & Guan, X. (2022). On large action space in EV charging scheduling optimization. *Science China. Information Sciences*, *65*(2), 122201. Advance online publication. DOI: 10.1007/s11432-020-3106-7

Jing, Y., Sun, H., Zhang, L., & Zhang, T. (2017). Variable speed control of wind turbines based on the quasi-continuous high-order sliding mode method. *Energies*, *10*(10), 1626. DOI: 10.3390/en10101626

Ji, Y. H., Jung, D. Y., Kim, J. G., Lee, J. H., Won, C. Y., & Yoo, D. W. (2011). A real maximum power point tracking method for mismatching compensation in PV array under partially shaded conditions. *IEEE Transactions on Power Electronics*, *26*(4), 1001–1009.

Johannsen, G. (2009). Human-machine interaction. *Control Systems. Robotics and Automation*, *21*, 132–162.

Johnson, S. C., Rhodes, J. D., & Webber, M. E. (2020). Understanding the impact of non-synchronous wind and solar generation on grid stability and identifying mitigation pathways. *Applied Energy*, *262*, 114492.

Joshu, S. D., & Talange, B. D. (2013, July). Fractional order modeling and control of depth system of AUV a case study. In *2nd International Conference on Advances in Computer Science and Engineering (CSE 2013)* (pp. 164-168). Atlantis Press.

Kadiyam, J., Mohan, S., Deshmukh, D., & Seo, T. (2021). Simulation-based semi-empirical comparative study of fixed and vectored thruster configurations for an underwater vehicle. *Ocean Engineering*, *234*, 109231.

Kalikatzarakis, M., Coraddu, A., Oneto, L., & Anguita, D. (2021). Optimizing fuel consumption in thrust allocation for marine dynamic positioning systems. *IEEE Transactions on Automation Science and Engineering*, *19*(1), 122–142.

Kang, T. (2020). BIM-based human machine interface (HMI) framework for energy management. *Sustainability*, *12*(21), 8861. Advance online publication. DOI: 10.3390/su12218861

Karami, M., Mariun, N. B., Ab-Kadir, M. Z. A., Misron, N., & Mohd Radzi, M. A. (2021). Motor current signature analysis-based non-invasive recognition of mixed eccentricity fault in line start permanent magnet synchronous motor. *Electric Power Components and Systems*, *49*(1-2), 133–145.

Katche, M. L., Makokha, A. B., Zachary, S. O., & Adaramola, M. S. (2023). A comprehensive review of maximum power point tracking (mppt) techniques used in solar pv systems. *Energies*, *16*(5), 2206.

Kaufmann, A., Schempp, T., Stoehr, I., Schmid, M., & Maier, T. (2019). The Development of an Adaptive HMI-From the Idea to the Prototype. *International Conference on Applied Human Factors and Ergonomics*, 142–150.

Kazemi, R., & Lee, S. C. (2024). Human Factors/Ergonomics (HFE) evaluation in the virtual reality environment: A systematic review. *International Journal of Human-Computer Interaction*, *40*(17), 4533–4549.

Kerdphol, T., Rahman, F. S., Mitani, Y., Hongesombut, K., & Küfeoğlu, S. (2017). Virtual inertia control-based model predictive control for microgrid frequency stabilization considering high renewable energy integration. *Sustainability*, *9*(5), 773.

Kerdphol, T., Rahman, F. S., Phunpeng, V., Watanabe, M., & Mitani, Y. (2018). Demonstration of Virtual Inertia Emulation Using Energy Storage Systems to Support Community-Based High Renewable Energy Penetration. *IEEE Global Humanitarian Technology Conference (GHTC)*, (pp. 1-7). San Jose, CA, USA.

Kerdphol, T., Rahman, F. S., Watanabe, M., Mitani, Y., Turschner, D., & Beck, H. P. (2019). Enhanced virtual inertia control based on derivative technique to emulate simultaneous inertia and damping properties for microgrid frequency regulation. *IEEE Access: Practical Innovations, Open Solutions*, *7*, 14422–14433.

Khalid, M. (2024). Smart grids and renewable energy systems: Perspectives and grid integration challenges. *Energy Strategy Reviews*, *51*, 101299.

Khalid, M., & Mohan, A. (2020, October). Performance analysis of vector controlled PMSM drive modulated with sinusoidal PWM and space vector PWM. *In 2020 IEEE International Power and Renewable Energy Conference* (pp. 1-6). IEEE.

Khan, M. W., Saad, S., Ammad, S., Rasheed, K., & Jamal, Q. (2024). Smart infrastructure and AI. In *AI in Material Science* (pp. 193–215). CRC Press.

Khan, P. W., & Byun, Y.-C. (2021). Blockchain-Based Peer-to-Peer Energy Trading and Charging Payment System for Electric Vehicles. *Sustainability, 13*(14), 7962. DOI: 10.3390/su13147962

Khaskheli, S., & Anvari-Moghaddam, A. (2024). Energy Trading in Local Energy Markets: A Comprehensive Review of Models, Solution Strategies, and Machine Learning Approaches. *Applied Sciences (Basel, Switzerland), 14*(24), 11510. DOI: 10.3390/app142411510

Khlaief, A. (2012). Contribution à la commande vectorielle sans capteur mécanique des machines synchrones à aimants permanents (MSAP) (Doctoral dissertation, Aix-Marseille Université).

Khov, M. (2009). Surveillance et diagnostic des machines synchrones à aimants permanents: Détection des courts-circuits par suivi paramétrique (Doctoral dissertation, Institut National Polytechnique de Toulouse-INPT).

Kiessling, J., Ström, E., & Tempone, R. (2021). Wind field reconstruction with adaptive random Fourier features. *Proceedings - Royal Society. Mathematical, Physical and Engineering Sciences, 477*(2252), 20210236. DOI: 10.1098/rspa.2021.0236

Kim, S. S. Y. (2024). Establishing appropriate trust in AI through transparency and explainability. *Extended Abstracts of the CHI Conference on Human Factors in Computing Systems*, 1–6.

Kiran, R. R., & Reddy, S. S. (2022). Human machine interface. *International Journal of Research Publication and Reviews, 3*(7), 3118–3120.

Kirsch, D. A. (1995). History of the Electric Automobile: Battery-Only Powered Cars by Ernest H. Wakefield. *Technology and Culture, 36*(3), 710–712.

Kjaer, S. B., Pedersen, J. K., & Blaabjerg, F. (2005). A review of single-phase grid-connected inverters for photovoltaic modules. *IEEE Transactions on Industry Applications, 41*(5), 1292–1306. DOI: 10.1109/TIA.2005.853371

Korneev, A., Niu, Y., & Ibrahim, A. (2024). Electric vehicles in the 21st century: Historical evolution, environmental impact, and safety challenges for sustainable mobility. *Khwarizmia, 2024*, 32–38.

Kottas, T. L., Boutalis, Y. S., & Karlis, A. D. (2006). New maximum power point tracker for PV arrays using fuzzy controller in close cooperation with fuzzy cognitive networks. *IEEE Transactions on Energy Conversion, 21*(3), 793–803.

Koutroulis, E., Kalaitzakis, K., & Voulgaris, N. C. (2001). Development of a microcontroller-based, photovoltaic maximum power point tracking control system. *IEEE Transactions on Power Electronics*, *16*(1), 46–54. DOI: 10.1109/63.903988

Krichen, M., Elbouchikhi, E., Benhadj, N., Chaieb, M., Benbouzid, M., & Neji, R. (2020). Motor current signature analysis-based permanent magnet synchronous motor demagnetization characterization and detection. *Machines (Basel)*, *8*(3), 35.

KS. (2023). A.K.; Kadiyam, J.; Mohan, S. Comparative performance investigations of the intervention-class underwater vehicle with different possible thruster configurations using eight identical thrusters. *Ocean Engineering*, *288*, 116147.

Kumar, T. (2025). Blockchain in EVs – Securing India's Electric Future with Cutting-Edge Tech. *Green Bharat News*. https://greenbharatnews.com/blockchain-in-evs

Kumar, V., Vyas, S. A. S., & Kumar, R. (2021). Blockchain Technology for Decentralized Energy Transaction in a Smart Grid. IEEE Smart Grid Bulletin. https://smartgrid.ieee.org/bulletins/november-2021/blockchain-technology-for-decentralized-energy-transaction-in-a-smart-grid

Kumar, A. (2024). A comprehensive review of an electric vehicle based on the existing technologies and challenges. *Energy Storage*, *6*(5), e70000.

Kumar, A., & Alaraj, M. (2021). Novel AI based energy management system for smart grid with RES integration. *IEEE Access: Practical Innovations, Open Solutions*, *9*, 162530–162542.

Kumar, P. S., Chandrasena, R. P. S., Ramu, V., Srinivas, G. N., & Babu, K. V. S. M. (2020). Energy management system for small scale hybrid wind solar battery based micro grid. *IEEE Access: Practical Innovations, Open Solutions*, *8*, 8336–8344. DOI: 10.1109/ACCESS.2020.2964538

Kumar, R., & Sharma, S. (2023). Quantum-resilient blockchain frameworks for secure energy trading in India. *Energy Policy*, *170*, 113–127. DOI: 10.1016/j.enpol.2022.114567

Kumar, R., Sharma, S., & Patel, P. (2023). Regulatory challenges and decentralized governance models for blockchain in energy markets. *Energy Policy*, *160*, 112–123. DOI: 10.1016/j.enpol.2022.112123

Kumar, R., & Singh, A. (2025). A blockchain consortium-based framework to enhance interoperability, security, and standardized energy trading mechanisms within smart grid ecosystems. *Energy Reports*, *11*, 1234–1245. DOI: 10.1016/j.egyr.2025.02.014

Kumar, R., Singh, M., Khan, S., Singh, J., Sharma, S., Kumar, H., & Aggarwal, V. (2023). A state-of-the-art review on the misalignment, failure modes and its detection methods for bearings. *Mapan, 38*(1), 265–274.

Kundu, S., Giri, A. K., & Singh, M. (2024). Adaptive control strategy for isolated renewable energy-based generation system with intermittency. *Electrical Engineering*. Advance online publication. DOI: 10.1007/s00202-024-01345-6

Lai, Y., Fan, L., & Zheng, W. (2024). Construction of a digital twin model for incremental aggregation of multi type load information in hybrid microgrids under integrity constraints. *Energy Informatics, 7*, 112. Advance online publication. DOI: 10.1186/s42162-024-00345-8

Lammouchi, Z., Naoui, M., Bekakra, Y., Khechekhouche, A., & Sbita, L. (2025). Model-free predictive control based on the execution time of active vectors using ultra-local model. *Arabian Journal for Science and Engineering, 50*(8), 5885–5898.

Lammouchi, Z., Naoui, M., Bekakra, Y., Khechekhouche, A., & Sbita, L. (2025). Model-Free Predictive Control based on the Execution Time of Active Vectors Using Ultra-Local Model. *Arabian Journal for Science and Engineering, 50*(8), 5885–5898.

Langer, M., Baum, K., & Schlicker, N. (2024). Effective Human Oversight of AI-Based Systems: A Signal Detection Perspective on the Detection of Inaccurate and Unfair Outputs. *Minds and Machines, 35*(1), 1–19. DOI: 10.1007/s11023-024-09701-0

Laux, J. (2024). Institutionalised distrust and human oversight of artificial intelligence: Towards a democratic design of AI governance under the European Union AI Act. *AI & Society, 39*, 2853–2866. DOI: 10.1007/s00146-023-01777-z PMID: 39640298

Lau, Y. Y., Wu, A. Y., & Yan, M. W. (2022). A way forward for electric vehicle in Greater Bay Area: Challenges and opportunities for the 21st century. *Vehicles (Basel), 4*(2), 420–432.

Lee-Pearce, M. L., Plowman, T. S., & Touchstone, D. (1998). Starbase-Atlantis, a school without walls: A comparative study of an innovative science program for at-risk urban elementary students. *Journal of Education for Students Placed at Risk, 3*(3), 223–235.

Liao, J.-T., Huang, H.-W., Yang, H.-T., & Li, D. (2021). Decentralized V2G/G2V scheduling of EV charging stations by considering the conversion efficiency of bidirectional chargers. *Energies, 14*(4), 962. Advance online publication. DOI: 10.3390/en14040962

Li, G., Jin, Y., Akram, M., & Chen, X. (2017). Research and current status of the solar photovoltaic water pumping system–a review. *Renewable & Sustainable Energy Reviews*, *79*, 440–458.

Li, L., Cao, S., Li, J., Nie, R., & Hou, L. (2021). Review of rotor balancing methods. *Machines (Basel)*, *9*(5), 89.

Lin, H.-Y., & Biggio, B. (2021). Adversarial machine learning: Attacks from laboratories to the real world. *Computer*, *54*(5), 56–60.

Liu, W., Teodorescu, R., Kerekes, T., & Dragicevic, T. (2022b, October). Advanced power synchronization control of modular multilevel converter in stiff grid. In *IECON 2022–48th Annual Conference of the IEEE Industrial Electronics Society* (pp. 1-6). IEEE.

Liu, D., Zhang, X., & Chi, K. T. (2022a). Effects of high level of penetration of renewable energy sources on cascading failure of modern power systems. *IEEE Journal on Emerging and Selected Topics in Circuits and Systems*, *12*(1), 98–106.

Liu, F., Duan, S., Liu, F., Liu, B., & Kang, Y. (2008). Variable step size INC MPPT method for PV systems. *IEEE Transactions on Industrial Electronics*, *55*(7), 2622–2628. DOI: 10.1109/TIE.2008.924035

Liu, F., Xu, W., Feng, Z., Yu, C., Liang, X., Su, Q., & Gao, J. (2025). Task Allocation and Path Planning Method for Unmanned Underwater Vehicles. *Drones (Basel)*, *9*(6), 411.

Liu, Z., Zhang, P., He, S., & Huang, J. (2021). A review of modeling and diagnostic techniques for eccentricity fault in electric machines. *Energies*, *14*(14), 4296.

Li, W.. (2016). Comparison study on SS and double-sided LCC compensation topologies for EV/PHEV wireless chargers. *IEEE Transactions on Vehicular Technology*, *65*(6), 4429–4439.

Lulhe, A. M., & Date, T. N. (2016). A technology review paper for drives used in electrical vehicle (EV) & hybrid electrical vehicles (HEV). *2015 International Conference on Control Instrumentation Communication and Computational Technologies (ICCICCT 2015)*, 632–636.

Lyden, S., & Haque, M. E. (2016). A simulated annealing global maximum power point tracking approach for PV modules under partial shading conditions. *IEEE Transactions on Power Electronics*, *31*(6), 4171–4181.

Majout, B., El Alami, H., Salime, H., Laabidine, N. Z., El Mourabit, Y., Motahhir, S., Bouderbala, M., Karim, M., & Bossoufi, B. (2022). A review on popular control applications in wind energy conversion system based on permanent magnet generator PMSG. *Energies, 15*(17), 6238. DOI: 10.3390/en15176238

Maka, A. O., Mehmood, M., & Chaudhary, T. N. (2023). Design, simulation and performance analysis of photovoltaic solar water pumping system. *Energy Harvesting and Systems, 10*(2), 287–299.

Malatji, M., Marnewick, A. L., & Von Solms, S. (2022). Cybersecurity capabilities for critical infrastructure resilience. *Information and Computer Security, 30*(2), 255–279.

Manwell, J. F., McGowan, J. G., & Rogers, A. L. (2009). *Wind energy explained: Theory, design and application* (2nd ed.). John Wiley & Sons.

Martinie, C., Navarre, D., Palanque, P., Barboni, E., & Steere, S. (2022). Engineering operations-based training. *Proceedings of the ACM on Human-Computer Interaction, 6*(EICS), 1–25.

Martyushev, N. V., Malozyomov, B. V., Sorokova, S. N., Efremenkov, E. A., & Qi, M. (2023). Mathematical modeling the performance of an electric vehicle considering various driving cycles. *Mathematics, 11*(11), 2586.

Marwa, M. E., Doaa, M. A., Aref, Y. E., Belal, A. A. Z., Mahmoud, S., & Abou, O. (2025). Multijunction solar cell parameter estimation based on metaheuristic algorithms. *Results in Engineering, 25*, 1–12.

Marzebali, M. H., Mazidi, M., & Mohiti, M. (2020). An adaptive droop-based control strategy for fuel cell-battery hybrid energy storage system to support primary frequency in stand-alone microgrids. *Journal of Energy Storage, 27*, 101127. Advance online publication. DOI: 10.1016/j.est.2019.101127

Masoum, M. A. S., Dehbonei, H., & Fuchs, E. F. (2002). Theoretical and experimental analyses of photovoltaic systems with voltage and current-based maximum power-point tracking. *IEEE Transactions on Energy Conversion, 17*(4), 514–522.

Masoum, M. A. S., & Sarvi, M. (2011). A new fuzzy-based combined variable step size INC and P&O maximum power point tracking algorithm for photovoltaic systems. In *Proceedings of the IEEE Power and Energy Society General Meeting* (pp. 1–8).

Mathew, R., & Panikkar, P. P. K. (2024). Inertial issues in renewable energy integrated systems and virtual inertia techniques. [IJPEDS]. *International Journal of Power Electronics and Drive Systems, 15*(1), 466.

Matsuda, Y., Yamazaki, Y., Oki, H., Takeda, Y., Sagawa, D., & Tanaka, K. (2021). Demonstration of Blockchain Based Peer to Peer Energy Trading System with Real-Life Used PHEV and HEMS Charge Control. *Energies*, *14*(22), 7484. DOI: 10.3390/en14227484

Mattiasson, C. (1999). The alarm system from the operator's perspective. *1999 International Conference on Human Interfaces in Control Rooms, Cockpits and Command Centres*, 217–221.

Maulidhia, A. N. A., Asfani, D. A., Priyadi, A., & Setiadi, H. (2022, November). Frequency stability analysis on optimization of virtual inertia control (VIC) capacitor energy storage (CES) controller settings using particle swarm optimization. In *2022 international conference on Computer engineering, network, and intelligent Multimedia (CENIM)* (pp. 359-364). IEEE.

Maurya, V. K., Kothari, R., Sachdev, P., & Rathore, N. S. (2024). Blockchain Technology for Secure Smart Grid Access Control. In Shrivastava, V., Bansal, J. C., & Panigrahi, B. K. (Eds.), *Power Engineering and Intelligent Systems. PEIS 2023. Lecture Notes in Electrical Engineering, 1097*. Springer., DOI: 10.1007/978-981-99-7216-6_28

Mazloumi, A., Poolad, A., Mokhtari, M. S., Altman, M. B., Abdelaziz, A. Y., & Elsisi, M. (2023). Optimal sizing of a photovoltaic pumping system integrated with water storage tank considering cost/reliability assessment using enhanced artificial rabbits optimization: A case study. *Mathematics*, *11*(2), 463.

McMahon, J., & Plaku, E. (2016). Mission and motion planning for autonomous underwater vehicles operating in spatially and temporally complex environments. *IEEE Journal of Oceanic Engineering*, *41*(4), 893–912.

Md Tonmoy Hossain, M. Z. (2024). Next generation power inverter for grid resilience: Technology review. *heliyon, 10*(21), e39596.

Meddeb, H., Abdellaoui, Z., & Ezzine, O. (2023). Human machine interface design of industrial automated machine using Simatic SCADA system. *International Journal of Industrial Engineering*, *30*(2), 435–450.

Meenakshi, V., Kaveri, V. V., Antony, A. S. M., Radhika, S., Kavitha, M., & Pushpavalli, M. (2024). IoT enhanced optimization of wind energy systems using LCL filter. In *Proceedings of the International Conference on Intelligent Computing and Control Systems (ICICCS)*. IEEE. https://doi.org/DOI: 10.1109/ACCAI61061.2024.10601921

Mehmood, M. U., Chun, D., Han, H., Jeon, G., & Chen, K. (2019). A review of the applications of artificial intelligence and big data to buildings for energy-efficiency and a comfortable indoor living environment. *Energy and Building*, *202*, 109383.

Mellit, A., & Kalogirou, S. A. (2014). MPPT-based artificial intelligence techniques for photovoltaic systems and its implementation into field programmable gate array chips: Review of current status and future perspectives. *Energy*, *70*, 1–18. DOI: 10.1016/j.energy.2014.03.070

Meng, L., & Xue, D. (2009, June). Design of an optimal fractional-order PID controller using multi-objective GA optimization. In *2009 Chinese Control and Decision Conference* (pp. 3849-3853). IEEE.

Miguel Angel, R. (2021). Quo Vadis Solar Energy Research? *Applied Sciences (Basel, Switzerland)*, *11*(30), 1–11.

Mijwil, M., Doshi, R., Hiran, K. K., Al-Mistarehi, A.-H., & Gök, M. (2022). Cybersecurity challenges in smart cities: An overview and future prospects. *Mesopotamian Journal of Cybersecurity*, *2022*, 1–4.

Mohamed, M. J., & Khashan, A. (2014, February). Comparison between PID and FOPID controllers based on particle swarm optimization. In *The Second Engineering Conference of Control, Computers and Mechatronics Engineering (ECCCM2, 2014)*.

Mohamed, S., Mohsen, Z., Miral, S., & Islam, M. (2024). Optimized and Sustainable PV Water Pumping System with Three-Port Converter, a Case Study: The Al-Kharijah Oasis. *Electricity*, *5*(2), 227–253.

Mohrehkesh, S., & Nadeem, T. (2011). Toward a wireless charging for battery electric vehicles at traffic intersections. *IEEE Conference on Intelligent Transportation Systems, Proceedings, ITSC*, 113–118.

Moniruzzaman, M., Islam, S. M. S., Bennamoun, M., & Lavery, P. (2017, September). Deep learning on underwater marine object detection: A survey. In *International Conference on Advanced Concepts for Intelligent Vision Systems* (pp. 150-160). Cham: Springer International Publishing.

Morgan, D., & Taylor, B. (2011). *U-Boat Attack Logs: A Complete Record of Warship Sinkings from Original Sources 1939-1945*. Seaforth Publishing.

Motahhir, S., El Hammoumi, A., & El Ghzizal, A. (2018). Photovoltaic system with quantitative comparative between an improved MPPT and existing INC and P&O methods under fast varying of solar irradiation. *Energy Reports*, *4*, 341–350.

Mourtzis, D., Angelopoulos, J., & Panopoulos, N. (2023). The future of the human–machine interface (HMI) in society 5.0. *Future Internet, 15*(5), 162.

Moutchou, R., & Abbou, A. (2021). Control of grid side converter in wind power-based PMSG with PLL method. *International Journal of Power Electronics and Drive Systems, 12*(4), 2191–2200. DOI: 10.11591/ijpeds.v12.i4.pp2191-2200

Muhsen, D., Khatib, T., & Nagi, F. (2017). A review of photovoltaic water pumping system designing methods, control strategies and field performance. *Renewable & Sustainable Energy Reviews, 68*, 70–86.

Naoui, M.. (2023). Intelligent control system for hybrid electric vehicle with autonomous charging. In Azar, A. T., & Koubaa, A. (Eds.), *Artificial intelligence for robotics and autonomous systems applications* (pp. 405–437). Springer., DOI: 10.1007/978-3-031-28715-2_13

Naoui, M., Aymen, F., Mouna, B. H., & Sbita, L. (2021). Brushless motor and wireless recharge system for electric vehicle design modeling and control. In *Handbook of research on modeling, analysis, and control of complex systems* (pp. 338–362). IGI Global Scientific Publishing.

Naoui, M., Flah, A., & Alqarni, M.. (2022). A new wireless charging system for electric vehicles using two receiver coils. *Ain Shams Engineering Journal, 13*(2). Advance online publication. DOI: 10.1016/j.asej.2021.08.012

Naoui, M., Flah, A., Altamimi, A., & Khan, Z. A. (2022). Power management and control of a hybrid electric vehicle based on photovoltaic, fuel cells, and battery energy sources. *Sustainability, 14*(5).

Naoui, M., Flah, A., & Ben Hamed, M. (2018). Wireless charging system for a mobile hybrid electric vehicle. *International Symposium on Advanced Electrical and Communication Technologies (ISAECT 2018)*, 1–5.

Naoui, M., Flah, A., Hamed, M. B., & Sbita, L. (2017). Review on autonomous charger for EV and HEV. *International Conference on Green Energy Conversion Systems (GECS)*, 1–6.

Naoui, M., Flah, A., & Sbita, L. (2023). Efficiency of intelligent control design for dual-stator machine on the EV traction system. *Arabian Journal for Science and Engineering*. Advance online publication. DOI: 10.1007/s13369-023-08070-7

Nasiri, M., Mobayen, S., Faridpak, B., Fekih, A., & Chang, A. (2020). Small-signal modeling of PMSG-based wind turbine for low voltage ride-through and artificial intelligent studies. *Energies, 13*(24), 6685. DOI: 10.3390/en13246685

Natarajan, S., Mathur, S., Sidheekh, S., Stammer, W., & Kersting, K. (2025). Human-in-the-loop or AI-in-the-loop? Automate or Collaborate? *Proceedings of the AAAI Conference on Artificial Intelligence, 39*(27), 28594–28600.

Nature Research. (2024). Blockchain-based energy consumption approaches in IoT. *Scientific Reports*. Advance online publication. DOI: 10.1038/s41598-024-77792-x PMID: 39543190

Ngamroo, I. (2017). Review of DFIG wind turbine impact on power system dynamic performances. *IEEJ Transactions on Electrical and Electronic Engineering, 12*(4), 401–411. DOI: 10.1002/tee.22379

Ng, K. W., & Tong, H. Y. (2024). Comparisons of Driving Characteristics between Electric and Diesel-Powered Bus Operations along Identical Bus Routes. *Sustainability, 16*(12), 4950.

Niforos, M., & Rojas, F. (2021). Blockchain governance and regulation as an enabler for market creation in emerging markets. *World Bank Policy Research Working Paper*, (9732). https://doi.org/DOI: 10.1596/1813-9450-9732

Ocaka, A., Briain, D. O., Davy, S., & Barrett, K. (2022). Cybersecurity threats, vulnerabilities, mitigation measures in industrial control and automation systems: a technical review. *2022 Cyber Research Conference-Ireland (Cyber-RCI)*, 1–8.

Oury, J. D., & Ritter, F. E. (2021). Cognition and Operator Performance. *Building Better Interfaces for Remote Autonomous Systems: An Introduction for Systems Engineers*, 37–62.

Özçiflikçi, O. E., Koç, M., Bahçeci, S., & Emiroğlu, S. (2024). Overview of PMSM control strategies in electric vehicles: A review. *International Journal of Dynamics and Control, 12*(6), 2093–2107.

Pandey, A., Dasgupta, N., & Mukerjee, A. K. (2023). A review on recent trends and developments in MPPT techniques for solar PV systems under partial shading conditions. *Renewable & Sustainable Energy Reviews, 172*, 113036. Advance online publication. DOI: 10.1016/j.rser.2022.113036

Park, I. H., Yoon, J. S., Sohn, J. H., & Lee, D. Y. (2024). Platform supporting intelligent human–machine interface (HMI) applications for smart machine tools. *International Journal of Precision Engineering and Manufacturing*. Advance online publication. DOI: 10.1007/s12541-024-00987-5

Patel, P., Reddy, S., & Verma, A. (2024). AI and blockchain convergence in decentralized energy systems: Insights from Indian pilot projects. *Renewable & Sustainable Energy Reviews, 182*, 113–132. DOI: 10.1016/j.rser.2024.114321

Patra, B., Nema, P., Khan, M. Z., & Khan, O. (2023). Optimization of solar energy using MPPT techniques and industry 4.0 modelling. *Sustainable Operations and Computers*, *4*, 22–28.

Paul, G. E. (2021). Modeling and simulation of human systems. *Handbook of human factors and ergonomics*, 704-735.

Pelekis, S., Karakolis, E., Lampropoulos, G., Mouzakitis, S., Markaki, O., Ntanos, C., & Askounis, D. (2024). Trustworthy artificial intelligence in the energy sector: Landscape analysis and evaluation framework. *2024 IEEE International Conference on Engineering, Technology, and Innovation (ICE/ITMC)*, 1–10.

Pfitzmann, A., & Hansen, M. (2010). A terminology for talking about privacy by data minimization: Anonymity, Unlinkability, Undetectability, Unobservability, Pseudonymity, and Identity Management. *University of Dresden.* Retrieved from http://dud.inf.tu-dresden.de/Anon_Terminology.shtml

Pham, T., Boone, A. P., & Ngo, M. K. (2023). Heuristic evaluation of supervisory control and data acquisition (SCADA) displays. *Proceedings of the Human Factors and Ergonomics Society Annual Meeting*, *67*(1), 102–107. DOI: 10.1177/21695067231193745

Pham, T., Boone, A. P., & Ngo, M. K. (2023). Heuristic Evaluation of Supervisory Control and Data Acquisition (SCADA) Displays. *Proceedings of the Human Factors and Ergonomics Society Annual Meeting*, *67*(1), 1767–1774.

Piccini, M. (2002). Human factors in the design of supervisory control systems and human–machine interfaces for highly automated complex systems. *Cognition Technology and Work*, *4*(4), 256–271.

Pietrzak, P., & Wolkiewicz, M. (2022). Machine learning-based stator current data-driven PMSM stator winding fault diagnosis. *Sensors (Basel)*, *22*(24), 9668. PMID: 36560037

Podder, T. K., & Sarkar, N. (2001). Fault-tolerant control of an autonomous underwater vehicle under thruster redundancy. *Robotics and Autonomous Systems*, *34*(1), 39–52.

Pop, T., Buzduga, C., Pentiuc, R. D., Ifrim, V. C., & Ungureanu, C. (2022, October). Mathematical modeling in MATLAB of a photovoltaic panel. []. IOP Publishing.]. *IOP Conference Series. Materials Science and Engineering*, *1256*(1), 012037.

Pradhan, A., & Panda, B. (2018). A simplified design and modeling of boost converter for photovoltaic sytem. *International Journal of Electrical and Computer Engineering*, *8*(1), 141.

Prevost, M.-C., Aubin, F., Gauthier, J., & Mailhot, R. (2007). Preventing human errors in power grid management systems through user-interface redesign. *2007 IEEE International Conference on Systems, Man and Cybernetics*, 626–631.

Radjai, T., Rahmani, L., Mekhilef, S., & Gaubert, J. P. (2014). Implementation of a modified perturb and observe MPPT algorithm with direct control based on a fuzzy duty cycle change estimator using dSPACE. *Solar Energy, 110*, 325–337.

Raghav, Y. Y., & Kait, R. (2024). Edge computing empowering distributed computing at the edge. In *Emerging Trends in Cloud Computing Analytics, Scalability, and Service Models* (pp. 67–83). IGI Global Scientific Publishing.

Rahimi, M. (2017). Modeling, control and stability analysis of grid-connected PMSG-based wind turbine assisted with diode rectifier and boost converter. *International Journal of Electrical Power & Energy Systems, 93*, 84–96. DOI: 10.1016/j.ijepes.2017.05.019

Rahman, M., Chowdhury, S., Shorfuzzaman, M., Hossain, M. K., & Hammoudeh, M. (2023). Peer-to-Peer Power Energy Trading in Blockchain Using Efficient Machine Learning Model. *Sustainability, 15*(18), 13640. DOI: 10.3390/su151813640

Rajkumar, V. S., Ştefanov, A., Presekal, A., Palensky, P., & Torres, J. L. R. (2023). Cyber attacks on power grids: Causes and propagation of cascading failures. *IEEE Access: Practical Innovations, Open Solutions, 11*, 103154–103176.

Rakesh, P., Shruti, J. R., Thippeswamy, I., Nithya, B. L., & Dheeraj, V. (2024). A Surrogate Approach to Explainable AI for Predictive Maintenance: Techniques and Applications. *2024 5th International Conference on Circuits, Control, Communication and Computing (I4C)*, 160–165.

Raney, A., Bendelac, S., Manville, K., Tan, M., & Yamaguchi, K. (2024). An AI red team playbook. *Assurance and Security for AI-Enabled Systems, 13054*, 71–97.

Rath, S., Nguyen, L. D., Sahoo, S., & Popovski, P. (2022). Self-healing secure blockchain framework in microgrids. *IEEE Transactions on Industrial Informatics, 18*(5), 3456–3465. DOI: 10.1109/TII.2022.3156789

Reed, D. A., Zabinsky, Z. B., & Boyle, L. N. (2011). A framework for optimizing civil infrastructure resiliency. *Structures Congress 2011*, 2104–2112.

Reijsbergen, D., Maw, A., Dinh, T. T. A., Li, W.-T., & Yuen, C. (2022). Securing Smart Grids Through an Incentive Mechanism for Blockchain-Based Data Sharing. *Energy Reports, 8*, 1320. DOI: 10.1016/j.egyr.2024.08.060

Renewable and Sustainable Energy Reviews. (2022). Smart contracts in energy systems: A systematic review of fundamental approaches and implementations. *Renewable and Sustainable Energy Reviews, 158*, 112013. https://doi.org/DOI: 10.1016/j.rser.2021.112013

Rethfeldt, C., & Jeinsch, T. (2024, September). Combined Actuator Allocation for Underwater Vehicles with Variable Buoyancy Systems Using QP-Based Optimization. In *OCEANS 2024-Halifax* (pp. 1–7). IEEE.

Romdhane, M., Naoui, M., & Mansouri, A. (2023). PMSM Inter-Turn Short Circuit Fault Detection Using the Fuzzy-Extended Kalman Filter in Electric Vehicles. *Electronics (Basel), 12*(18), 3758.

Romdhane, M., Naoui, M., & Mansouri, A. (2023). PMSM inter-turn short circuit fault detection using the fuzzy-extended Kalman filter in electric vehicles. *Electronics (Basel), 12*(18), 3758. Advance online publication. DOI: 10.3390/electronics12183758

Rosero, J., Romeral, L., Rosero, E., & Urresty, J. (2009, February). Fault Detection in dynamic conditions by means of Discrete Wavelet Decomposition for PMSM running under Bearing Damage. *In 2009 Twenty-Fourth Annual IEEE Applied Power Electronics Conference and Exposition* (pp. 951-956). IEEE.

Ruiz, J. R. R., Rosero, J. A., Espinosa, A. G., & Romeral, L. (2009). Detection of demagnetization faults in permanent-magnet synchronous motors under nonstationary conditions. *IEEE Transactions on Magnetics, 45*(7), 2961–2969.

Safari, A., & Mekhilef, S. (2011). Incremental conductance MPPT method for PV systems. In *Proceedings of the IEEE Applied Power Electronics Conference* (pp. 1147–1152).

Safari, A., & Mekhilef, S. (2011). Simulation and hardware implementation of incremental conductance MPPT with direct control method using cuk converter. *IEEE Transactions on Industrial Electronics, 58*(4), 1154–1161. DOI: 10.1109/TIE.2010.2048834

Sahuquet, A., Meunier, S., Cherni, J. A., Charpentier, A., Vezin, T., Darga, A., & Quéval, L. (2024). Photovoltaic pumping tests: A novel supervision method for photovoltaic water pumping systems. *Heliyon, 10*(21). PMID: 39553633

Sajid, A., Abbas, H., & Saleem, K. (2016). Cloud-assisted IoT-based SCADA systems security: A review of the state of the art and future challenges. *IEEE Access : Practical Innovations, Open Solutions, 4*, 1375–1384.

Salah, L., Franck, P., & Alain, G. (2007). Higher order sliding mode control based on integral sliding mode. *Automatica*, *43*(3), 531–537.

Salas, V., Olias, E., Barrado, A., & Lázaro, A. (2006). Review of the maximum power point tracking algorithms for stand-alone photovoltaic systems. *Solar Energy Materials and Solar Cells*, *90*(11), 1555–1578.

Sallán, J., Villa, J. L., Llombart, A., & Sanz, J. F. (2009). Optimal design of ICPT systems applied to electric vehicle battery charge. *IEEE Transactions on Industrial Electronics*, *56*(6), 2140–2149.

Salvatti, G. A., Carati, E. G., Cardoso, R., da Costa, J. P., & Stein, C. M. O. (2020). Electric vehicles energy management with V2G/G2V multifactor optimization of smart grids. *Energies*, *13*(5), 1191. Advance online publication. DOI: 10.3390/en13051191

Sandrini, G., Chindamo, D., Gadola, M., Candela, A., & Magri, P. (2024). Exploring the impact of vehicle lightweighting in terms of energy consumption: Analysis and simulation on real driving cycle. *Energies*, *17*(24), 6398.

Sankhwar, P. (2024). Application of permanent magnet synchronous motor for electric vehicle. *Application of Permanent Magnet Synchronous Motor for Electric Vehicle*, *4*(2), 1–6.

Sarkar, N., Podder, T. K., & Antonelli, G. (2002). Fault-accommodating thruster force allocation of an AUV considering thruster redundancy and saturation. *IEEE Transactions on Robotics and Automation*, *18*(2), 223–233.

Sawyer, S., & Tapia, A. (2005). The sociotechnical nature of mobile computing work: Evidence from a study of policing in the United States. *International Journal of Technology and Human Interaction*, *1*(3), 1–14.

Schillai, S. M., Turnock, S. R., Rogers, E., & Phillips, A. B. (2019). Experimental analysis of low-altitude terrain following for hover-capable flight-style autonomous underwater vehicles. *Journal of Field Robotics*, *36*(8), 1399–1421.

Segalas, J. (n.d.). Ethical Governance of AI: An Integrated Approach via Human-in-the-Loop. *Machine Learning*.

Sera, D., Teodorescu, R., & Rodriguez, P. (2006). Improved MPPT algorithms for rapidly changing environmental conditions. In *Proceedings of the IEEE Power Electronics and Applications* (pp. 1420–1425).

Sera, D., Teodorescu, R., & Rodriguez, P. (2008). Optimized maximum power point tracker for fast-changing environmental conditions. *IEEE Transactions on Industrial Electronics*, *55*(7), 2629–2637. DOI: 10.1109/TIE.2008.924015

Serpanos, D. (2022). False Data Injection Attacks on Sensor Systems. *Proceedings of the 41st IEEE/ACM International Conference on Computer-Aided Design*, 1–5.

Shafidah, S., Mohd, N. H., Lin, X., & Kyungkon, K. (2025). Predicting high-performance perovskite solar cells using AI-based machine learning models. *Materials Today Sustainability*, *18*, 1–14.

Shafqat, A., & Taosif, I. (2025). Sensorless induction motor drive using coupled inductor based three phase inverters for solar pumping applications. *Results in Engineering*, *25*, 1–15.

Shahab, M. A., Srinivasan, B., & Srinivasan, R. (2023). Enhancing Human Machine Interface design using cognitive metrics of process operators. [). Elsevier.]. *Computer-Aided Chemical Engineering*, *52*, 3513–3518.

Sharma, R., Sahay, K., & Singh, S. (2023). Control aspects for PMSG-based grid-connected wind energy conversion system – A review. *SSRG International Journal of Electrical and Electronics Engineering*, *10*(12), 72–86. DOI: 10.14445/23488379/IJEEE-V10I12P108

Shayeghi, H., Pourjafar, S., & Sedaghati, F. (2021). A buck-boost converter: Design, analysis and implementation suggested for renewable energy systems. *Iranian Journal of Electrical and Electronic Engineering*, *17*(2), 1862. DOI: 10.22068/IJEEE.17.2.1862

Shobug, M. A., Chowdhury, N. A., Hossain, M. A., Sanjari, M. J., Lu, J., & Yang, F. (2024). Virtual inertia control for power electronics-integrated power systems: Challenges and prospects. *Energies*, *17*(11), 2737.

Siew, C. Y., Ong, S. K., & Nee, A. Y. C. (2021). Improving maintenance efficiency and safety through a human-centric approach. *Advances in Manufacturing*, *9*(1), 104–114.

Singh, A., Mehta, R., & Kumar, V. (2024). Quantum-resistant blockchains for decentralized energy markets: Global perspectives. *Applied Energy*, *345*, 125–142. DOI: 10.1016/j.apenergy.2023.123456

Singh, G., Kour, V., & Singh, L. (2017). Design and performance analysis of PID controller for automatic generation control of an autonomous power system. *International Journal of Engineering Research & Technology (Ahmedabad)*.

Singirikonda, S., & Pedda, O. Y. (2023). Investigation on performance evaluation of electric vehicle batteries under different drive cycles. *Journal of Energy Storage*, *63*, 106966.

SNAME, T. (1950). Nomenclature for treating the motion of a submerged body through a fluid. *The Society of Naval Architects and Marine Engineers, Technical and Research Bulletin*, (1950), 1-5.

Soliman, M. S., Belkhier, Y., Ullah, N., Achour, A., Alharbi, Y. M., Alahmadi, A. A. A., Abeida, H., & Khraisat, Y. S. H. (2021). Supervisory energy management of a hybrid battery/PV/tidal/wind sources integrated in DC-microgrid energy storage system. *Energy Reports*, *7*, 7728–7740. DOI: 10.1016/j.egyr.2021.11.056

Soni, J., & Bhattacharjee, K. (2023). Optimization of MPPT Controller for Standalone Photovoltaic Systems. In *Perspectives on Social Welfare Applications' Optimization and Enhanced Computer Applications* (pp. 231–263). IGI Global.

Sridevi, S. (2014). User interface design. *International Journal of Computer Science and Information Technology Research*, *2*(2), 415–426.

Stanton, N. A., Chambers, P. R. G., & Piggott, J. (2001). Situational awareness and safety. *Safety Science*, *39*(3), 189–204.

Stone, D., Jarrett, C., Woodroffe, M., & Minocha, S. (2005). *User interface design and evaluation*. Elsevier.

Stoneman, R. (2008). *Alexander the Great: A life in legend*. Yale University Press.

Stowers, K., Oglesby, J., Sonesh, S., Leyva, K., Iwig, C., & Salas, E. (2017). A framework to guide the assessment of human–machine systems. *Human Factors*, *59*(2), 172–188. PMID: 28324673

Strobhar, D. A. (1995). Evolution of operator decision making. *ISA Transactions*, *34*(4), 405–409.

Stuck, R. E., Tomlinson, B. J., & Walker, B. N. (2022). The importance of incorporating risk into human-automation trust. *Theoretical Issues in Ergonomics Science*, *23*(4), 500–516.

Subudhi, B., & Pradhan, R. (2013). A comparative study on maximum power point tracking techniques for photovoltaic power systems. *IEEE Transactions on Sustainable Energy*, *4*(1), 89–98. DOI: 10.1109/TSTE.2012.2202294

Swiegers, W., & Enslin, J. H. R. (1998). An integrated maximum power point tracker for photovoltaic panels. In *Proceedings of the IEEE International Symposium on Industrial Electronics* (pp. 40–44).

Tanakitkorn, K., & Phoemsapthawee, S. (2022). Impacts of thruster configurations on the task performance of an unmanned surface vehicle. *Ocean Engineering, 256*, 111465.

Tan, R., Krishna, V. B., Yau, D. K. Y., & Kalbarczyk, Z. (2016). Impact of Integrity Attacks on Real-Time Pricing in Smart Grids. *IEEE Transactions on Smart Grid, 7*(4), 1772–1781. DOI: 10.1109/TSG.2016.2532563

Tarusan, S. A. A.. (2020). A review of direct torque control development in various multilevel inverter applications. *International Journal of Power Electronics and Drive Systems, 11*(3), 1675–1688.

Tehrani, K. A., Capitaine, T., Barrandon, L., Hamzaoui, M., Rafiei, S. M. R., & Lebrun, A. (2011, August). Current control design with a fractional-order PID for a three-level inverter. In *Proceedings of the 2011 14th European Conference on Power Electronics and Applications* (pp. 1-7). IEEE.

Tey, K. S., Mekhilef, S., Seyedmahmoudian, M., Horan, B., Oo, A. T., & Stojcevski, A. (2018). Improved differential evolution-based MPPT algorithm using SEPIC for PV systems under partial shading conditions and load variation. *IEEE Transactions on Industrial Informatics, 14*(10), 4322–4333.

The Energy and Resources Institute (TERI). (2025). Blockchain for Peer-to-Peer Energy Trading: A Case Study. *TERI Energy and Environment Data Diary and Yearbook, 2025*, 45–50. https://www.teriin.org/publications/Blockchain-for-Peer-to-Peer-Energy-Trading

Tiwari, R. (2023). Explainable AI (XAI) and its applications in building trust and understanding in AI decision making. *International J. Sci. Res. Eng. Manag, 7*, 1–13.

Triatmaja, A. K., Budiastuti, P., & Rismarinandyo, M. Y. (2024). Development of an IoT-based PLC trainer: Bridging the practical divide in industrial automation education. *International Journal of Educational Management and Innovation, 5*(1), 39–52.

Udayakumar, P. (2023). Design and deploy security for infrastructure, data, and applications. In *Design and Deploy a Secure Azure Environment: Mapping the NIST Cybersecurity Framework to Azure Services* (pp. 75–148). Springer.

Udo, W. S., Kwakye, J. M., Ekechukwu, D. E., & Ogundipe, O. B. (2024). Smart grid innovation: Machine learning for real-time energy management and load balancing. *International Journal of Smart Grid Applications, 22*(4), 405–423.

Ulbig, A., Borsche, T. S., & Andersson, G. (2014). Impact of low rotational inertia on power system stability and operation. *IFAC Proceedings Volumes, 47*(3), 7290-7297.

Ullah, Z., Lodhi, B. A., & Hur, J. (2020). Detection and identification of demagnetization and bearing faults in PMSM using transfer learning-based VGG. *Energies, 13*(15), 3834.

Un-Noor, F.. (2017). A comprehensive study of key electric vehicle (EV) components, technologies, challenges, impacts, and future direction of development. *Energies, 10*(8).

Usman, A., Doiphode, N. T., & Rajpurohit, B. S. (2019, September). Stator winding faults investigation in permanent magnet synchronous motor using motor signatures: Part I. In 2019 *international conference on electrical drives & power electronics (EDPE)* (pp. 160-168). IEEE.

Usman, A., & Rajpurohit, B. S. (2020). Modeling and classification of stator interturn fault and demagnetization effects in BLDC motor using rotor back-EMF and radial magnetic flux analysis. *IEEE Access: Practical Innovations, Open Solutions, 8*, 118030–118049.

Vafaeipour, M., Zolfani, S., Varzandeh, M., Derakhti, A., & Eshkalag, M. (2014). Assessment of regions priority for implementation of solar projects in Iran: New application of a hybrid multi-criteria decision making approach. *Energy Conversion and Management, 86*, 653–663.

Varshney, N., Sreetharan, V., Vennila, C., Ahila, R., Mane, D. T., Kshirsagar, K., & Pragathi, B. (2025). Blockchain Technology in Energy Management Systems: Enhancing Security and Transparency.

Vasiliev, A. S. (2016). Controlled Forming of Machine Components Operating Characteristics. *Procedia Engineering, 150*, 975–979.

Vasudevan, K. R., Ramachandaramurthy, V. K., Babu, T. S., & Pouryekta, A. (2020). Synchronverter: A comprehensive review of modifications, stability assessment, applications and future perspectives. *IEEE Access: Practical Innovations, Open Solutions, 8*, 131565–131589.

Vasudevan, S., Piazza, A., & Carr, M. (2022). Qualitative factors in organizational cyber resilience. *2022 International Conference on Cyber Resilience (ICCR)*, 1–5.

Veerachary, M., Senjyu, T., & Uezato, K. (2006). MPPT schemes for PV fed matrix converter. *IEEE Transactions on Industrial Electronics, 53*(4), 1175–1184.

Vega, E. P., Chocron, O., & Benbouzid, M. (2016). A flat design and a validated model for an AUV reconfigurable magnetic coupling thruster. *IEEE/ASME Transactions on Mechatronics, 21*(6), 2892–2901.

Verma, D., Farr, J., & Johannesen, L. H. (2003). System training metrics and measures: A key operational effectiveness imperative. *Systems Engineering*, *6*(4), 238–248.

Vieira, G., & Zhang, J. (2021). Peer-to-Peer Energy Trading in a Microgrid Leveraged by Smart Contracts. arXiv. https://arxiv.org/abs/2201.04944

Villa-Ávila, E.. (2025). Improving V2G systems performance with low-pass filter and fuzzy logic for PV power smoothing in weak low-voltage networks. *Applied Sciences (Basel, Switzerland)*, *15*(4), 1952.

Vionis, P., & Kotsilieris, T. (2024). The potential of blockchain technology and smart contracts in the energy sector: A review. *Applied Sciences (Basel, Switzerland)*, *14*(1), 253. DOI: 10.3390/app14010253

Vlachos, A. G., & Biskas, P. N. (2010). Balancing supply and demand under mixed pricing rules in multi-area electricity markets. *IEEE Transactions on Power Systems*, *26*(3), 1444–1453.

Vukic, Z., & Kuljaca, O. (2002). *Lectures on PID controllers*. Faculty of Electrical Engineering and Computing University of Zagreb.

Wagner, B., Kuebler, J., & Zalnieriute, M. (2025). Humans in the loop: Exploring the challenges of human participation in automated decision-making systems. *Frontiers in Political Science*, *7*, 1611563.

Wais, P. (2017). A review of Weibull functions in wind sector. *Renewable & Sustainable Energy Reviews*, *70*, 1099–1107. DOI: 10.1016/j.rser.2016.12.014

Wakefield, E. H. (1998). *History of the electric automobile: hybrid electric vehicles* (Vol. 187). SAE International.

Wali, A., & Alshehry, F. (2024). A survey of security challenges in cloud-based SCADA systems. *Computers*, *13*(4), 97.

Wang, C., Mei, D., Wang, Y., Yu, X., Sun, W., Wang, D., & Chen, J. (2022). Task allocation for Multi-AUV system: A review. *Ocean Engineering*, *266*, 112911.

Wang, J.. (2020). Review on multi-power sources dynamic coordinated control of hybrid electric vehicle during driving mode transition process. *International Journal of Energy Research*, *44*(8), 6128–6148.

Wang, L., Jiao, S., Xie, Y., Mubaarak, S., Zhang, D., Liu, J., Jiang, S., Zhang, Y., & Li, M. (2021). A permissioned blockchain-based energy management system for renewable energy microgrids. *Sustainability*, *13*(3), 1317. DOI: 10.3390/su13031317

Wang, M. S., Chen, S. C., & Shih, C. H. (2018). Speed control of brushless DC motor by adaptive network-based fuzzy inference. *Microsystem Technologies, 24*(1), 33–39.

Wang, N., Chau, S. C.-K., & Zhou, Y. (2021). Privacy-preserving energy storage sharing with blockchain and secure multi-party computation. *Energy, 225*, 120–130. DOI: 10.1016/j.energy.2021.120130

Wang, Q., & Chang, L. (2004). An intelligent maximum power extraction algorithm for inverter-based variable speed wind turbine systems. *IEEE Transactions on Power Electronics, 19*(5), 1242–1249. DOI: 10.1109/TPEL.2004.833459

Wang, S., Taha, A. F., Wang, J., Kvaternik, K., & Hahn, A. (2019). Energy crowdsourcing and peer-to-peer energy trading in blockchain-enabled smart grids. *Applied Energy, 235*, 1078–1090. DOI: 10.1016/j.apenergy.2018.11.031

Wang, S., Yang, M., Wang, Y., Yang, S., Lan, S., & Zhang, X. (2021, July). Optimization of Flight Parameters for Petrel-L Underwater Glider. *IEEE Journal of Oceanic Engineering, 46*(3), 817–828.

Wang, X., & Suh, C. S. (2017). Precision concurrent speed and position tracking of brushed dc motors using nonlinear time-frequency control. *Journal of Vibration and Control, 23*(19), 3266–3291.

Wang, Y., Huang, Y., Li, Y., & Ren, C. (2024). Optimization of thruster configuration and control allocation for a spherical magnetic coupling thrusters actuated underwater robot. *Ocean Engineering, 309*, 118445.

Weschenfelder, F., Leite, G. D. N. P., da Costa, A. C. M., de Castro Vilela, O., Ribeiro, C. M. S., Cavalcante, A. C., & Ochoa, A. A. V. (2020). A review on the complementarity between grid-connected solar and wind power systems. *Journal of Cleaner Production, 257*, 120617. Advance online publication. DOI: 10.1016/j.jclepro.2020.120617

Windarko, N. A., Fu, J. H., & Choi, J. H. (2010). Maximum power point tracking of photovoltaic system using adaptive neuro-fuzzy inference system. In *Proceedings of the IEEE International Conference on Sustainable Energy Technologies* (pp. 1–6). IEEE.

Wireko-Brobby, A., Hu, Y., Wang, G., Gong, C., Lang, W., & Zhang, Z. (2023). Analysis of the sources of error within PMSM-based electric powertrains—A review. *IEEE Transactions on Transportation Electrification, 10*(3), 6370–6406.

Wu, H., Chen, Y., Yang, Q., Yan, B., & Yang, X. (2024). A review of underwater Robot localization in confined spaces. *Journal of Marine Science and Engineering, 12*(3), 428.

Wu, X., Xiao, L., Sun, Y., Zhang, J., Ma, T., & He, L. (2022). A survey of human-in-the-loop for machine learning. *Future Generation Computer Systems*, *135*, 364–381.

Xiao, W., Dunford, W. G., Palmer, P. R., & Capel, A. (2006). Real-time identification of optimal operating points in photovoltaic power systems. *IEEE Transactions on Industrial Electronics*, *53*(4), 1017–1026.

Xiao, Y., Lin, X., Lei, Y., Gu, Y., Tang, J., Zhang, F., & Qian, B. (2024). Blockchain-Assisted Secure Energy Trading in Electricity Markets: A Tiny Deep Reinforcement Learning-Based Stackelberg Game Approach. *Electronics (Basel)*, *13*(18), 3647. DOI: 10.3390/electronics13183647

Xu, D., Yan, Z., Xie, H., Lu, L., Xiao, Z., & Han, B. (2024). Demand-driven optimization method for thruster configuration of underwater vehicles considering capability boundary and system efficiency. *Ocean Engineering*, *310*, 118749.

Xu, S., Wu, Y., & Wang, C. (2019). Study on human–machine interface design of nuclear power plant control room. *International Conference on Man-Machine-Environment System Engineering*, 631–639.

Yang, A.-M., Li, S.-S., Ren, C.-H., Liu, H.-X., Han, Y., & Liu, L. (2018). Situational awareness system in the smart campus. *IEEE Access : Practical Innovations, Open Solutions*, *6*, 63976–63986.

Yang, B., Linen, Z., Xiaoshun, Z., Hongchun, S., Tao, Y., Haofei, L., Lin, J., & Liming, S. (2019). Novel bio-inspired memetic salp swarm algorithm and application to MPPT for PV systems considering partial shading condition. *Journal of Cleaner Production*, *215*, 1203–1222.

Yang, Q., Wang, H., Wang, T., Zhang, S., Wu, X., & Wang, H. (2021). Blockchain-Based Decentralized Energy Management Platform for Residential Distributed Energy Resources in A Virtual Power Plant. arXiv. https://arxiv.org/abs/2105.00174

Yang, Z.. (2023). Review on optimal planning of new power systems with distributed generations and electric vehicles. *Energy Reports*, *9*, 501–509. DOI: 10.1016/j.egyr.2022.11.168

Yap, K. Y., Sarimuthu, C. R., & Lim, J. M. Y. (2019). Virtual inertia-based inverters for mitigating frequency instability in grid-connected renewable energy system: A review. *Applied Sciences (Basel, Switzerland)*, *9*(24), 5300.

You, S., Zhao, Y., Mandich, M., Cui, Y., Li, H., Xiao, H., & Zhang, Y. (2020, November). A review on artificial intelligence for grid stability assessment. In *2020 IEEE International Conference on Communications, Control, and Computing Technologies for Smart Grids (SmartGridComm)* (pp. 1-6). IEEE.

Youssef, R., Hamida, M. A., Mohamed, N., Lassaad, S., Djamel, T., & Yasmine, A. Single-Stage Standalone Photovoltaic Water Pumping System Using Predictive Torque Control (PTC) of Induction Machine. Electric Power Components and Systems, 1-17.

Youssef, R.. (2024). Comparative study of the MPPT control for the photovoltaic water pumping system between FSS-P&O and VSS-P&O. *WSEAS Transactions on Power Systems*, *19*, 229–241.

Youssef, R., Hamida, M. A., Mohamed, N., Lassaad, S., Djamel, T., & Yasmine, A. (2023). Single-stage standalone photovoltaic water pumping system using predictive torque control (PTC) of induction machine. *Electric Power Components and Systems*, •••, 1–17.

Youssef, R., Hamida, M. A., Mohamed, N., Lassaad, S., Djamel, T., & Yasmine, A. (2023). Single-Stage Standalone Photovoltaic Water Pumping System Using Predictive Torque Control (PTC) of Induction Machine. *Electric Power Components and Systems*, •••, 1–17. DOI: 10.1080/15325008.2023.2278624

Yuan, M., Chen, J., Hu, Y., Feng, S., Xie, M., Mohammadi, G., Xing, Z., & Quigley, A. (2024). Towards Human-AI Synergy in UI Design: Enhancing Multi-Agent Based UI Generation with Intent Clarification and Alignment. *ArXiv Preprint ArXiv:2412.20071*.

Zaman, T., & Hossain, A. (2021). *HMI Design: An Analysis of A Good Display for Seamless Integration between User Understanding and Automatic Controls*. Journal American Society for Engineering Education.

Zandi, G., Yaacob, N. A., Tajuddin, M., & Nik Abdul Rahman, N. K. (2024). Artificial intelligence and the evolving cybercrime paradigm: Current threats to businesses. *Journal of Information Technology Management*, *16*(4), 162–170.

Zeng, Z., Li, Y., Cao, Y., Zhao, Y., Zhong, J., Sidorov, D., & Zeng, X. (2020). Blockchain Technology for Information Security of the Energy Internet: Fundamentals, Features, Strategy and Application. *Energies*, *13*(4), 881. DOI: 10.3390/en13040881

Zhang, L., Li, X., & Wang, J. (2025). Decentralized EV Charging Network with Tokenized Access and Blockchain Technology. *ScienceDirect*. https://www.sciencedirect.com/science/article/pii/S2773186325001021

Zhang, G., Chen, S. C.-I., & Yue, X. (2024). Blockchain Technology in Carbon Trading Markets: Impacts, Benefits, and Challenges—A Case Study of the Shanghai Environment and Energy Exchange. *Energies*, *17*(13), 3296. DOI: 10.3390/en17133296

Zhang, H.. (2018). Six-plate capacitive coupler to reduce electric field emission in large air-gap capacitive power transfer. *IEEE Transactions on Power Electronics, 33*(1), 665–675.

Zhang, L., Harnefors, L., & Nee, H. P. (2010a). Modeling and control of VSC-HVDC links connected to island systems. *IEEE Transactions on Power Systems, 26*(2), 783–793.

Zhang, L., Harnefors, L., & Nee, H. P. (2010b). Interconnection of two very weak AC systems by VSC-HVDC links using power-synchronization control. *IEEE Transactions on Power Systems, 26*(1), 344–355.

Zhang, L., Wang, Y., & Li, J. (2025). Towards Blockchain Interoperability: A Comprehensive Survey on Cross-Chain Solutions. *Computer Networks, 202*, 108–123. DOI: 10.1016/j.comnet.2025.04.015

Zhang, Q., & Feng, M. (2019). Fast fault diagnosis method for Hall sensors in brushless DC motor drives. *IEEE Transactions on Power Electronics, 34*(3), 2585–2596.

Zhang, X., Beram, S. M., Haq, M. A., Wawale, S. G., & Buttar, A. M. (2022). Research on algorithms for control design of human–machine interface system using ML. *International Journal of System Assurance Engineering and Management, 13*(Suppl 1), 462–469.

Zhang, X., Li, Y., & Wang, Z. (2023). AI-integrated blockchain for energy transaction optimization. *Journal of Cleaner Production, 404*, 136–151. DOI: 10.1016/j.jclepro.2023.136789

Zhang, X., Li, Y., & Wang, Z. (2024). Blockchain for energy markets: A comprehensive survey. *Renewable & Sustainable Energy Reviews, 135*, 110–122. DOI: 10.1016/j.rser.2020.110122

Zhang, Y., & Wang, L. (2023). Exploring the effects of blockchain scalability limitations on performance and user behavior in blockchain-based shared manufacturing systems: An experimental approach. *Applied Sciences (Basel, Switzerland), 13*(7), 4251. DOI: 10.3390/app13074251

Zhao, C.. (2020). Design and analysis of an engine-start control strategy for a single-shaft parallel hybrid electric vehicle. *Energy, 202*, 117621.

Zhao, J., Guan, X., Li, C., Mou, Q., & Chen, Z. (2020). Comprehensive evaluation of inter-turn short circuit faults in PMSM used for electric vehicles. *IEEE Transactions on Intelligent Transportation Systems, 22*(1), 611–621.

Zhao, J., Wang, L., Xu, L., Dong, F., Song, J., & Yang, X. (2021). Uniform demagnetization diagnosis for permanent-magnet synchronous linear motor using a sliding-mode velocity controller and an ALN-MRAS flux observer. *IEEE Transactions on Industrial Electronics*, *69*(1), 890–899.

Zhao, Y., & Behdad, S. (2024). State-of-health estimation for sustainable electric vehicle batteries using temporal-enhanced self-attention graph neural networks. *Journal of Energy Storage*, *146*(June), 1–9.

Zhao, Z., Liu, Y., Zeng, Z., Chen, Z., & Zhou, H. (2023). Privacy-preserving electricity theft detection based on blockchain. *IEEE Transactions on Industrial Informatics*, *19*(5), 3456–3465. DOI: 10.1109/TII.2023.3145678

Zhong, Q. C., & Weiss, G. (2010). Synchronverters: Inverters that mimic synchronous generators. *IEEE Transactions on Industrial Electronics*, *58*(4), 1259–1267.

Zhou, B., Meng, L., Yang, D., Ma, Z., & Xu, G. (2019). A Novel VSG-Based Accurate Voltage Control and Reactive Power Sharing Method for Islanded Microgrids. *Sustainability*, *11*(23), 6666.

Żywiec, J., Tchórzewska-Cieślak, B., & Sokolan, K. (2024). Assessment of Human Errors in the Operation of the Water Treatment Plant. *Water (Basel)*, *16*(17), 2399.

About the Contributors

Naoui Mohamed was born in Nefta, Tunis, in 1991. He received the degree in electrical engineering from the University of Gabès, Tunisia, in 2015, and the Ph.D. degree from the Department of Electrical Engineering, in 2020. From 2016 to 2018, he worked as a Professional Engineer in electrical and automatic engineering. He is currently an Associate Professor in electrical engineering with the University of Gabès (ENIG). He has academic experience of three years. He has published over 100 research articles in reputed journals, international conferences, and book chapters. His research interests include electric vehicles, power systems, and renewable energy.

Sbita Lassaad was born in Hammam Lif, Tunisia, in 1962. He received the B.E. degree in electrical engineering from the University of Tunis, Tunis, Tunisia, in 1985, and the D.E.A. and Thesis degrees in electrical engineering from the École nationale supérieure d'ingénieurs de Tunis, Tunis, in 1987 and 1997, respectively. In 1988, he joined the Department of Electrical Engineering, National School of Engineering of Sfax, University of Sfax, as a Professor Assistant, and the Department of Electrical Engineering, National Engineering School of Gabes, University of Gabés, Gabés, Tunisia, in 1991, where he became an Associate Professor, in 1998, and a Professor, in 2009.

***

N. Rajavel, is a principal of Selvamm Arts and Science College (Autonomous), Namakkal. He completed his Ph.D. at Bharathiyar University (Coimbatore) and his M.S. (Education) at Alagappa University. He has started his teaching career since 1999. He has been an elected senate member of Periyar University, Salem, since 2007. The specialized subject areas of Dr. N. Rajavel are marketing, Human Resource Management and Finance. He is recognized to guide for MBA, M.Phil., and Ph.D scholars. He has guided more than 100 MBA candidates and 12 M.Phil. candidates. He has presented various research articles on various topics at 32 National and International Conferences to his credit. He is one of the Authors for five books, namely Marketing, Production & Materials Management, Service Marketing, Human Resource Management, and Principles of Management Practices. He has published 12 research articles in National and International Journals and received the Best Programme Officer Award, Best Citizens of India Award, and Pride of India Award from various national agencies.

Naoui Mohamed was born in Nefta, Tunis, in 1991. He received the degree in electrical engineering from the University of Gabès, Tunisia, in 2015, and a Ph.D. degree from the Department of Electrical Engineering, in 2020. From 2016 to 2018, he worked as a Professional Engineer in electrical and automatic engineering. He is currently an Associate Professor in electrical engineering with the University of Gabès (ENIG). He has academic experience of three years. He has published over 80 research articles in reputed journals, international conferences, and book chapters. His research interests include electric vehicles, power systems, and renewable energy.

R Maheswari has completed her Graduation Bachelor of Computer Applications from Kongu Arts and Science College, Erode and secured University Gold Medal. She did her Post Graduation Master of Business Administration at Kongu Arts and Science College, Erode. Secured University 8th Rank. NET Qualified faculty. Completed PhD in Management from Bharathiar University, Coimbatore. She has teaching experience of 11 years and presently working as Assistant Professor at Kongu Engineering College, Erode. She also has Industrial Experience of 2 years and Entrepreneurial Experience of 3 years. She holds 2 patents. She is author of various book chapters and has published articles in various academic journals. She hascleared NISM Equity and Derivatives trader certification and Mutual Funds Distributor certification. She has conducted ManagementDevelopmentFacultyforvariousorganizations across the state. She is a MotivationalSpeakerandTrainerforvariousorganizationandinstitutions.Posted articles in her Blog and Facebook Page. Certified Trainer at Lions Club International and JCI. She has completed NeuroLinguisticProgram(NLP)–Basic level and Certified Adolescencestudents' handler.

SC Vetrivel is a faculty member in the Department of Management Studies, Kongu Engineering College (Autonomous), Perundurai, Erode Dt. Having experience in Industry 20 years and Teaching 16 years. Awarded with Doctoral Degree in Management Sciences in Anna University, Chennai. He has organized various workshops and Faculty Development Programmes. He is actively involved in research and consultancy works. He acted as a resource person to FDPs & MDPs to various industries like, SPB ltd, Tamilnadu Police, DIET, Rotary school and many. His areas of interest include Entrepreneurship, Business Law, Marketing and Case writing. Articles published more than 100 International and National Journals. Presented papers in more than 30 National and International conferences including IIM Bangalore, IIM Kozhikode, IIM Kashipur and IIM Indore. He was a Chief Co-ordinator of Entrepreneurship and Management Development Centre (EMDC) of Kongu Engineering College, he was instrumental in organizing various Awareness Camps, FDP, and TEDPs to aspiring entrepreneurs which was funded by NSTEDB – DST/GoI

V.P. Arun is a driven and accomplished professional with a diverse educational background and extensive hands-on experience across various industries. Graduating with honors, Arun earned his Master of Business Administration (M.B.A) with a specialization in Human Resources and Marketing from the renowned Sona School of Management in Salem in 2018, where he excelled academically with an impressive 8.3 Cumulative Grade Point Average (CGPA).Throughout his academic journey, Arun displayed an unwavering commitment to learning and personal growth, actively seeking opportunities to expand his knowledge and skills beyond the confines of traditional education. He sought practical experiences to complement his theoretical understanding, such as a 45-day summer internship focused on conducting a feasibility study for R-Doc Sustainability in the market. Additionally, Arun broadened his horizons through a 7-day industrial visit to Malaysia and Singapore, immersing himself in diverse cultural and professional environments. Arun's academic pursuits were further enriched by his involvement in hands-on projects, including a comprehensive study on Employee Job Satisfaction at Roots Cast Private Limited. His professional trajectory includes serving as a Growth Officer at Parle Agro Private Limited, where he played a crucial role in driving sales growth and market expansion. Arun is a dedicated professional, currently employed at JKKN Engineering College in the Management Studies Department. He holds expertise in his field and has contributed significantly to academia through the publication of two journal articles. Arun's work focuses on advancing knowledge and understanding in management studies, making valuable contributions to the academic community.

Wafa Hayder is a Tunisian researcher and innovator specializing in renewable energy, energy systems optimization, and green hydrogen production. She earned her Ph.D. in 2020, focusing on the optimization of photovoltaic systems through advanced approaches such as Particle Swarm Optimization (PSO) and Artificial Neural Networks (ANN). Her research has significantly improved the efficiency of Maximum Power Point Tracking (MPPT), particularly under complex conditions such as partial shading and variable atmospheric factors. During her doctoral journey, she collaborated with prestigious institutions, including Aalborg University (Denmark) under the supervision of Dr. Dezso Sera, and the Polytechnic University of Milan (Italy) with Dr. Emanuele Ogliari. These international collaborations strengthened her expertise in hybrid methods and intelligent optimization of energy systems. Recognized for her ability to bridge academic research, technological innovation, and industrial application, Wafa Hayder represents a new generation of scientists committed to building a sustainable energy future.

Index

A

B

C

D

E

F

G

H

I

T

U

V

W

www.ingramcontent.com/pod-product-compliance
Lightning Source LLC
LaVergne TN
LVHW062332300926
849161LV00007B/1663

* 9 7 9 8 3 3 7 3 5 6 2 1 1 *